Molecular Modeling of Proteins

METHODS MOLECULAR BIOLOGY™

John M. Walker, SERIES EDITOR

457. **Membrane Trafficking**, edited by *Ales Vancura, 2008*
456. **Adipose Tissue Protocols**, *Second Edition*, edited by *Kaiping Yang, 2008*
455. **Osteoporosis**, edited by *Jennifer J. Westendorf, 2008*
454. **SARS- and Other Coronaviruses: Laboratory Protocols**, edited by *Dave Cavanagh, 2008*
453. **Bioinformatics, Volume 2:** *Structure, Function, and Applications*, edited by *Jonathan M. Keith, 2008*
452. **Bioinformatics, Volume 1:** *Data, Sequence Analysis, and Evolution*, edited by *Jonathan M. Keith, 2008*
451. **Plant Virology Protocols: From Viral Sequence to Protein Function**, edited by *Gary Foster, Elisabeth Johansen, Yiguo Hong, and Peter Nagy, 2008*
450. **Germline Stem Cells**, edited by *Steven X. Hou and Shree Ram Singh, 2008*
449. **Mesenchymal Stem Cells: Methods and Protocols**, edited by *Darwin J. Prockop, Douglas G. Phinney, and Bruce A. Brunnell, 2008*
448. **Pharmacogenomics in Drug Discovery and Development**, edited by *Qing Yan, 2008*
447. **Alcohol: Methods and Protocols**, edited by *Laura E. Nagy, 2008*
446. **Post-translational Modification of Proteins: Tools for Functional Proteomics, Second Edition**, edited by *Christoph Kannicht, 2008*
445. **Autophagosome and Phagosome**, edited by *Vojo Deretic, 2008*
444. **Prenatal Diagnosis**, edited by *Sinhue Hahn and Laird G. Jackson, 2008*
443. **Molecular Modeling of Proteins**, edited by *Andreas Kukol, 2008*
442. **RNAi: Design and Application**, edited by *Sailen Barik, 2008*
441. **Tissue Proteomics: Pathways, Biomarkers, and Drug Discovery**, edited by *Brian Liu, 2008*
440. **Exocytosis and Endocytosis**, edited by *Andrei I. Ivanov, 2008*
439. **Genomics Protocols, Second Edition**, edited by *Mike Starkey and Ramnanth Elaswarapu, 2008*
438. **Neural Stem Cells: Methods and Protocols, Second Edition**, edited by *Leslie P. Weiner, 2008*
437. **Drug Delivery Systems**, edited by *Kewal K. Jain, 2008*
436. **Avian Influenza Virus**, edited by *Erica Spackman, 2008*
435. **Chromosomal Mutagenesis**, edited by *Greg Davis and Kevin J. Kayser, 2008*
434. **Gene Therapy Protocols: Volume 2:** *Design and Characterization of Gene Transfer Vectors*, edited by *Joseph M. LeDoux, 2008*
433. **Gene Therapy Protocols: Volume 1:** *Production and In Vivo Applications of Gene Transfer Vectors*, edited by *Joseph M. LeDoux, 2008*
432. **Organelle Proteomics**, edited by *Delphine Pflieger and Jean Rossier, 2008*
431. **Bacterial Pathogenesis: Methods and Protocols**, edited by *Frank DeLeo and Michael Otto, 2008*
430. **Hematopoietic Stem Cell Protocols**, edited by *Kevin D. Bunting, 2008*
429. **Molecular Beacons: Signalling Nucleic Acid Probes, Methods and Protocols**, edited by *Andreas Marx and Oliver Seitz, 2008*
428. **Clinical Proteomics: Methods and Protocols**, edited by *Antonio Vlahou, 2008*
427. **Plant Embryogenesis**, edited by *Maria Fernanda Suarez and Peter Bozhkov, 2008*
426. **Structural Proteomics: High-Throughput Methods**, edited by *Bostjan Kobe, Mitchell Guss, and Huber Thomas, 2008*
425. **2D PAGE: Volume 2: Sample Preparation and Fractionation**, edited by *Anton Posch, 2008*
424. **2D PAGE: Volume 1:** *Sample Preparation and Fractionation*, edited by *Anton Posch, 2008*
423. **Electroporation Protocols: Preclinical and Clinical Gene Medicine**, edited by *Shulin Li, 2008*
422. **Phylogenomics**, edited by *William J. Murphy, 2008*
421. **Affinity Chromatography: Methods and Protocols, Second Edition**, edited by *Michael Zachariou, 2008*
420. **Drosophila: Methods and Protocols**, edited by *Christian Dahmann, 2008*
419. **Post-Transcriptional Gene Regulation**, edited by *Jeffrey Wilusz, 2008*
418. **Avidin-Biotin Interactions: Methods and Applications**, edited by Robert J. *McMahon, 2008*
417. **Tissue Engineering, Second Edition**, edited by *Hannsjörg Hauser and Martin Fussenegger, 2007*
416. **Gene Essentiality: Protocols and Bioinformatics**, edited by *Svetlana Gerdes and Andrei L. Osterman, 2008*
415. **Innate Immunity**, edited by *Jonathan Ewbank and Eric Vivier, 2007*
414. **Apoptosis in Cancer: Methods and Protocols**, edited by *Gil Mor and Ayesha Alvero, 2008*

Molecular Modeling of Proteins

Andreas Kukol

School of Life Sciences
University of Hertfordshire
Hertfordshire, UK

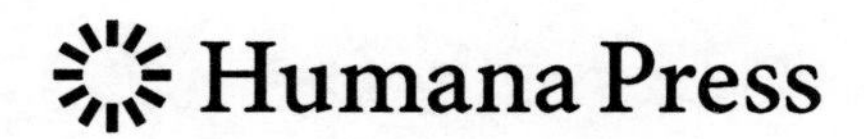

Editor
Andreas Kukol
School of Life Sciences
University of Hertfordshire
College Lane
Hatfield, Hertfordshire AL10 9AB, UK
a.kukol@herts.ac.uk

Series Editor
John M. Walker
School of Life Sciences
University of Hertfordshire
College Lane
Hatfield, Hertfordshire AL10 9AB, UK

Cover illustrations (clockwise from upper right): representative structures from principle components analysis (PCA; *see* complete caption for Fig. 3 in Chapter 5); molecular assembly formed by the homotetrameric facilitator of glycerol transport, GlpF, embedded in a fully hydrated POPE lipid bilayer (*see* complete caption for Fig. 4 in Chapter 7); TM domain of GpA formed by a homodimer of α-helices, embedded in a lipid membrane mimetic (*see* complete caption for Fig. 5A in Chapter 7); conformational change of the prion protein obtained from a molecular dynamics trajectory enhanced by conformational flooding (*see* complete caption for Fig. 4C in Chapter 5).

ISBN: 978-1-58829-864-5 e-ISBN: 978-1-59745-177-2
DOI: 10.1007/978-1-59745-177-2
ISSN: 1064-3745

Library of Congress Control Number: 2007942569

Printed on acid-free paper

9 8 7 6 5 4 3 2 1

springer.com

Preface

Although two decades ago molecular modeling and simulation of biomolecules were in the realm of specialists with access to supercomputers, ongoing improvements in force fields and powerful software readily available to the academic community have stimulated a great interest among bioscientists who are primarily interested in investigating biological or chemical problems. This development has been accompanied by a decrease in the price/performance ratio of hardware that enables us to carry out meaningful simulations on desktop workstations or small clusters of workstations. For example, all-atom models of a protein in a lipid membrane with water molecules included can be simulated for a few nanoseconds.

The purpose of *Molecular Modeling of Proteins* is to enable nonspecialists, first, to grasp the scope of methods available and, second, to apply methods easily to their own problems. Although software packages in molecular modeling are accompanied by good manuals, the first-time user may easily be frustrated over a problem that requires only a small tweak of an input file to solve. Thus, most chapters contain, apart from a thorough introduction, step-by-step instructions and notes on troubleshooting and hints about how to avoid pitfalls.

The first part of the book describes the methodologies of molecular modeling including a chapter about normal modes and essential dynamics. This part contains, apart from practical hints and tips, a thorough treatment of the underlying theories. The next part focuses on free energy calculations, followed by various chapters about the molecular modeling of membrane proteins. A later part contains chapters about protein structure determination by comparative protein modeling as well as modeling based on experimental data. A further part is devoted to the conformational changes of proteins, and protein folding and unfolding and misfolding in prion diseases. The last part contains several chapters about applications to drug design. The topics have been chosen to represent the latest developments in the field, albeit highly relevant to biochemical and biomedical problems. Although this book is directed at the modeling of proteins, the techniques described are equally applicable to other biomolecules, such as DNA or carbohydrates, provided the adequate force fields are used. The chapters are written by internationally well-established investigators; they include leading developers of popular simulation packages or force fields.

Molecular Modeling of Proteins is directed to scientists in chemistry, biochemistry, biology, biophysics, and bioinformatics working in industry and academia, who are interested in applying the techniques described to their own research. Additionally, the book forms a valuable resource for educators who wish to teach courses to university students or professionals about molecular modeling.

Andreas Kukol

Contents

Contributors

Andrew J. Beevers
Department of Biological Sciences, University of Warwick, Coventry, UK

Philip C. Biggin
Department of Biochemistry, University of Oxford, Oxford, UK

Peter J. Bond
Department of Biochemistry, University of Oxford, Oxford, UK

A.M.J.J. Bonvin
Bijvoet Center for Biomolecular Research, Utrecht University, Utrecht, the Netherlands

Christophe Chipot
Equipe de dynamique des assemblages membranaires, Universite Henri Poincare, Vandœuvre-les-Nancy cedex, France

Shubhra Ghosh Dastidar
Genome Center, University of California, Davis, CA

Bert L. de Groot
Max Planck Institute for Biophysical Chemistry, Göttingen, Germany

Michael W. Deem
Departments of Bioengineering and Physics and Astronomy, Rice University, Houston, TX

Yong Duan
Genome Center, University of California, Davis, CA

David J. Earl
Department of Chemistry, University of Pittsburgh, Pittsburgh, PA

Yao Fan
Department of Chemistry and Supercomputer Institute, University of Minnesota, Minneapolis, MN

Michael Feig
Department of Biochemistry & Molecular Biology, Michigan State University, East Lansing, MI

B. Fischer
Institut für Nanotechnologie, Forschungszentrum Karlsruhe, Karlsruhe, Germany

G. Fuentes
Bijvoet Center for Biomolecular Research, Utrecht University, Utrecht, the Netherlands

Jiali Gao
Department of Chemistry and Supercomputer Institute, University of Minnesota, Minneapolis, MN

Olgun Guvench
Department of Pharmaceutical Sciences, University of Maryland, Baltimore, MD

Steven Hayward
School of Computing Sciences and School of Biological Sciences, University of East Anglia, Norwich, UK

Andreas Kukol
School of Life Sciences, University of Hertfordshire, Hatfield, UK

Matthew C. Lee
Hogan and Hartson, Los Angeles, CA

Hongxing Lei
Genome Center, University of California, Davis, CA

Marc F. Lensink
Center for Structural Biology and Bioinformatics, Université Libre de Bruxelles, Brussels, Belgium

Marguerita Lim-Wilby
BioSolveIT GmbH, Sankt Augustin, Germany

Yen-lin Lin
Department of Chemistry and Supercomputer Institute, University of Minnesota, Minneapolis, MN

Erik R. Lindahl
Department of Biochemistry and Biophysics, Stockholm University, Stockholm, Sweden

Haiguang Liu
Genome Center, University of California, Davis, CA

Gerald H. Lushington
Molecular Graphics and Modeling Laboratory, University of Kansas, Lawrence, KS

Shuhua Ma
Department of Chemistry and Supercomputer Institute, University of Minnesota, Minneapolis, MN

Alexander D. MacKerell, Jr.
Department of Pharmaceutical Sciences, University of Maryland, Baltimore, MD

Dan T. Major
Department of Chemistry and Supercomputer Institute, University of Minnesota, Minneapolis, MN

Edyta B. Małolepsza
Faculty of Chemistry, Warsaw University, Warsaw, Poland

H. Merlitz
Department of Physics and ITPA, Xiamen University, Xiamen, China

Garrett M. Morris
Department of Molecular Biology, The Scripps Research Institute, La Jolla, CA

A.D.J. van Dijk
Bijvoet Center for Biomolecular Research, Utrecht University, Utrecht, the Netherlands

Wolfgang Wenzel
Institut für Nanotechnologie, Forschungszentrum Karlsruhe, Karlsruhe, Germany

David S. Wishart
Departments of Computing Science and Biological Sciences, University of Alberta, Edmonton, Canada

Kin-Yiu Wong
Department of Chemistry and Supercomputer Institute, University of Minnesota, Minneapolis, MN

Hyung-June Woo
Department of Chemistry, University of Nevada, Reno, NV

Wei Zhang
Genome Center, University of California, Davis, CA

Part I
Methodology

Chapter 1
Molecular Dynamics Simulations

Erik R. Lindahl

Summary Molecular simulation is a very powerful toolbox in modern molecular modeling, and enables us to follow and understand structure and dynamics with extreme detail—literally on scales where motion of individual atoms can be tracked. This chapter focuses on the two most commonly used methods, namely, energy minimization and molecular dynamics, that, respectively, optimize structure and simulate the natural motion of biological macromolecules. The common theoretical framework based on statistical mechanics is covered briefly as well as limitations of the computational approach, for instance, the lack of quantum effects and limited timescales accessible. As a practical example, a full simulation of the protein lysozyme in water is described step by step, including examples of necessary hardware and software, how to obtain suitable starting molecular structures, immersing it in a solvent, choosing good simulation parameters, and energy minimization. The chapter also describes how to analyze the simulation in terms of potential energies, structural fluctuations, coordinate stability, geometrical features, and, finally, how to create beautiful ray-traced movies that can be used in presentations.

Keywords: Energy minimization · Equilibration · Force field · Molecular dynamics · Position restraints · Protein · Secondary structure · Simulation · Solvent · Trajectory analysis

1 Introduction

Biomolecular dynamics occur over a wide range of scales in both time and space, and the choice of approach to study them depends on the question asked. Molecular simulation is far from the only theoretical method; when the aim is to predict, e.g., the structure and/or function of proteins rather than studying the folding process, the best tool is normally bioinformatics that detect related proteins from amino acid sequence similarity; and, for computational drug design, often it is much more productive to use statistical methods such as quantitative structure–activity

From: Methods in Molecular Biology, vol. 443, Molecular Modeling of Proteins
Edited by Andreas Kukol © Humana Press, Totowa, NJ

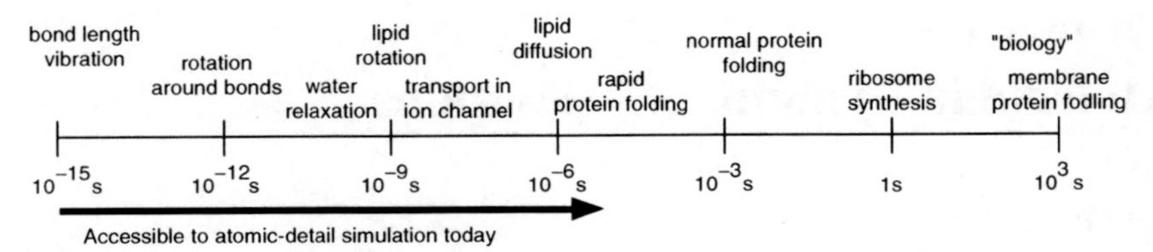

Fig. 1 Range of time scales for dynamics in biomolecular systems. Although the individual time steps of molecular dynamics is 1 to 2 fs, parallel computers make it possible to simulate on a microsecond scale, and distributed computing techniques can sample even slower processes, almost reaching milliseconds

relationship (QSAR) instead of spending billions of CPU hours to simulate binding of thousands of compounds.

The most important point of simulations is that they provide a way to test whether theoretical models predict experimental observations. As an example, simulations of ion channels cannot compete with experiments when it comes to measuring ion currents, but they have been useful to explain *why* some ions pass whereas others are blocked. Similarly, simulations can provide detail not accessible through experiments, for instance, pressure distributions inside membranes. Further, structural refinement and energy minimizations are regularly used to improve both experimental and predicted protein structures, and drug design is moving toward more accurate models, even including large-scale simulations for free energy screening.

Ideally, the time-dependent Schrödinger equation should be able to predict all properties of any molecule with arbitrary precision *ab initio*. However, as soon as more than a handful of particles are involved, it is necessary to introduce approximations. For most biomolecular systems, we, therefore, choose to work with empirical parameterizations of models instead; for instance, classic Coulomb interactions between pointlike atomic charges rather than a quantum description of the electrons. These models are not only orders of magnitude faster, but because they have been parameterized from experiments, they also perform better when it comes to reproducing observations on a microsecond scale (Fig. 1), rather than extrapolating quantum models 10 orders of magnitude. The first molecular dynamics simulation was performed as late as 1957 [1], although it was not until the 1970s that it was possible to simulate water [2] and biomolecules [3].

2 Theory

Macroscopic properties measured in an experiment are not direct observations, but averages over billions of molecules representing a statistical mechanics *ensemble*. This has deep theoretical implications, which are covered in great detail in the literature [4, 5], but, even from a practical point of view, there are important consequences. 1) It is not sufficient to work with individual structures, but systems have to be expanded to generate a representative ensemble of structures at the given experimental conditions, e.g., temperature and pressure. 2) Thermodynamic equilibrium properties related to free energy, such as binding constant, solubilities, and relative stability, cannot be calculated directly from individual simulations, but require more elaborate techniques covered in later chapters. 3) For equilibrium properties (in contrast to kinetic), the aim is to examine the ensemble of structures, and *not* necessarily to reproduce individual atomic trajectories!

The two most common ways to generate statistically faithful equilibrium ensembles are *Monte Carlo* and *molecular dynamics simulations*; the latter also has the advantage of accurately reproducing kinetics of non-equilibrium properties such as diffusion or folding times. When a starting configuration is very far from equilibrium, large forces can cause the simulation to crash or distort the system, and, in this case, it is necessary to start with *energy minimization* of the system before the molecular dynamics simulation. In addition, energy minimizations are commonly used to refine low-resolution experimental structures.

All classic simulation methods rely on more or less empirical approximations called *force fields* [6–9] to calculate interactions and evaluate the potential energy of the system as a function of pointlike atomic coordinates. A force field consists of both the set of equations used to calculate the potential energy and forces from particle coordinates, as well as a collection of parameters used in the equations. For most purposes, these approximations work well, but they cannot reproduce quantum effects such as bond formation or breaking.

All common force fields subdivide potential functions into two classes. *Bonded interactions* cover covalent bond-stretching, angle-bending, torsion potentials when rotating around bonds, and out-of-plane “improper torsion” potentials, all which are normally fixed throughout a simulation—see Fig. 2. The remaining *nonbonded interactions* consist of Lennard-Jones repulsion and dispersion as well as Coulomb electrostatics. These are typically computed from neighborlists updated every 5 to 10 steps.

Given the potential and force (negative gradient of potential) for all atoms, the coordinates are updated for the next step. For energy minimization, the *steepest descent* algorithm simply moves each atom a short distance in the direction of

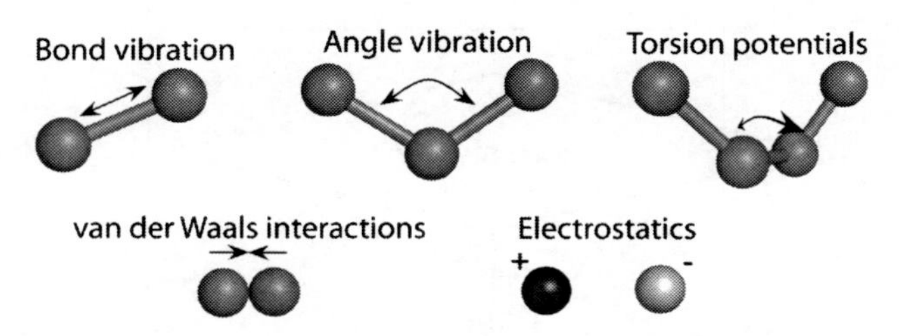

Fig. 2 Examples of interaction functions in modern force fields. Bonded interactions include covalent bond-stretching, angle-bending, torsion rotation around bonds and out-of-plane or “improper” torsions (not shown). Nonbonded interactions are based on neighborlists and consist of Lennard-Jones attraction and repulsion as well as Coulomb electrostatics

decreasing energy, while molecular dynamics is performed by integrating Newton's equations of motion [10]:

$$\mathbf{F}_i = -\frac{\partial V(\mathbf{r}_1, \ldots, \mathbf{r}_N)}{\partial \mathbf{r}_i}$$

$$m_i \frac{\partial^2 \mathbf{r}_i}{\partial t^2} = \mathbf{F}_i$$

The updated coordinates are then used to evaluate the potential energy again, as shown in the flowchart of Fig. 3.

Typical biomolecular simulations use *periodic boundary conditions* to avoid surface artifacts, so that a water molecule that exits to the right reappears on the left; if the box is sufficiently large, the molecules will not interact significantly with their

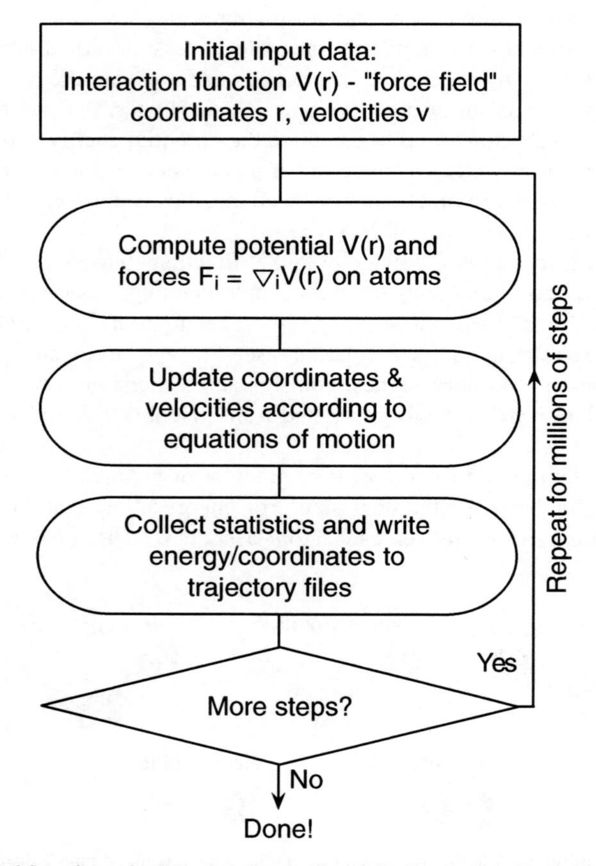

Fig. 3 Simplified flowchart of a typical molecular dynamics simulation. The basic idea is to generate structures from a natural ensemble by calculating potential functions and integrating Newton's equations of motion; these structures are then used to evaluate equilibrium properties of the system. A typical time step is on the order of 1 or 2 fs!

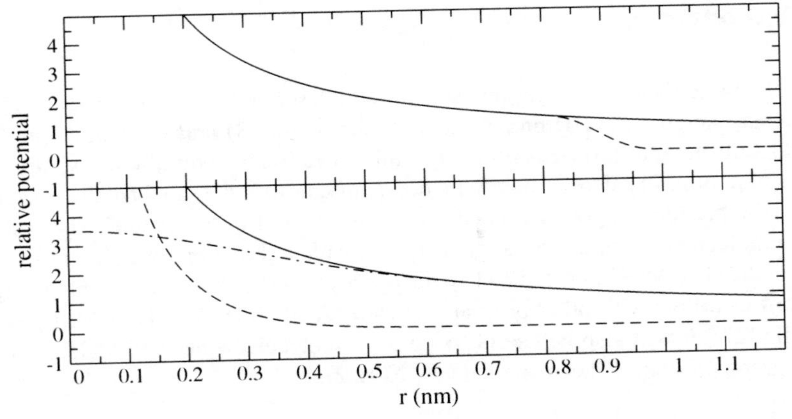

Fig. 4 Alternatives to a sharp cut-off for nonbonded coulomb interactions. *Top:* By switching off the interaction (*dashed*) before the cut-off, the force will be the exact derivative of potential, but the derivative (and, thus, the force) will unnaturally *increase* just before the cutoff. *Bottom:* PME is an amazing algorithm in which the coulomb interaction (*solid*) is divided into a short-range term that is evaluated within a cut-off (*dashed*), and a long-range term that can be solved exactly in reciprocal space with Fourier transforms (*dot-dash*)

periodic copies. This is intimately related to the nonbonded interactions, which ideally should be summed over all neighbors in the resulting infinite periodic system. Simple cut-offs can work for Lennard-Jones interactions that decay very rapidly, but, for Coulomb interactions, a sudden cut-off can lead to large errors. One alternative is to "switch off" the interaction before the cut-off, as shown in Fig. 4, but a better option is to use particle mesh Ewald summation (PME) to calculate the infinite electrostatic interactions by splitting the summation into short- and long-range parts [11]. For PME, the cut-off only determines the balance between the two parts, and the long-range part is treated by assigning charges to a grid that is solved in reciprocal space through Fourier transforms.

Cut-offs and rounding errors can lead to drifts in energy, which will cause the system to heat up during the simulation. To control this, the system is normally coupled to a thermostat that scales velocities during the integration to maintain room temperature. Similarly, the total pressure in the system can be adjusted through scaling the simulation box size, either isotropically or separately in x, y, and z dimensions.

The single most demanding part of simulations is the computation of nonbonded interactions, because millions of pairs have to be evaluated for each time step. Extending the time step is, thus, an important way to improve simulation performance, but, unfortunately, errors are introduced in bond vibrations already at 1 fs. However, in most simulations, the bond vibrations are not of interest per se, and can be removed entirely by introducing *bond constraint* algorithms such as SHAKE [12] or LINCS [13]. Constraints make it possible to extend time steps to 2 fs, and fixed-length bonds are likely better approximations of the quantum mechanical grounds state than harmonic springs.

3 Methods

With the basic theory covered, this section will describe how to: 1) choose and obtain a starting structure, 2) prepare it for a simulation, 3) create a simulation box, 4) add solvent water, 5) perform energy minimization, 6) equilibrate the structure with simulation, 7) perform the production simulation, and 8) analyze the trajectory data. To reproduce it, you will need access to a Unix/Linux machine (see **Note 1**) with a molecular dynamics package installed. Although the options and files below refer to the GROMACS program [14], the description should be reasonably straightforward to follow with other programs such as AMBER [15], CHARMM [16], or NAMD [17]. It will also be useful to have the molecular viewer PyMOL [18] and Unix graph program Grace installed (see **Note 2**).

3.1 Obtaining a Starting Structure

Lysozyme is a 164-residue protein with antibiotic effect first described by Alexander Fleming [19], and one of the first biomolecular structures to be determined [20]. There are plenty of lysozyme structures in the Protein Data Bank (PDB; http://www.pdb.org), but many are bound to special compounds or determined at special conditions such as high pressure. Choose the entry 1LYD with 2-Å resolution [21], and download it as 1LYD.pdb (see **Note 3**). Figure 5 shows a cartoon

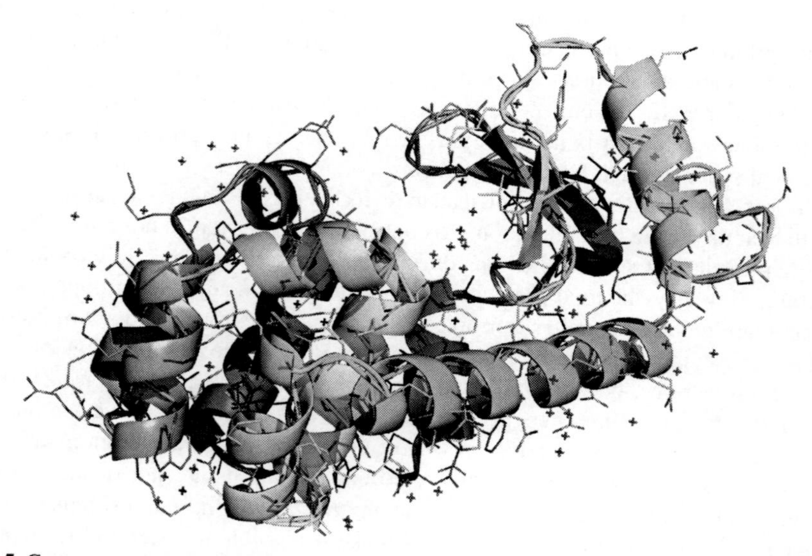

Fig. 5 Cartoon representation of the lysozyme structure 1LYD from PDB, with side chains shown as sticks. Including hydrogens, the protein contains almost 2,900 atoms. Ray-traced image generated with PyMOL

representation of this structure; the small crosses are crystal water oxygen atoms visible in the x-ray experiment (see **Note 4**).

3.2 Preparation of Input Data

In addition to the coordinates and velocities that change each step, simulations also need a static description of all atoms and interactions in the system, called *topology*. In GROMACS, this is created from the PDB structure by the program pdb2gmx, which also adds all of the hydrogen atoms that are not present in x-ray structures. For this example, we will work with the OPLS-AA force field, the TIP3P [22] water model (see **Note 5**), and accept the default choices for all residue protonation states, termini, disulfide bridges, etc. The command to use is then:

```
pdb2gmx –f 1LYD.pdb –water tip3p
```

You will be prompted for the force field (select OPLS), and the command will produce three files: `conf.gro` contains coordinates with hydrogens, `topol.top` is the topology, and `posre.itp` contains a list of position restraints that will be used shortly. For all of these programs, you can use the `–h` flag for help and a detailed list of options (see **Note 6**).

3.3 Creating a Simulation Box

The default box is taken from the PDB crystal cell, but a simulation in water requires something larger. The box size is a trade-off, however: volume is proportional to the box side cubed, and more water means the simulation is slower. The easiest option is to place the solute in the center of a cube, with greater than 0.5 nm to the box sides. The drawback with this is that a cube wastes volume in the corners—the ideal case would a sphere, but, as mentioned in the theory section, we also require periodic boundary conditions, which excludes spheres. There are, however, periodic cells, such as a *truncated octahedron* or *rhombic dodecahedron* that are more spherical than a cube (see **Note 7**). This is far from trivial to see in three dimensions, but Fig. 6 shows how a *hexagonal* cell similarly is more efficient than a square in two dimensions (very useful for membrane simulations). The box creation is accomplished with:

```
editconf –f conf.gro –bt dodecahedron –d 0.5 –o box.gro
```

where the distance (`–d`) flag automatically centers the protein in the box, and the new conformation is written to the file `box.gro` (see **Note 8**).

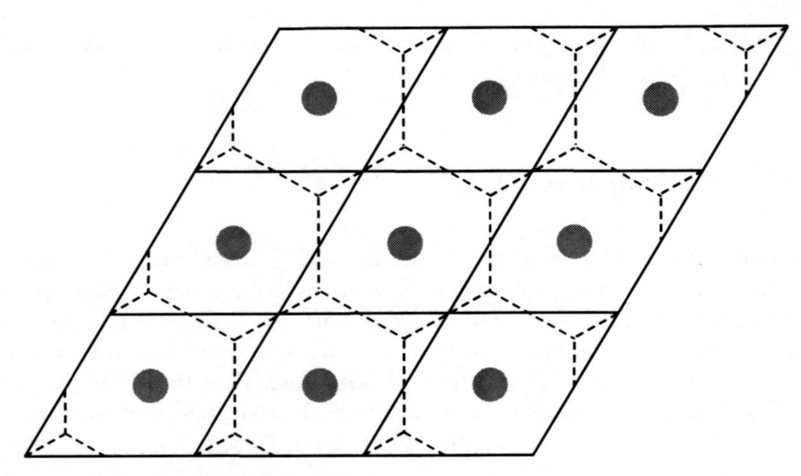

Fig. 6 Two-dimensional example of how a hexagonal box leads to lower volume than a square box, with the same separation distance

3.4 Adding Solvent Water

The last step before the simulation is to add water in the box to solvate the protein. This is performed by using a small pre-equilibrated system of water coordinates that is repeated over the box, with overlapping water molecules removed. The lysozyme system will require roughly 6,000 water molecules, which increases the number of atoms significantly (from 2,900 to more than 20,000). GROMACS does not use a special pre-equilibrated system for TIP3P water because water coordinates can be used with any model—the actual parameters are stored in the topology and force field. In GROMACS, a suitable command to solvate the new box would be:

```
genbox −cp box.gro −cs spc216.gro −p topol.top \
−o solvated.gro
```

The backslash means that the entire command should be written on a single line. Solvent coordinates (`−cs`) are taken from an SPC water system [23], and the `−p` flag adds the new water to the topology file. The resulting system is illustrated in Fig. 7 (see **Note 9**).

3.5 Energy Minimization

The added hydrogens and broken hydrogen bond network in water would lead to very large forces and structure distortion if molecular dynamics was started immediately. To remove these forces, it is necessary to first run a short energy minimization. The aim is not to reach any local energy minimum, therefore, 5,000 steps of *steepest descent* (as mentioned in the theory section) works very well as a stable rather than

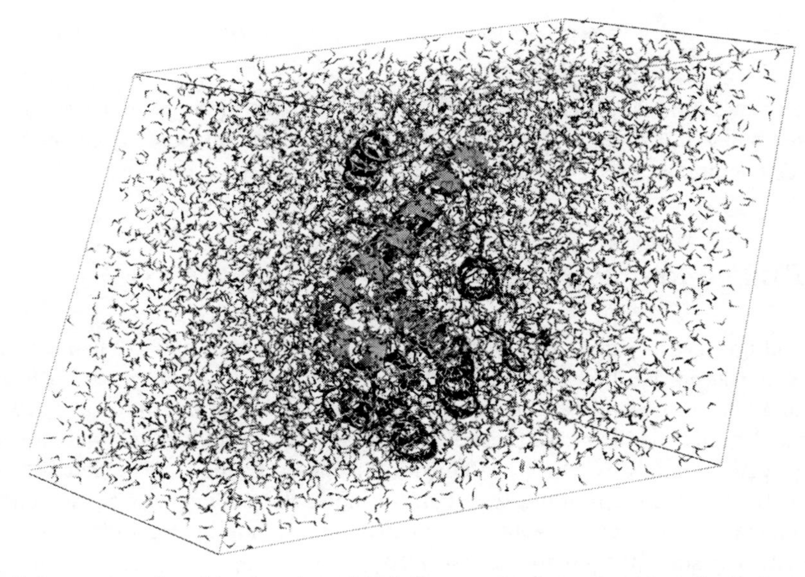

Fig. 7 Lysozyme solvated in water in a triclinic box representing a rhombic dodecahedron (30% lower volume than a cube)

maximally efficient minimization. Nonbonded interactions and other settings are specified in a parameter file (em.mdp); it is only necessary to specify parameters where we deviate from the default value, for example (also included on the CD-ROM):

```
 ------em.mdp------
integrator  = steep
nsteps      = 5000
nstlist     = 10
rlist       = 1.0
coulombtype = pme
rcoulomb    = 1.0
vdw-type    = cut-off
rvdw        = 1.0
nstenergy   = 10
 ------------------
```

Note 10 contains a more detailed description of these settings. GROMACS uses a separate preprocessing program, grompp, to collect parameters, topology, and coordinates into a single run input file (em.tpr) from which the simulation is started (this makes it easier to move it to a separate supercomputer). These two commands are:

```
grompp -f em.mdp -p topol.top -c solvated.gro -o em.tpr
mdrun -v -deffnm em
```

The `-deffnm` is a smart shortcut that uses "`em`" as the base filename for all options, but with different extensions. The minimization takes approximately 10 minutes to complete (see **Note 11**).

3.6 Position-Restrained Equilibration

To avoid unnecessary distortion of the protein when the molecular dynamics simulation is started, we first perform a 100-ps equilibration run in which all heavy protein atoms are restrained to their starting positions (using the file `posre.itp` generated earlier) while the water is relaxing around the structure. As covered in the theory section, bonds will be constrained to enable 2-fs time steps. Other settings are identical to energy minimization, but, for molecular dynamics, we also control the temperature and pressure with the Berendsen weak coupling algorithm [24] (see **Note 12**). The settings used are (see **Note 13**):

```
------pr.mdp------
integrator       = md
nsteps           = 50000
dt               = 0.002
constraints      = all-bonds
nstlist          = 10
rlist            = 1.0
coulombtype      = pme
rcoulomb         = 1.0
vdw-type         = cut-off
rvdw             = 1.0
tcoupl           = Berendsen
tc-grps          = protein non-protein
tau-t            = 0.1 0.1
ref-t            = 298 298
Pcoupl           = Berendsen
tau-p            = 1.0
compressibility  = 5e-5 5e-5 5e-5 0 0 0
ref-p            = 1.0
nstenergy        = 100
define           = -DPOSRES
------------------
```

For a small protein such as lysozyme, 100 ps (50,000 steps) should be more than enough for the water to equilibrate around the protein, but, in a large membrane system, the slow lipid motions can require several nanoseconds of relaxation. The only way to know for certain is to watch the potential energy, and to extend the equilibration until it has converged. Running this equilibration in GROMACS, you execute:

```
grompp -f pr.mdp -p topol.top -c em.gro -o pr.tpr
mdrun -v -deffnm pr
```

This mdrun invocation will take 1 to 2 h to finish on a normal workstation.

3.7 Production Runs

The difference between equilibration and production run is minimal: the position restraints and pressure coupling are turned off, we decide how often to write output coordinates to analyze (say, every 1,000 steps), and start a significantly longer simulation. How long depends on what you are studying, and that should be decided before starting any simulations! For decent sampling, the simulation should be at least 10 times longer than the phenomena you are studying, which, unfortunately, sometimes conflicts with reality and available computer resources. We will perform a 10-ns simulation (5 million steps), which should take approximately 1 week on a modern workstation. If you are not that patient, you can choose a shorter simulation just to get an idea of the concepts, and the analysis programs in the next section can read the simulation output trajectory as it is being produced.

```
------run.mdp------
integrator  = md
nsteps      = 5000000
dt          = 0.002
constraints = all-bonds
nstlist     = 10
rlist       = 1.0
coulombtype = pme
rcoulomb    = 1.0
vdw-type    = cut-off
rvdw        = 1.0
tcoupl      = Berendsen
tc-grps     = protein non-protein
tau-t       = 0.1 0.1
```

```
ref-t       = 298 298
nstxout     = 100000
nstvout     = 100000
nstxtcout   = 1000
nstenergy   = 1000
------------------
```

Storing full precision coordinates and velocities every 100,000 steps enables restart if runs crash (power outage, full disk, etc.). The analysis only uses the compressed coordinates stored every 10,000 steps. Perform the production run as:

```
grompp -f run.mdp -p topol.top -c pr.gro -o run.tpr
mdrun -v -deffnm run
```

The production run will use roughly 200 MB of hard disk space for the output data.

3.8 Trajectory Analysis

3.8.1 Deviation from X-Ray Structure

One of the most important fundamental properties to analyze is whether the protein is stable and close to the experimental structure. The standard way to measure this is the root mean square displacement (RMSD) of all heavy atoms with respect to the x-ray structure. GROMACS has a finished program to do this, as:

```
g_rms -s em.tpr -f run.xtc
```

Note that the reference structure here is taken from the input before energy minimization. The program will prompt for both the fit group, and the group for which to calculate RMSD—choose "`Protein-H`" (protein except hydrogens) for both. The output will be written to `rmsd.xvg`, and, if you installed the Grace program, you will directly get a finished graph with:

```
xmgrace rmsd.xvg
```

The RMSD is also illustrated in Fig. 8. It increases rapidly in the first part of the simulation, but stabilizes around 0.19 nm, roughly the resolution of the x-ray structure. The difference is partly caused by limitations in the force field, but also because atoms in the simulation are moving and vibrating around an equilibrium structure. A better measure can be obtained by first creating a running average structure (see **Note 14**) from the simulation and comparing the running average with the x-ray structure, which gives a more realistic RMSD around 0.16 nm (see **Note 15**).

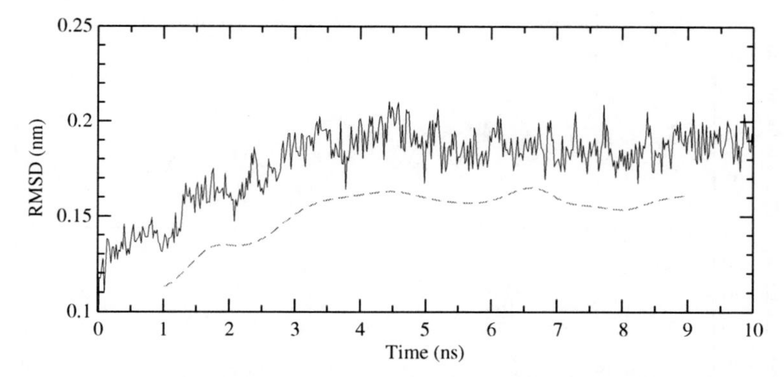

Fig. 8 Instantaneous RMSD of all heavy atoms in lysozyme during the simulation (*solid*), relative to the crystal structure. To a large extent, atoms are vibrating around an equilibrium, therefore the RMSD of a 1-ns running average structure (*dashed*) is a better measure

3.8.2 Comparing Fluctuations with Temperature Factors

Vibrations around the equilibrium are not random, but depend on local structure flexibility. The root mean square fluctuation (RMSF) of each residue is straightforward to calculate over the trajectory, but, more important, they can be converted to *temperature factors* that are also present for each atom in a PDB file. Once again, there is a program that will do the entire job:

```
g_rmsf -s run.tpr -f run.xtc -o rmsf.xvg -oq bfac.pdb
```

You can use the group "C-alpha" to get one value per residue. Figure 9 displays both the residue RMSF from the simulation (`xmgrace rmsf.xvg`), as well as the calculated and experimental temperature factors. The overall agreement is very good, which lends further credibility to the accuracy and stability of the simulation.

3.8.3 Secondary Structure

Another measure of stability is the protein secondary structure. This can be calculated for each frame with a program such as DSSP [25]. If the DSSP program is installed and the environment variable DSSP points to the binary (see **Note 16**), the GROMACS program `do_dssp` can create time-resolved secondary structure plots. Because the program writes output in a special xpm (X pixmap) format, you probably also need the GROMACS program xpm2ps to convert it to postscript:

```
do_dssp -s run.tpr -f run.xtc
xpm2ps -f ss.xpm -o ss.eps
```

Use the group "`protein`" for the calculation. Figure 10 shows the resulting output in grayscale, with some unused formatting removed. The DSSP secondary

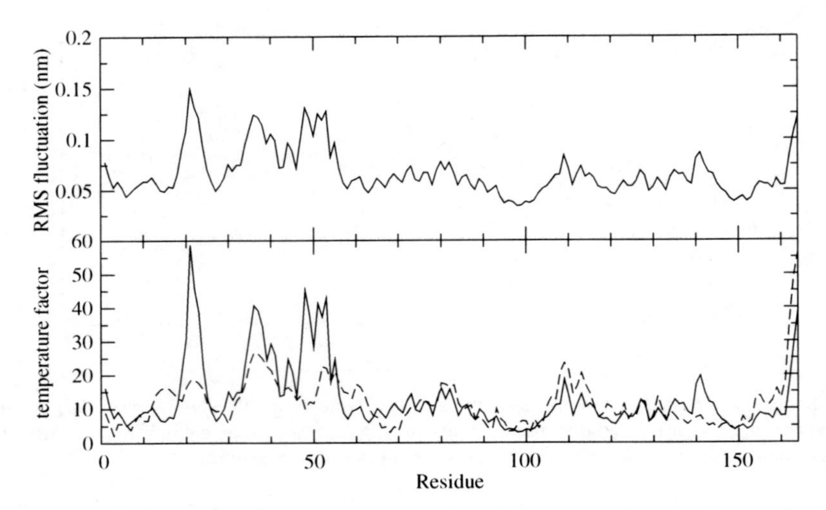

Fig. 9 *Top*: RMSF of residue coordinates in the simulation. *Bottom*: The fluctuations can be converted to x-ray temperature factors (*solid*), which agree well with the experimental B-factors from the PDB file (*dashed*)

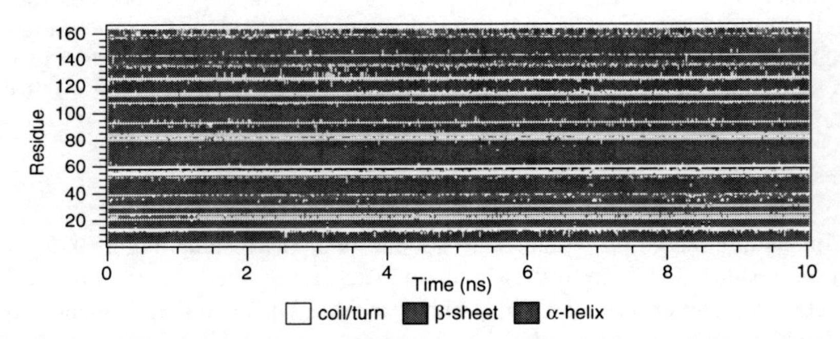

Fig. 10 Local secondary structure in lysozyme as a function of time during the simulation, according to the DSSP definition. Note how helices sometimes are unrolled slightly at the start and end, but the overall structure is very stable over 10 ns

structure definition is tight, therefore, it is normal for residues to fluctuate around the well-defined state, in particular at the ends of helices or sheets. For a (long) protein-folding simulation, a DSSP plot would show how the secondary structures form during the simulation.

3.8.4 Distance and Hydrogen Bonds

With basic properties accurately reproduced, we can use the simulation to analyze more specific details. As an example, lysozyme seems to be stabilized by hydrogen bonds between the residues GLU22 and ARG137, therefore, how much does this

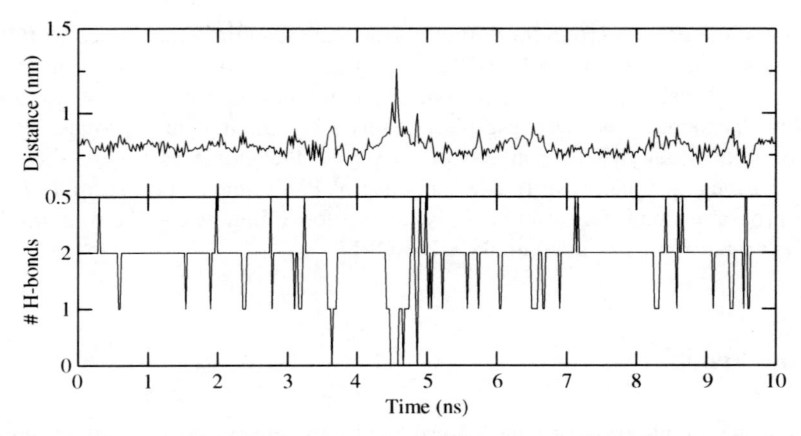

Fig. 11 *Top:* Distance between GLU22 and ARG137 residues as a function of simulation time. *Bottom:* Number of hydrogen bonds between GLU22 and ARG137

fluctuate in the simulation, and are the hydrogen bonds intact? To determine this, first create an index file with these groups as:

```
make_ndx -f run.gro
```

At the prompt, create a group for GLU22 with "`r 22`," ARG137 with "`r 137`," and then "`q`" to save an index file as `index.ndx`. The distance and number of hydrogen bonds can now be calculated with the commands:

```
g_dist -s run.tpr -f run.xtc -n index.ndx -o dist.xvg
g_hbond -s run.tpr -f run.xtc -n index.ndx -num hbnum.xvg
```

In both cases, you should select the two groups you just created. Figure 11 shows the results merged into a single plot. Two hydrogen bonds are present almost throughout the simulation, and, in a few frames, there is even a third bond formed, likely because we included the backbone atoms in the residue groups (see **Note 17**).

3.8.5 Making a Movie

A normal movie uses approximately 30 frames/s, therefore a 10-s movie requires 300 simulation trajectory frames. To make a smooth movie, the frames should not be more 1- to 2-ps apart, or it will seem to shake nervously (see **Note 18**). Export a short trajectory from the first 500 ps in PDB format (readable by PyMOL) as:

```
trjconv -s run.tpr -f run.xtc -e 500.0 -o movie.pdb
```

Choose the protein group for output rather than the entire system (see **Note 19**). If you open this trajectory in PyMOL as "`PyMOL movie.pdb`," you can immediately play it using the VCR-style controls on the bottom right, adjust visual settings in the menus, and even use photorealistic ray tracing for all images in the movie. With MacPyMOL, you can directly save the movie as a quicktime file; and, on Linux, you can save it as a sequence of PNG images for assembly in another program. Rendering a movie only takes a few minutes, and the final product `lysozyme.mov` is included on the CD-ROM.

4 Conclusions

This chapter should provide a basic introduction to general simulations. An important lesson is that high-quality simulations require a lot of care from the user—just as with experimental techniques, the entire result can be ruined by a single sloppy step. Further, even simulations using empirical force fields are still very limited in the range of timescales accessible, but recent techniques based on distributed computing and Markovian state models have been able to probe dynamics in the millisecond range without extending individual simulations to those scales [30]. Although simulations are advancing rapidly because of the continuous development of faster computers, the field has also been plagued by (published) simulations that have not advanced our knowledge either of simulation methods or biomolecules. Instead of just starting a simulation and hoping for something to happen, you should decide beforehand what you want to study, estimate the timescales necessary or see whether it can be accomplished with more advanced methods (e.g., free energy calculations), and not start simulations until you are fairly confident both regarding sampling, analysis required, and the force field accuracy. Used with caution, molecular dynamics is an amazingly powerful tool, and a great complement to experiments.

5 Notes

1. Computer hardware changes extremely rapidly, but, nevertheless, there are a couple of general recommendations that have been roughly constant during the last couple of years. First, no matter which molecular dynamics program you would like to use, they have all been developed for Unix/Linux, so give up the idea of running it under the Windows operating system—it is simply not worth the hassle. Standard Intel or AMD PC hardware provides the best price to performance ratio, for instance, a Dell or HP workstation. If you are serious about simulations, it is worth investing in a dual dual-core machine (total of 4 processors) to run in parallel later, and the performance will essentially be proportional to the speed of the processors. Most vendors will offer a bundled

expensive Linux distribution, but the best option is really to order the machine without an operating system and the download the free CentOS ("community enterprise OS") from http://www.centos.org. This Linux distribution is built from the same sources as RedHat Linux (completely legal, it is open source) and there is both free and commercial support available if you need it. Accessories such as graphics card, large amounts of memory, and hard disks do not affect your performance, therefore you can go very cheap there. If you are hesitant about installing Linux, purchase an Apple Macintosh instead (MacPro for high performance). Apple OS X is a real Unix operating system underneath, therefore you can open a terminal window for your simulation work while still having commercial support and a nice user interface—and, in addition, they make *great* workstations when it comes to creating movies! The free *gcc* C compiler is very good (for OS X it is part of the free developer tools), but if any programs you use require a Fortran compiler you should invest $100 or so in the commercial Intel *ifort* compiler for x86—it can be 50 to 100% faster than the free *gfortran*, which adds up in the long run. Intel also has a good C compiler (*icc*) and mathematical libraries (*mkl*). Most of the molecular simulation packages are free or at least very cheap for academic use, therefore you can try several of them, although we have to point you to the program documentation for details on how to compile, install, and use them.

2. GROMACS is freely available from http://www.gromacs.org. It should be very easy to install using the step-by-step instructions, and, for most common platforms, there are finished binary packages (installation might require root access, however). PyMOL is distributed from http://www.PyMOL.org, with binaries for Windows, Linux, and Mac OS X. The MacPyMOL version requires a license after a trial period, but is very much recommended for the better movie export capabilities. Unfortunately, the Grace package is not as trivial to install. The distribution site http://plasma-gate.weizmann.ac.il/Grace/ only provides source code, so you might want to use Google to search for a binary for your platform. Linux RPMs can often be found at http://www.rpmfind.net. Grace uses Motif X11 library, but it compiles with the open source clone, LessTif, http://www.lesstif.org.
3. For this tutorial, most of the other lysozyme structures would have also been acceptable, but some of the more esoteric bound compounds can be difficult to model automatically, both in GROMACS and in other programs. It is often a good idea to look at the structure in PyMOL, and read the text information at the top of the PDB file to see whether there are any special issues. For 1LYD, the header mentions that residues 162 to 164 were not visible in the electron density map, and they have been modeled in. If large parts of the protein are inaccurate, it might be better to choose a different structure.
4. Sometimes people remove the crystal water to replace it with their own solvent later, but this is usually a bad idea. The reason why they are visible is that these waters are tightly bound to the structure and often form salt bridges, therefore, if they are discarded, the structure might distort before the new solvent has a chance to equilibrate in these positions. Keep the crystal water!

5. Water is a very special liquid, and actually very difficult to model accurately. However, biomolecular simulations usually focus on the protein, DNA, etc., and, thus, normally prefer cheap and simple approximate solvent models to the most accurate one. The most common such models are SPC [23] (used with the GROMOS96 force field) and TIP3P [22] (OPLS and Amber force fields), which both represent the water as an entirely rigid molecule with three sites (one oxygen and two hydrogens). There are a couple of modified models, such as SPC/E, that improve bulk properties, but the standard models are often preferred for interface systems such as membranes. TIP4P [26] is a smart model with a fourth interaction site offset from the oxygen, and still reasonably cheap computationally (recommended), whereas TIP5P [27], with five interaction sites, is too expensive for most simulations.
6. `pdb2gmx` can be somewhat picky with the input structures, but that is usually a good thing—it will, for instance, not accept proteins with missing heavy atoms. If that happens, the best option is to find a better structure, and if that is not possible, you can try to build the missing parts with a program such as Modeller (http://salilab.org/modeller/). However, if you have to build more than a handful of residues, it is doubtful that the resulting structure is accurate enough to simulate. For 1LYD, `pdb2gmx` will also issue a warning regarding net charge, but that is fine. In general, all GROMACS program try to do both double and triple checking of your input, so if you do not get any warning, you can be fairly confident regarding the correctness of your input.
7. The volume of a rhombic dodecahedron is approximately 71% of a cube with the same spacing, and, for a truncated octahedron, it is 77%. These differences can seem small, but 30% is very significant when simulations use weeks of supercomputer time, and it is a free lunch after all! If you are working with a program that does not support these shapes, it is also acceptable to use a cubic box.
8. All GROMACS programs that write coordinates support a number of different output formats. The default format is `.gro`, simply because it also has support for velocities, but if you want a PDB file to view, e.g., in PyMOL, you simply change the output file extension to `.pdb`.
9. To compensate for the +8 charge on lysozyme, we could also add eight chloride counter ions at this stage. The GROMACS program `genion` can replace selected waters with positive/negative ions, but for brevity we will skip this step here—ions can take a long time to equilibrate.
10. We choose a standard cut-off of 1.0 nm, both for the neighborlist generation and the coulomb and Lennard-Jones interactions. `nstlist=10` means it is updated at least every 10 steps, but for energy minimization it will usually be every step. Energies and other statistical data are stored every 10 steps (`nstenergy`), and we have chosen the more expensive PME for electrostatic interactions. The treatment of nonbonded interactions frequently borders on religion. One camp advocates that standard cutoffs are fine, another camp swears by switched-off interactions, whereas the third camp would not even consider anything but PME. One argument in this context is that 'true' interactions should conserve

energy, which is violated by sharp cut-offs, because the force is no longer the exact derivative of the potential. On the other hand, just because an interaction conserves energy does not mean that it describes nature accurately. In practice, the difference is most pronounced for systems that are very small or with large charges, but the key lesson is really that it is a trade off. PME is great, but also clearly slower than cut-offs. Longer cut-offs are always better than short ones (but slower), and although switched interactions improve energy conservation, they introduce artificially large forces. Using PME is the safe option, but if that is not fast enough, it is worth investigating reaction-field or cut-off interactions. It is also a good idea to check and follow the recommended settings for the force field used.

11. The `mdrun` program will write several output files: `em.edr` is an "energy file" with statistical data (energies, temperature, pressure, etc.). `em.trr` is a trajectory with full coordinates and velocities of the system during the run, and `em.log` is a log file. Depending on the parameters (disabled here), it might also write a compressed trajectory with low-precision coordinates only, `em.xtc`.
12. Berendsen weak coupling provides the most efficient control of both temperature and pressure, at the cost of violating the statistical mechanics ensemble. When that is important, Nose-Hoover thermostats [28] and Parinello-Rahman barostats [29] are better choices, but also more sensitive during equilibration. Berendsen coupling guarantees exponential relaxation to the correct temperature and pressure, with the provided time constraints. For temperature, 0.1 ps is a reasonable choice, whereas pressure scaling should be an order of magnitude slower (1 ps). In principle, we need to know the compressibility for pressure scaling, but, because it only affects the relaxation time, we can use water values, roughly 5×10^{-5} bar^{-1}.
13. For molecular dynamics simulations, the integrator has now been changed to "md," and all bonds are constrained to enable 2-fs time steps. Temperature coupling has been enabled for protein and water separately (to avoid heating the water more than the protein or vice versa), with a 298 K reference temperature. The compressibility is really a symmetric tensor, and, by setting the last three elements (off-diagonal) to 0, we disable any box shear deformation. The last line causes `grompp` to include the position restraint file `posre.itp` generated by `pdb2gmx`, which turns on position restraints.
14. The easiest way to create a running average in GROMACS is to use the `g_filter` program. The command "`g_filter -nf 50 -all -s run.tpr -f run.xtc -ol lowpass.xtc`" will create a low-pass version of the trajectory (cosine averaging over 50 frames), which then can be used as modified input file to the `g_rms` program.
15. If the RMSD is significantly higher than this, or continuously increasing, there is likely something very wrong. Start over with the PDB file, read the headers carefully, and make sure the starting structure is accurate. In the next step, check the different energy terms and RMSD change both during minimization and position restraints. You can also use the `-posrefc` flag with `pdb2gmx` to increase the strength of the position restraints, and extend the equilibration run.

16. The DSSP program can be obtained from http://swift.cmbi.ru.nl/gv/dssp/. It is free for academic users, but requires a signed license to be submitted. Compile the binary and install it, e.g., in /usr/local/bin, and set the environment variable with a command such as "`export DSSP=/usr/local/bin/dssp`" (bash shell). If you do not have access to the real DSSP program, you can download the GROMACS source and try to build our homegrown version in src/contrib/my_dssp, but be warned that it might not produce identical results.
17. Modern force fields no longer use special hydrogen bond interactions, partly because it is not necessary and partly because it is difficult to track formation and breaking of hydrogen bonds separately. "Hydrogen bonds" are, therefore, defined from geometric criteria, typically, that the distance between the donor and acceptor atoms should be smaller than 0.35 nm, and the angle donor–acceptor hydrogen should be below 30 degrees.
18. To visualize slower phenomena such as protein folding, you can use `g_filter` to smooth out motions in longer trajectories. In some cases, this can lead to strange artifacts, e.g., when averaging torsion rotation around a bond, but it is usually better than taking raw trajectory frames with too large spacing.
19. PyMOL loads all frames of the trajectory into memory, therefore, if the water molecules are included, it will likely run out of memory when creating graphical representations for more than 20,000 atoms repeated in 250 frames. Trajectories restricted to the protein part can, thus, be much longer.

References

1. Alder, B.J. and Wainwright, T.E. (1957) Phase transition for a hard sphere system. *J. Chem. Phys.* **27**, 1208–1209.
2. Rahman, A. and Stillinger, F.H. (1971) Molecular dynamics study of liquid water. *J. Chem. Phys.* **55**, 3336–3359.
3. McCammon, J.A. and Karplus, M. (1977) Internal motions of antibody molecules. *Nature* **268**, 765–766.
4. Allen, M.P. and Tildesley, D.J. (1989) *Computer Simulation of Liquids*. Clarendon Press, New York, NY.
5. Frenkel, D. and Smit, B. (2001) *Understanding Molecular Simulation*. Academic Press, New York, NY.
6. Kaminski, G.A., Friesner, R.A., Tirado-Rives, J., and Jorgensen, W.L. (2001) Evaluation and reparametrization of the OPLS-AA force field for proteins via comparison with accurate quantum chemical calculations on peptides. *J. Phys. Chem. B* **105**, 6474–6487.
7. MacKerell, A.D. Jr, et al. (1998) All-atom empirical potential for molecular modeling and dynamics Studies of proteins. *J. Phys. Chem. B* **102**, 3586–3616.
8. Oostenbrink, C., Villa, A., Mark, A.E., and van Gunsteren, W.F. (2004) A biomolecular force field based on the free enthalpy of hydration and solvation: the GROMOS force-field parameter sets 53A5 and 53A6. *J. Comput. Chem.* **25**, 1656–1676.
9. Wang, J., Cieplak, P., and Kollman, P.A. (2000) How well does a restrained electrostatic potential (RESP) model perform in calculating conformational energies of organic and biological molecules? *J. Comput. Chem.* **21**, 1049–1074.

10. Chandler, D. (1987) *Introduction to Modern Statistical Mechanics*. Oxford University Press, New York, NY.
11. Essman, U., Perera, L., Berkowitz, M., Darden, T., Lee, H., and Pedersen, L.G. (1995) A smooth particle mesh Ewald method. *J. Chem. Phys.* **103**, 8577–8593.
12. Ryckaert, J.P, Ciccotti, G., and Berendsen, H.J.C. (1977) Numerical ;ntegration of the cartesian equations of motion of a system with constraints; molecular dynamics of n-alkanes. *J. Comp. Phys.* **23**, 327–341.
13. Hess, B., Bekker, H., Berendsen, H.J.C., and Fraaije, J.G.E.M. (1998) LINCS: A linear constraint solver for molecular simulation. *J. Comput. Chem.* **18**, 1463–1472.
14. Lindahl, E., Hess, B., and van der Spoel, D. (2001) GROMACS 3.0: A package for molecular simulation and trajectory analysis. *J. Mol. Model.* **7**, 306–317.
15. Case, D.A., et al. (2005) The Amber biomolecular simulation programs. *J. Comput. Chem.* **26**, 1668–1688.
16. Brooks, B.R., et al. (1983) CHARMM: A program for macromolecular energy, minimization, and dynamics calculations. *J. Comput. Chem.* **4**, 187–217.
17. Phillips, J.C., et al. (2005). Scalable molecular dynamics with NAMD. *J. Comput. Chem.* **26**, 1781–1802.
18. DeLano, W.L. (2002) The PyMOL Molecular Graphics System. DeLano Scientific, San Carlos, CA. http://www.PyMOL.org.
19. Fleming, A. (1922) On a remarkable bacteriolytic element found in tissues and secretions. *Proc. Royal Soc. Ser. B* **93**, 306–317.
20. Blake, C.C., Koenig, D.F., Mair, G.A., North, A.C., Phillips, D.C., and Sarma, V.R. (1965) Structure of hen egg-white lysozyme. A three-dimensional Fourier synthesis at 2 Ångstrom resolution. *Nature*, **206**, 757–761.
21. Rose, D.R., Phipps, J., Michniewicz, J., Birnbaum, G.I., Ahmed, F.R., Muir, A., Anderson, W.F., and Narang, S. (1988) Crystal structure of T4-lysozyme generated from synthetic coding DNA expressed in Escherichia coli. *Protein Eng.* **2**, 277–282.
22. Jorgensen, W.L., Chandrasekhar J., Madura, J.D., Impey, R.W., and Klein, M.L. (1983) Comparison of simple potential functions for simulating liquid water. *J. Chem. Phys.* **79**, 926–935.
23. Berendsen, H.J.C, Postma, J.P.M., and van Gunsteren, W.F. (1981) Interaction models for water in relation to protein hydration, in *Intermolecular Forces* (Pullman, B., ed.), D. Reidel Publishing Company, Dordrecht, Germany, pp. 331–342.
24. Berendsen, H.J.C., Postma, J.P.M., DiNola, A., and Haak, J.R. (1984) Molecular dynamics with coupling to an external bath. *J. Chem. Phys.* **81**, 3684–3690.
25. Kabsch, W. and Sanders, C. (1983) Dictionary of protein secondary structure: Pattern recognition of hydrogen-bonded and geometrical features, *Biopolymers* **22**, 2577–2637.
26. Jorgensen, W.L. and Madura, J.D. (1985) Temperature and size dependence for Monte Carlo simulations of TIP4P water. *Mol. Phys.* **56**, 1381–1392.
27. Mahoney, M.W. and Jorgensen, W.L. (2000). A five-site model for liquid water and the reproduction of the density anomaly by rigid, nonpolarizable potential functions. *J. Chem. Phys.* **112**, 8910–8922.
28. Nosé, S. (1984) A molecular dynamics method for simulations in the canonical ensemble. *Mol. Phys.* **52**, 255–268.
29. Parrinello, M. and Rahman, A. (1981) Polymorphic transitions in single crystals: A new molecular dynamics method. *J. Appl. Phys.* **52**, 7182–7190.
30. Kasson, P., Kelley, N., Singhal, N., Vrjlic, M., Brunger, A., and Pande, V.S. (2006) Ensemble molecular dynamics yields submillisecond kinetics and intermediates of membrane fusion. *Proc. Natl. Acad. Sci.* **103**, 11916–11921.

Chapter 2
Monte Carlo Simulations

David J. Earl and Michael W. Deem

Summary A description of Monte Carlo methods for simulation of proteins is given. Advantages and disadvantages of the Monte Carlo approach are presented. The theoretical basis for calculating equilibrium properties of biological molecules by the Monte Carlo method is presented. Some of the standard and some of the more recent ways of performing Monte Carlo on proteins are presented. A discussion of the estimation of errors in properties calculated by Monte Carlo is given.

Keywords: Markov chain · Metropolis algorithm · Monte Carlo · Protein simulation · Stochastic methods

1 Introduction

The term Monte Carlo generally applies to all simulations that use stochastic methods to generate new configurations of a system of interest. In the context of molecular simulation, specifically, the simulation of proteins, Monte Carlo refers to importance sampling, which we describe in Sect. 2, of systems at equilibrium. In general, a Monte Carlo simulation will proceed as follows: starting from an initial configuration of particles in a system, a Monte Carlo move is attempted that changes the configuration of the particles. This move is accepted or rejected based on an *acceptance criterion* that guarantees that configurations are sampled in the simulation from a statistical mechanics ensemble distribution, and that the configurations are sampled with the correct weight. After the acceptance or rejection of a move, one calculates the value of a property of interest, and, after many such moves, an accurate average value of this property can be obtained. With the application of statistical mechanics, it is possible to calculate the equilibrium thermodynamic properties of the system of interest in this way. One important necessary condition for Monte Carlo simulations is that the scheme used must be *ergodic*, namely, every point that is accessible in configuration space should be able to be reached from any other point in configuration space in a finite number of Monte Carlo moves.

From: Methods in Molecular Biology, vol. 443, Molecular Modeling of Proteins
Edited by Andreas Kukol © Humana Press, Totowa, NJ

1.1 Advantages of Monte Carlo

Unlike molecular dynamics simulations, Monte Carlo simulations are free from the restrictions of solving Newton's equations of motion. This freedom allows for cleverness in the proposal of moves that generate trial configurations within the statistical mechanics ensemble of choice. Although these moves may be nontrivial, they can lead to huge speedups of up to 10^{10} or more in the sampling of equilibrium properties. Specific Monte Carlo moves can also be combined in a simulation allowing the modeler great flexibility in the approach to a specific problem. In addition, Monte Carlo methods are generally easily parallelizable, with some techniques being ideal for use with large CPU clusters.

1.2 Disadvantages of Monte Carlo

Because one does not solve Newton's equations of motion, no dynamical information can be gathered from a traditional Monte Carlo simulation. One of the main difficulties of Monte Carlo simulations of proteins in an explicit solvent is the difficulty of conducting large-scale moves. Any move that significantly alters the internal coordinates of the protein without also moving the solvent particles will likely result in a large overlap of atoms and, thus, the rejection of the trial configuration. Simulations using an implicit solvent do not suffer from these drawbacks, and, therefore, coarse-grained protein models are the most popular systems where Monte Carlo methods are used. There is also no general, good, freely available program for the Monte Carlo simulation of proteins because the choice of which Monte Carlo moves to use, and the rates at which they are attempted, vary for the specific problem one is interested in, although we note that a Monte Carlo module has recently been added to CHARMM [1].

2 Theoretical Basis for Monte Carlo

The aim of a Monte Carlo simulation is the accurate calculation of equilibrium thermodynamic and physical properties of a system of interest. Let us consider calculating the average value of some property, $\langle A \rangle$. This could be calculated by evaluating the following:

$$\langle A \rangle = \frac{\int d\mathbf{r}^N \exp\left[-\beta U\left(\mathbf{r}^N\right)\right] A\left(\mathbf{r}^N\right)}{\int d\mathbf{r}^N \exp\left[-\beta U\left(\mathbf{r}^N\right)\right]}, \tag{1}$$

where $\beta = 1/k_{\mathrm{B}}T$, U is the potential energy, and $\mathbf{r}^N$ denotes the configuration of an N particle system (i.e., the positions of all N particles). Now, the probability density of finding the system in configuration $\mathbf{r}^N$ is:

$$\rho\left(\mathbf{r}^N\right) = \frac{\exp\left[-\beta U\left(\mathbf{r}^N\right)\right]}{\int \mathrm{d}\mathbf{r}^N \exp\left[-\beta U\left(\mathbf{r}^N\right)\right]}, \tag{2}$$

where the denominator in Eq. 2 is the configurational integral. If one can randomly generate N_{MC} points in configuration space according to Eq. 2, then Eq. 1 can be expressed as:

$$\langle A \rangle \approx \frac{1}{N_{MC}} \sum_{i=1}^{N_{MC}} A\left(\mathbf{r}_i^N\right). \tag{3}$$

After equilibration of our system of interest, errors in $\langle A \rangle$ scale as $1/\sqrt{N_{MC}}$, as discussed in Sect. 4. The Monte Carlo methods that we outline in Sect. 3 are responsible for generating the N_{MC} points in configuration space in a simulation.

A Monte Carlo algorithm consists of a group of Monte Carlo moves that generate a Markov chain of states. This Markov process has no history dependence, in the sense that new configurations are generated with a probability that depends only on the current configuration and not on any previous configurations. If our system is currently in state m, then the probability of moving to a state n is defined as π_{mn}, where π is the transition matrix. Let us introduce a probability vector, ρ, that defines the probability that the system is in a particular state; ρ_i is the probability of being in state i. The initial probability vector, for a randomly chosen starting configuration, is $\rho^{(0)}$ and the probability vector for subsequent points in the simulation is $\rho^{(j)} = \rho^{(j-1)}\pi$. The equilibrium, limiting distribution of the Markov chain, ρ^*, results from applying the transition matrix an infinite number of times, $\rho^* = \lim_{N_{MC}\to\infty} \rho^{(0)}\pi^{N_{MC}}$. We note:

$$\rho^* = \rho^*\pi. \tag{4}$$

If we are simulating in the canonical ensemble, then ρ^* is reached when ρ_i is equal to the Boltzmann factor (Eq. 4) for all states. Thus, once a system has reached equilibrium, any subsequent Monte Carlo moves leave the system in equilibrium. The initial starting configuration of the system should not matter as long as the system is simulated for a sufficient number of Monte Carlo steps, N_{MC}. For the simulation to converge to the limiting distribution, the Monte Carlo moves used must satisfy the balance condition and they must result in ergodic sampling [2]. If the transition matrix satisfies Eq. 4, then the balance condition is met. For the stricter condition of detailed balance to be satisfied, the net flux between two states must be zero at equilibrium, i.e.:

$$\rho_m \pi_{mn} = \rho_n \pi_{nm}. \tag{5}$$

For all of the Monte Carlo moves we present in Sect. 3, the balance or detailed balance conditions hold.

Now, how is the transition matrix chosen, such that balance or detailed balance is satisfied? In other words, how do we propose a Monte Carlo move and correctly choose whether to accept or reject it? As an example, let us consider the Metropolis acceptance criterion. The transition matrix between states m and n can be written as:

$$\pi_{mn} = \alpha_{mn} p_{mn}, \tag{6}$$

where α_{mn} is the probability of proposing a move between the two states and p_{mn} is the probability of accepting the move. Assuming that $\alpha_{mn} = \alpha_{nm}$, as is the case for many but not all Monte Carlo moves, then Metropolis proposed the following scheme: if the new state (n) is lower in energy than the old state (m), then accept the move (i.e., $p_{mn} = 1$ and $\pi_{mn} = \alpha_{mn}$ for $\rho_n \geq \rho_m$), or, if the new state is higher in energy than the old state, then accept the move with $p_{mn} = \exp\left(-\beta\left[U(n) - U(m)\right]\right) < 1$; i.e., $\pi_{mn} = \alpha_{mn}\left(\rho_n/\rho_m\right)$ for $\rho_n < \rho_m$, where the energy of state i is $U(i)$. If the old and new states are the same, then the transition matrix is given by $\pi_{mm} = 1 - \sum_{n \neq m} \pi_{mn}$. The Metropolis *acceptance criterion* [3,4] can be summarized as:

$$p_{mn} = \min\left\{1, \exp\left(-\beta\left[U(n) - U(m)\right]\right)\right\}. \tag{7}$$

In a computer simulation, for the case where $U(n)$ is greater than $U(m)$, one generates a pseudorandom number with a value between 0 and 1, and if this number is less than p_{mn}, then the trial move is accepted.

Monte Carlo simulations can be conducted in several different statistical mechanics ensembles, and the distribution that we sample from depends on the ensemble. These ensembles include, but are not limited to, the canonical ensemble (constant number of particles N, volume V, and temperature T), the isobaric–isothermal ensemble (constant N, pressure P, and T), the grand canonical ensemble (constant chemical potential μ, V, T), and the semigrand canonical ensemble. For simulations of proteins, the modeler is unlikely to stray too often from the canonical ensemble where the partition function is:

$$Q(N, V, T) \equiv \prod_{i=1}^{N} \frac{1}{\Lambda_i^3 N!} \int d\mathbf{r}^N \exp\left[-\beta U\left(\mathbf{r}^N\right)\right], \tag{8}$$

where Λ_i is the thermal de Broglie wavelength, $\Lambda_i = \sqrt{h^2/\left(2\pi m_i k_B T\right)}$ and m_i is the mass of particle i. A thorough description of Monte Carlo simulations in other ensembles can be found elsewhere [5,6].

3 Monte Carlo Methods for Protein Simulation and Analysis

3.1 Standard Monte Carlo Moves

For a molecular system, there are several standard Monte Carlo moves that one can use to explore conformational degrees of freedom. The simplest trial move is to select at random an atom, i, in the system, and propose a change in its Cartesian coordinates:

$$x_i^{new} = x_i^{old} + \Delta(\chi - 0.5), \quad (9a)$$

$$y_i^{new} = y_i^{old} + \Delta(\chi - 0.5), \quad (9b)$$

$$z_i^{new} = z_i^{old} + \Delta(\chi - 0.5), \quad (9c)$$

where χ is a pseudorandom number between 0 and 1 that is different for each axis at each attempted move, and Δ sets the maximum displacement. After moving the atom, one calculates the energy of the new trial structure, and the new structure is accepted or rejected based on the Metropolis criterion, Eq. 7. For molecules such as proteins, these moves are often not very efficient, because a change in the coordinates of one atom changes the bond lengths between the selected atom and all the atoms to which it is connected. This move will also change any bond angles and torsional angles that the atom is involved in, and, therefore, acceptance of the move is unlikely for all but the very smallest of moves, especially because proteins often have a number of stiff or rigid bonds (or bond angles or torsional angles). As an alternative scheme, one can select individual bond lengths, bond angles, or torsional angles in the molecule to change at random, while keeping all other bond lengths, bond angles, and torsional angles fixed. For bond lengths, bond angles, and torsional angles that are effectively rigid, one can impose constraints on their movement. Torsional angle (or pivot) moves can result in significant changes in the structure of the molecule, and, thus, can be problematic when the protein is surrounded by a solvent. This is because any significant change in the conformation of the protein will result in an overlap with the solvent with a high energetic penalty, and, thus, the rejection of the move. Pivot moves can be very effective in implicit solvent simulations, however.

When considering multiple molecules, one can also use Monte Carlo moves that translate or rotate entire molecules. For flexible, nonlinear molecules such as proteins, the method of quarternions should be used for rotation moves. For translation moves, the center of mass of the molecule in question is simply changed, in the same way as for a single atom move (Eqs. 9a–9c).

3.2 Configurational-Bias Monte Carlo

Configurational-bias Monte Carlo moves fall under the more general category of biased Monte Carlo moves. In these moves, the Monte Carlo trial move is biased toward the proposal of reasonable configurations and, to satisfy detailed balance, the acceptance rules used must be altered because α, in Eq. 6, will no longer be symmetric. Configurational-bias Monte Carlo was originally proposed for the simulation of chain molecules [7] and was later adapted for the simulation of biomolecules [8]. The idea behind the approach is to delete a section of a molecule and to then "regrow" the section. The regrowth of the molecule is biased such that energetically reasonable conformations are likely to be produced by the method used. Biasing during the regrowth can occur by, for example, restricting torsional angles to reasonable bounds and, if one is regrowing a central section of a molecule, biasing the regrowth toward an area of the simulation box. A configurational-bias Monte Carlo algorithm is composed of the following main steps. First, a section of a molecule is deleted. Second, a trial conformation for the regrowth is generated using a Rosenbluth scheme, and the Rosenbluth weight for the new configuration, $W(n)$, is calculated. Third, the Rosenbluth weight for the old configuration, $W(m)$, is calculated by "retracing" the original conformation. Finally, the trial configuration is accepted with the probability:

$$p_{mn} = \min\left\{1, \frac{W(n)}{W(m)}\right\}. \tag{10}$$

We do not provide details here for determining the Rosenbluth weight, or the generation of trial orientations for the regrowth, of peptide and protein molecules within configurational-bias Monte Carlo methods, because they are lengthy and nontrivial. We instead refer the reader to the following excellent references in which the theory is fully described [6, 7].

3.3 Rebridging and Fixed End Moves

Rebridging, concerted rotation, or end-bridging Monte Carlo moves have been developed to explore the conformational degrees of freedom of internal, backbone portions of polymers, peptides, and proteins [9–12]. An example of a rebridging move for a peptide molecule that causes a local change in conformation but leaves the positions of the rest of the molecule fixed, is shown in Fig. 1. Torsional angles ϕ_0 and

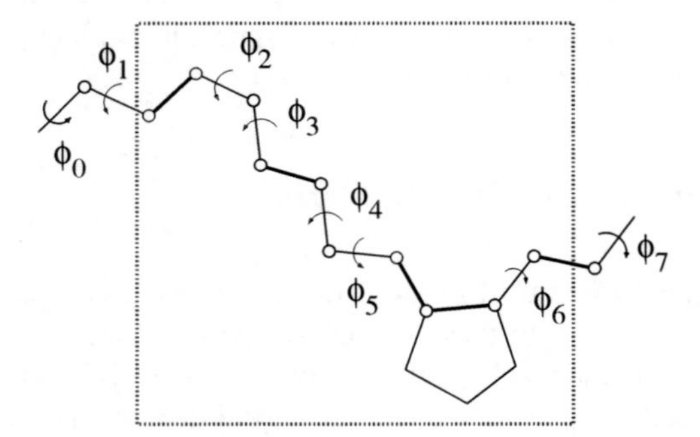

Fig. 1 Driver angles ϕ_0 and ϕ_7 are changed, breaking the connectivity of the molecule in the enclosed region, in an analytic rebridging move. The backbone segment is rebridged in the enclosed region, and the *solid lines* represent pi bonds or rigid molecular fragments within which no rotation is possible. Reprinted from [12] with permission. Copyright American Institute of Physics (1999)

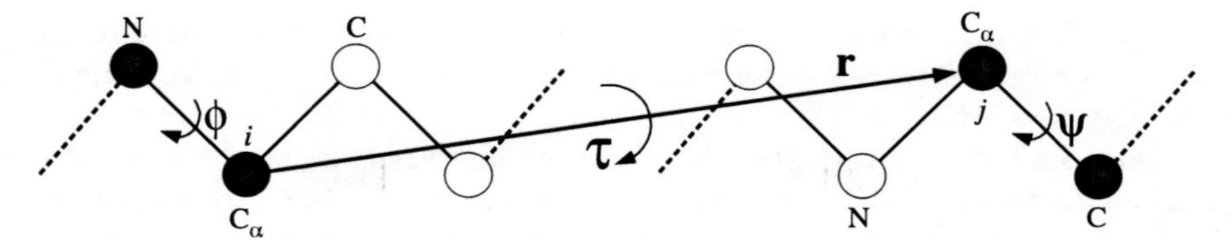

Fig. 2 Fixed end move for biomolecules. The atoms between i and j are rotated about $\mathbf{r}$ by an angle τ

ϕ_7 are rotated by $\Delta\phi_0$ and $\Delta\phi_7$, causing a change in the units between 0 and 6, and breaking the connectivity of the molecule. In this move, rigid units only, where bond lengths and angles are fixed at a prescribed or equilibrium value, are considered to reduce the situation to a geometric problem, and it is possible to determine all of the solutions that reinsert the backbone in a valid, connected way. Where there are side chains attached to the backbone, they are considered to be rigidly rotated. To satisfy detailed balance, solutions for new and old configurations are determined, and the maximum number of solutions can be shown to be limited to 16. One of these solutions is randomly selected, and the new conformation is accepted or rejected based on a modified acceptance criterion.

An alternative to the rebridging move is a fixed end move, shown in Fig. 2, in which a rigid rotation, τ, along the C_α ends of an arbitrary length of the backbone is attempted. This move changes two torsional angles, ϕ and ψ, as well as the bond angle at each end of the segment, but keeps the other internal angles and torsional angles in the molecule fixed. The rigid rotation, τ, is chosen such that the bond angles at the end of the segment do not vary by more than 10° from their natural range. The relations between the torsional angles, the bond angles, and τ are nonlinear, but it has been shown that trial moves can be proposed that satisfy detailed balance and are effective in equilibrating peptide molecules [13].

3.4 Hybrid Monte Carlo

A hybrid Monte Carlo move involves running a molecular dynamics simulation for a fixed length of time [14]. The configuration at the end of this molecular dynamics run is accepted or rejected based on the Metropolis criterion (Eq. 7), where $U(m)$ is the potential energy before the molecular dynamics run, and $U(n)$ is the potential energy at the end of the run [15]. Thus, molecular dynamics can be used to propose a Monte Carlo move that involves the collective motion of many particles. An advantage that this method has over conventional molecular dynamics simulations is that large time steps may be used in the hybrid Monte Carlo move, because there is no need to conserve energy during the molecular dynamics run. As long as the molecular dynamics algorithm that is used is time reversible and area preserving, then it is acceptable for use in a hybrid Monte Carlo move. Indeed, several very suitable multiple-time-step molecular dynamics algorithms now exist that are ideal for use in hybrid Monte Carlo moves, including those proposed by Martyna, Tuckerman, and coworkers [16].

At the start of each new hybrid MC move, particle velocities can be assigned at random from a Maxwell distribution, or alternatively they can be biased toward particle velocities from previous, successful hybrid Monte Carlo steps, with, of course, a modified acceptance criterion. The total length of time one uses for the molecular dynamics run is at the discretion of the simulator. However, the general rule of not using too long or short a run applies, because long runs that are not accepted are wasteful of CPU time, and short runs do not advance the configuration through phase space quickly.

3.5 Parallel Tempering

In the parallel tempering method [17], one simulates M replicas of an original system of interest. These replicas are typically each at a different temperature. The general idea of the method is that the higher-temperature replicas are able to access large volumes of phase space, whereas lower-temperature systems may become trapped in local energy minima during the timescales of a typical computer simulation. By allowing configuration swaps between different (typically adjacent in temperature) replicas, the lower-temperature systems can access a representative set of low-energy regions of phase space. The partition function for the M replica system in the canonical ensemble is:

$$Q \equiv \prod_{i=1}^{M} \frac{q_i}{N!} \int d\mathbf{r}_i^N \exp\left[-\beta_i U\left(\mathbf{r}_i^N\right)\right], \tag{11}$$

where $q_i = \prod_{j=1}^{N} \left(2\pi m_j k_B T_i / h^2\right)^{3/2}$ comes from integrating out the momenta, and $\beta_i = 1/k_B T_i$ is the reciprocal temperature of the replica. If the probability of performing a swap move is equal for all conditions, then exchanges of configurations between replicas i and j are accepted with the probability:

$$p_{ij} = \min\left\langle 1, \exp\left\{+\left[\beta_i - \beta_j\right]\left[U\left(\mathbf{r}_i^N\right) - U\left(\mathbf{r}_j^N\right)\right]\right\}\right\rangle \tag{12}$$

A rather complex issue when using parallel tempering is how many replicas one should use and what temperatures each replica should be set at. As with most Monte Carlo moves, there are no hard and set rules. The highest temperature must be high enough for the simulation to pass over all of the energy barriers in phase space in a manageable computational time, and there must be a sufficient number of replicas to allow a reasonable probability of acceptance for parallel tempering moves, allowing each system to sample both high and low temperatures. To maximize the efficiency of parallel tempering simulations, recent theoretical [18] and numerical studies [19] have recommended "tuning" the temperature intervals such that moves between replicas are accepted with a probability of 20 to 25%. An adaptive approach based on the total "round-trip" time between high and low temperature replicas seems to be well founded for tuning the temperature intervals [20].

With appropriate scheduling, parallel tempering is an ideal approach to use with large CPU clusters [21], and its effectiveness in the simulation of biomolecules, as well as in solid state and material modeling, is well proven [22].

3.6 Density of States Methods

The canonical partition function can also be written as:

$$Q(N, V, T) \equiv \sum_{E_i} g(N, V, E_i) \exp[-\beta E_i], \tag{13}$$

where g is the density of states. Wang and Landau [23] proposed a Monte Carlo scheme whereby one performs a random walk in energy space, while keeping a running estimate of the inverse of the density of states as a function of the energy, or, alternatively, collecting the configurational temperature as a function of the energy and determining the density of states by integrating the inverse temperature. Although density of states methods do not satisfy detailed balance, it has been proven that they asymptotically satisfy balance [24]. During the simulation, the density of states is continuously modified to produce a flat histogram in energy space, preventing the simulation from becoming stuck in low-energy minima. de Pablo and coworkers have used the density of states method, combined with hybrid Monte Carlo and pivot moves, to fold small protein molecules successfully [25].

3.7 Monte Carlo as an Optimization Tool

Monte Carlo methods that satisfy the balance condition are essential if one is interested in calculating properties at equilibrium. If one is, instead, only interested in optimizing a property of interest, for example, determining a minimum energy structure, then there is no such restriction, and one has greater flexibility in the choice of moves one can use. Simulated annealing, in which a simulation begins at a high temperature and is sequentially reduced, is one such method that does not satisfy balance but can be highly effective in overcoming energetic barriers.

In protein structure prediction, Monte Carlo methods can be particularly effective as an optimization tool. For example, Baker and coworkers have had great success in the high-resolution de novo structure prediction of small proteins [26]. In their method, they use a combination of multiscale force fields, Monte Carlo torsional angle moves, side-chain optimization using rotamer representations, and gradient-based minimization of their all-atom energy function. Similar methods can be used in the refinement of structures from nuclear magnetic resonance (NMR) and x-ray data [27]. Saven and coworkers have also shown that biased Monte Carlo moves and parallel tempering-based simulations can be particularly useful in protein sequence design efforts, in which one attempts to define amino acid sequences that fold to a particular tertiary structure [28].

4 Statistical Errors

Although we may wish that the results from our computer simulations could provide exact values for a property that we are interested in, this is never the case, because we are unable to simulate our system for an infinite length of time. One must, therefore, estimate the error in the calculated value of the property. One way to do this is to use the method of block averaging. Suppose that we have n samples of a fluctuating property of interest, A, that were taken when our system was in equilibrium. The average value of A is estimated as:

$$\langle A \rangle \approx \bar{A} \equiv \frac{1}{n} \sum_{i=1}^{n} A_i . \tag{14}$$

If all of our samples were uncorrelated, then our estimate of the variance would be:

$$\sigma^2 (A) = \left\langle A^2 \right\rangle - \langle A \rangle^2 \approx \frac{1}{n} \sum_{i=1}^{n} \left[A_i - \bar{A} \right]^2 . \tag{15}$$

However, in a simulation, our samples are correlated. To account for this fact, one may compute averages for blocks of data, and use these block averages to estimate the variance. Where we have n' blocks of data, the variance of our new set of data is:

$$\sigma^2 \left(A' \right) = \left\langle A'^2 \right\rangle - \left\langle A' \right\rangle^2 = \frac{1}{n'} \sum_{i=1}^{n'} \left[A'_i - \bar{A} \right]^2 , \tag{16}$$

where A'_i is the block average of set i, and the average for the whole set of data remains the same. As the size of the block used increases, the correlation between the blocks will decrease. Thus, as the block size becomes sufficiently large, $\sigma^2 \left(A' \right) / \left(n' - 1 \right)$ should be constant and provide an estimate of the error of our ensemble average of A:

$$\varepsilon^2 (A) \approx \frac{\sigma^2 \left(A' \right)}{n' - 1} , \tag{17}$$

and an accurate measure of our statistical error ε, is now available.

Another way of calculating errors in a quantity is through the bootstrap method. From our original n samples, one produces N new sets of data, each containing n

samples. The samples that comprise each new set of data are picked at random from the original data set, and, therefore, it is probable that there will be repeat elements in the new data sets. These new data sets provide an estimate of the probability distribution of our property of interest, and, thus, provide another estimate of the variance and error in the simulation results.

5 Conclusions

Having presented a large number of Monte Carlo moves in Sect. 3, we conclude by discussing how one can combine these moves in an algorithm that satisfies balance. Most Monte Carlo algorithms are comprised of several different Monte Carlo moves that are effective at relaxing different degrees of freedom in a molecule or system. How to choose which Monte Carlo moves to include in a combined algorithm, and the probability of attempting each of them, is problem specific.

To satisfy detailed balance, the choice of what type of Monte Carlo move to attempt at a particular time is chosen on a probabilistic basis, and each of the specific Monte Carlo moves satisfies detailed balance individually. This scheme presents a problem when using methods such as parallel tempering, in which it is convenient to synchronize replicas in Monte Carlo move number and attempt swaps after a set number of Monte Carlo steps. This latter approach satisfies the balance condition, however, and therefore, it is perfectly acceptable. Also acceptable, because it satisfies balance, is sequential updating of a system, in which the moves are performed in a defined sequence, rather than the moves chosen at random.

Most Monte Carlo simulations to date are performed by software custom written by research groups. The most typical language for these Monte Carlo simulations is C. Both the Allen and Tildesley book [6] and the Frenkel and Smit book [5] offer web sites with useful routines [29, 30]. C language code for analytical rebridging [12] is available [31]. A web site by Mihaly Mezei lists several useful links related to Monte Carlo simulation of proteins [32].

References

1. Hu, J., Ma, A., and Dinner, A. R. (2006) Monte Carlo simulations of biomolecules: The MC module in CHARMM. *J. Comput. Chem.* **27**, 203–216.
2. Manousiouthakis, V. I. and Deem, M. W. (1999) Strict detailed balance is unnecessary in Monte Carlo simulation. *J. Chem. Phys.* **110**, 2753–2756.
3. Metropolis, N., Rosenbluth, A. W., Rosenbluth, M. N., Teller, A. N., and Teller, E. (1953) Equation of state calculations by fast computing machines. *J. Chem. Phys.* **21**, 1087–1092.
4. Metropolis, N. (1987) The beginning of the Monte Carlo method. *Los Alamos Science*, **12**, 125–130.
5. Frenkel, D. and Smit, B. (2002) *Understanding Molecular Simulation: From Algorithms to Applications.* Academic Press, San Diego, CA.

6. Allen, M. P. and Tildesley, D. J. (1987) *Computer Simulation of Liquids*. Clarendon Press, Oxford.
7. Siepmann, J. I. and Frenkel, D. (1992) Configurational-bias Monte Carlo: A new sampling scheme for flexible chains. *Mol. Phys.* **75**, 59–70.
8. Deem, M. W. and Bader, J. S. (1996) A configurational bias Monte Carlo method for linear and cyclic peptides. *Mol. Phys.* **87**, 1245–1260.
9. Dodd, L. R., Boone, T. D., and Theodorou, D. N. (1993) A concerted rotation algorithm for atomistic Monte Carlo simulation of polymer melts and glasses. *Mol. Phys.* **78**, 961–996.
10. Mavrantzas, V. G., Boone, T. D., Zevropoulou, E., and Theodorou, D. N. (1999) End-bridging Monte Carlo: A fast algorithm for atomistic simulation of condensed phases of long polymer chains. *Macromolecules* **32**, 5072–5096.
11. Wu, M. G. and Deem, M. W. (1999) Efficient Monte Carlo methods for cyclic peptides. *Mol. Phys.* **97**, 559–580.
12. Wu, M. G. and Deem, M. W. (1999) Analytical rebridging Monte Carlo: Application to cis/trans isomerization in proline-containing, cyclic peptides. *J. Chem. Phys.* **111**, 6625–6632.
13. Betancourt, M. R. (2005) Efficient Monte Carlo moves for polypeptide simulations. *J. Chem. Phys.* **123**, 174905.
14. Duane, S., Kennedy, A., Pendleton, B. J., and Roweth, D. (1987) Hybrid Monte Carlo. *Phys. Rev. Lett.* **195**, 216–222.
15. Mehlig, B., Heermann, D. W., and Forrest, B. M. (1992) Hybrid Monte Carlo method for condensed matter systems. *Phys. Rev. B* **45**, 679–685.
16. Tuckerman, M. E., Berne, B. J., and Martyna, G. J. (1992) Reversible multiple time scale molecular dynamics. *J. Chem. Phys.* **97**, 1990–2001.
17. Geyer, C. J. and Thompson, E. A. (1995) Annealing Markov-Chain Monte Carlo with applications to ancestral inference, *J. Am. Stat. Assn.* **90**, 909–920.
18. Kone, A. and Kofke, D. A. (2005) Selection of temperature intervals for parallel tempering simulations, *J. Chem. Phys.* **122**, 206101.
19. Rathore, N., Chopra, M., and de Pablo, J. J. (2005) Optimal allocation of replicas in parallel tempering simulations, *J. Chem. Phys.* **122**, 024111.
20. Katzgraber, H. G., Trebst, S., Huse, D. A., and Troyer, M. (2006) Feedback-optimized parallel tempering Monte Carlo, *J. Stat. Mech.: Exp. & Theory* P03018.
21. Earl, D. J. and Deem, M. W. (2004) Optimal allocation of replicas to processors in parallel tempering simulations. *J. Phys. Chem. B* **108**, 6844–6849.
22. Earl, D. J. and Deem, M. W. (2005) Parallel tempering: theory, applications, and new perspectives, *Phys. Chem. Chem. Phys.* **7**, 3910–3916.
23. Wang, F. and Landau, D. P. (2001) Efficient, multiple-range random walk algorithm to calculate the density of states, *Phys. Rev. Lett.* **86**, 2050–2053.
24. Earl, D. J. and Deem, M. W. (2005) Markov chains of infinite order and asymptotic satisfaction of balance: Application to the adaptive integration method, *J. Phys. Chem. B* **109**, 6701–6704.
25. Rathore, N., Knotts IV, T. A., and de Pablo, J. J. (2003) Configurational temperature density of states simulations of proteins, *Biophys. J.* **85**, 3963–3968.
26. Bradley, P., Misura, K. M. S., and Baker, D. (2005) Toward high-resolution de novo structure prediction for small proteins, *Science* **309**, 1868–1871.
27. Meiler, J. and Baker, D. (2003) Rapid protein fold determination using unassigned NMR data, *Proc. Natl. Acad. Sci. USA* **100**, 15404–15409.
28. Yang, X. and Saven, J. G. (2005) Computational methods for protein design sequence variability: biased Monte Carlo and replica exchange, *Chem. Phys. Lett.* **401**, 205–210.
29. http://www.ccl.net/cca/software/SOURCES/FORTRAN/allen-tildesley-book/index.shtml
30. http://molsim.chem.uva.nl/frenkel_smit/index.html
31. http://www.mwdeem.rice.edu/rebridge/
32. http://fulcrum.physbio.mssm.edu/~mezei/

Chapter 3
Hybrid Quantum and Classical Methods for Computing Kinetic Isotope Effects of Chemical Reactions in Solutions and in Enzymes

Jiali Gao, Dan T. Major, Yao Fan, Yen-lin Lin, Shuhua Ma, and Kin-Yiu Wong

Summary A method for incorporating quantum mechanics into enzyme kinetics modeling is presented. Three aspects are emphasized: 1) combined quantum mechanical and molecular mechanical methods are used to represent the potential energy surface for modeling bond forming and breaking processes, 2) instantaneous normal mode analyses are used to incorporate quantum vibrational free energies to the classical potential of mean force, and 3) multidimensional tunneling methods are used to estimate quantum effects on the reaction coordinate motion. Centroid path integral simulations are described to make quantum corrections to the classical potential of mean force. In this method, the nuclear quantum vibrational and tunneling contributions are not separable. An integrated centroid path integral–free energy perturbation and umbrella sampling (PI-FEP/UM) method along with a bisection sampling procedure was summarized, which provides an accurate, easily convergent method for computing kinetic isotope effects for chemical reactions in solution and in enzymes. In the ensemble-averaged variational transition state theory with multidimensional tunneling (EA-VTST/MT), these three aspects of quantum mechanical effects can be individually treated, providing useful insights into the mechanism of enzymatic reactions. These methods are illustrated by applications to a model process in the gas phase, the decarboxylation reaction of *N*-methyl picolinate in water, and the proton abstraction and reprotonation process catalyzed by alanine racemase. These examples show that the incorporation of quantum mechanical effects is essential for enzyme kinetics simulations.

Keywords: Combined QM/MM · Dual-level potential · Enzyme kinetics · Kinetic isotope effects · Path integral simulations · PI-FEP/UM · Solvent effects

From: Methods in Molecular Biology, vol. 443, Molecular Modeling of Proteins
Edited by Andreas Kukol © Humana Press, Totowa, NJ

1 Introduction

The remarkable ability that enzymes accelerate the rates of chemical reactions has fascinated chemists and biochemists for nearly a century since the identification of proteins as the primary biological catalysts [1]. In the absence of structural information, Pauling proposed that enzymes specifically bind the transition state more strongly than the reactant state, consequently lowering the free energy of activation [2]. This remains a key concept in our understanding of enzyme catalysis [3]. Now, detailed energy calculations can be carried out to provide an understanding of the mechanism of enzymes at the atomistic level [4–6], thanks to the development of modern transition state theory (TST), the advent of computer simulation methods, and the availability of three-dimensional structures of enzymes and enzyme–substrate complexes. Quantum mechanics is essential for modeling enzyme mechanism and kinetics, and it contributes to these calculations in three specific ways [7, 8]: 1) electronic structural theory provides the necessary potential energy surface (PES) for the enzyme system to adequately treat the bond forming and breaking processes, 2) quantum mechanical (QM) treatment of vibrational motions allows the more accurate estimation of the rate constant for enzyme reactions, and 3) inclusion of nuclear tunneling provides further insight in the understanding of enzyme mechanism and the transition state through quantitative computation of kinetic isotope effects (KIE). In this chapter, we present some of the methods that have been developed in our laboratories for studying enzymatic reactions.

We begin our discussion by considering the general kinetic equation for an enzyme reaction:

$$E + S \underset{k_{-1}}{\overset{k_1}{\rightleftharpoons}} ES \xrightarrow{k_{cat}} EP \longrightarrow E + P$$

where the symbols E, S, and P are the enzyme, substrate, and product, respectively, and ES and EP are substrate and product complexes with the enzyme, respectively. Although the bimolecular rate constant k_{cat}/K_M, where $K_M = (k_{-1} + k_{cat})/k_1$, is the key kinetic parameter for the specific biological function of the enzyme [9], the rate reduction in the catalyze reaction, k_{cat}, relative to that of the corresponding uncatalyzed process in aqueous solution, k_{aq}, is of special interest in understanding the origin of enzyme catalysis. The experimental approach is best illustrated by the systematic studies of Wolfenden, who pioneered the concept of comparing the rate constant of the catalyzed reaction with that of the same reaction in the absence of the enzyme in water [9, 10]. Furthermore, site-directed mutagenesis experiments help to identify residues that make important contributions to catalysis, but the experimental findings do not always shed light on the mechanism of the enzyme action, for a variety of reasons, including the possibility of altering the mechanism of the actual catalyzed reaction. Computer simulations provide a powerful and a complementary tool for elucidating the mechanism of enzyme reactions and this involves the computation and comparison of the catalyzed (k_{cat}) and uncatalyzed (k_{aq}) rate constant.

The theoretical framework in the present discussion is TST, which yields the expression of the classical mechanical (CM) rate constant [11, 12]. However, evaluation of the QM rate constant is of particular interest. Although a number of methods have been developed, we describe a method to make QM corrections to classical trajectories. This is especially attractive because it is efficient to use classical molecular dynamics simulations to sample enzyme configurations. Thus, for a unimolecular reaction, the forward rate constant is written as:

$$k \equiv k_{qm} = \gamma \cdot k_{TST}, \tag{1}$$

where k_{qm} and k_{TST} are the QM and TST rate constants, respectively, and γ is the generalized transmission coefficient [5], which includes the classical dynamic recrossing factor, Γ, and the quantum correction factor, κ:

$$\gamma = \kappa \cdot \Gamma. \tag{2}$$

The classical transmission coefficient accounts for the dynamical correction to TST, and the time-dependent transmission coefficient can be calculated by the reaction flux method [13, 14]:

$$\Gamma(t) = N < \dot{z}(0) H[z(t) - z^{\neq}] >_{z^{\neq}}, \tag{3}$$

where z is the reaction coordinate, $\dot{z}(0)$ is the time derivative of z at time $t = 0$, $z^{\neq}$ is the value of the reaction coordinate at the transition state, N is a normalization factor, H is a step function such that it is 1 when the trajectory is in the product side and 0 otherwise, and the brackets, $\langle \cdots \rangle_{z^{\neq}}$, specify an ensemble average over transition state configurations. Although it is important to consider dynamic effects [15, 16], it has been noted that the "sobering fact for the theorist" is that the solvent contribution to the free energy of activation often has much greater influence on the computed rate constant because of its exponential dependence [17]. Here, we assume that the Γ factor is unity and is identical for the CM and QM rate constant; interested readers are directed to our recent article on this subject [8].

The quantum correction factor is then defined as follows:

$$\kappa = \frac{k_{qm}}{k_{TST}} = e^{-\beta\left(\Delta F^{\neq}_{qm} - \Delta F^{\neq}_{TST}\right)}. \tag{4}$$

In Eq. 4, $\beta = 1/k_B T$ with k_B being Boltzmann's constant and T the temperature; $\Delta F^{\neq}_{qm}$ and $\Delta F^{\neq}_{TST}$ are, respectively, the quantum and the classical free energy of activation. The different methods applied to enzymatic reactions to incorporate nuclear quantum effects differ in the specific approximations to estimate the free energy difference in Eq. 4.

In classical dynamics, the TST rate constant is the rate of one-way flux through the transition state dividing surface [11, 17]:

$$k_{TST} = < \dot{z} >_{z^{\neq}} e^{-\beta w(z^{\neq})} / \int_{-\infty}^{z^{\neq}} dz\, e^{-\beta w(z)}, \tag{5}$$

where $w(z)$ is the potential of mean force (PMF) along the reaction coordinate z. The frequency for passage through the transition state is given by the average velocity of the reaction coordinate z' at the transition state, $z = z^{\neq}$. Alternatively, Eq. 5 can be written as:

$$k_{TST} = \frac{1}{\beta h} e^{-\beta \Delta F^{\neq}_{TST}}, \tag{6}$$

where h is Planck's constant and $\Delta F^{\neq}_{TST}$ is the molar standard state free energy of activation, defined as [18]:

$$\Delta F^{\neq}_{TST} = w_{\mathrm{CM}}(z^{\neq}) - [w_{\mathrm{CM}}(z_{\mathrm{R}}) + F^{\mathrm{R}}_{\mathrm{CM}}(z)] + C(z), \tag{7}$$

where z_{R} is the value of the reaction coordinate at the reactant state, $F^{\mathrm{R}}_{\mathrm{CM}}(z)$ corresponds to the free energy of the mode in the reactant (R) state that correlates with the progress coordinate z, and $C(z)$ is a small correction term that is caused by the Jacobian of the transformation from a locally rectilinear reaction coordinate to the curvilinear reaction coordinate z.

The classical PMF is defined as follows:

$$w(z) = -\frac{1}{\beta} \ln \int dz' d\mathbf{q} \delta[z - z'] e^{-\beta V(z', \mathbf{q})}, \tag{8}$$

where $\mathbf{q}$ represents all degrees of freedom of the system except that corresponding to the reaction coordinate, and $V(z', \mathbf{q})$ is the potential energy function. Computationally, the PMF $w(z)$ can be obtained by umbrella sampling [19] from Monte Carlo and molecular dynamics simulations [20–23].

There is no unique way of separating the exact CM rate constant into the dynamical correction factor, Γ, and the TST rate constant, k_{TST}. Both quantities can be determined from computer simulations, in which the solvent is in thermal equilibrium along the reaction coordinate, z. Thus, solvation affects both $\Delta F^{\neq}_{TST}$ and Γ, and these two quantities are not independent of each other, but they are related by the choice of the reaction coordinate, z [7, 17]. Consequently, in analyzing computational results, it is important to examine the effect of using a specific reaction coordinate on the computed PMF [24–26].

The rest of this chapter focuses on discussion of treatment of the PES and methods for computing free energies of activation and for incorporating nuclear quantum effects in enzyme reactions.

2 Methods

Almost all enzyme reactions can be well described by the Born-Oppenheimer approximation, in which the sum of the electronic energy and the nuclear repulsion provides a potential energy function, or PES, governing the interatomic motions. Therefore, the molecular modeling problem breaks into two parts: the PES and the dynamics simulations.

2.1 Potential Energy Surface

The potential energy function describes the energetic changes as a function of the variations in atomic coordinates, including thermal fluctuations and rearrangements of the chemical bonds. The accuracy of the potential energy function used to carry out molecular dynamics simulations directly affects the reliability of the computed $\Delta F^{\neq}_{TST}$ and its nuclear quantum correction [7,27]. The accuracy can be achieved by the use of analytical functions fitted to reproduce key energetic, structural, and force constant data, from either experiments or high-level *ab initio* calculations. Molecular mechanical (MM) potentials or force fields [28, 29], however, are not general for chemical reactions, and require reparameterization of the empirical parameters for every new reaction, which severely limits applicability. More importantly, often, little information is available in regions of the PES other than the stationary reactant and product states and the saddle point (transition state). On the other hand, combined QM and MM (QM/MM) potentials offer the advantages of both computational efficiency and accuracy for all regions of the PES [30, 31].

2.1.1 Empirical Potentials

The application of MM methods to modeling molecule–solvent interactions in uncatalyzed chemical reactions in solution was pioneered by Chandrasekhar and Jorgensen in their classic study of a model S_N2 reaction in water [32, 33]. Their study involved three key steps: 1) defining a reaction path, 2) determining potential functions that reproduce experimental or *ab initio* results along the entire reaction path in the gas phase, and 3) performing free energy simulations. This procedure remains valuable for studying chemical reactions in solution and in enzymes [34]. Yang and coworkers further developed and applied this approach in a number of calculations of enzymatic reactions, using the reaction path and charges derived from combined QM/MM energy minimizations (the QM/MM method is explained as type 3 below) and density functional theory (DFT) [35, 36].

The most widely used potential energy function for modeling enzyme reactions is the empirical valence bond (EVB) model developed by Warshel and coworkers [37–39]. The form of the potential function is derived on the basis of a two-state valence bond theory for the chemical bond, such as that in a hydrogen molecule. Thus, it has a flavor of quantum mechanics. In applications to enzyme reactions, two empirical functions represent the reactant (H_{11}) and product (H_{22}) configurations, called effective diabatic states. They are represented by a force field and the bonding term is treated by a Morse potential [38, 39]. The EVB potential function is given by:

$$U_{EVB} = 0.5\left[(H_{11} + H_{22}) + \sqrt{(H_{11} - H_{22})^2 - 4\varepsilon_{12}^2}\right]. \tag{9}$$

In Eq. 9, the coupling between the two potentials or resonance integral ε_{12} is treated empirically [40], typically by an exponential function. Although, in principle, many valence bond states can be included in constructing an EVB potential, and sometimes this is done, most applications to enzymatic reactions have used a simple two-state procedure [39, 41]. The successful application of the EVB method relies on the parameterization of the resonance integral ε_{12} to the barrier height of the specific reaction and of the diagonal constant $\Delta\varepsilon$ (embedded in H_{22}) to the free energy of reaction [38, 39]. The parameterization process has been typically carried out for the uncatalyzed reaction in aqueous solution [6, 42], and, then, the study of enzymatic reactions is performed to estimate $\Delta\Delta F^{\neq}(aq \rightarrow enzyme)$ using these parameters. This method has been reviewed in a number of publications by Aqvist and Warshel [43, 44].

The assumption that the atomic partial charges are invariant, although it is not an inherent restriction of the EVB model, is a major shortcoming in practical applications because these charges do polarize and vary as the geometry of the substrate and the positions of the rest of the system fluctuate during the chemical process. Although the free energy barrier can be parameterized and its change in the enzyme can be computed to reproduce experimental results, there is no rigorous justification for its representing other regions of the PES, casting doubt on results that require a knowledge of this information, such as tunneling and KIE. QM models that define the Lewis resonance structures based on block-localized wave functions have been developed [25, 45], and these effective diabatic states have been used in effective valence bond treatment at the Hartree-Fock level, coupled with configuration interaction (CI) theory, which showed a remarkable charge polarization within each diabatic state [46–48].

Truhlar and coworkers developed a multiconfiguration molecular mechanics (MCMM) method that involves a systematic parameterization for the off-diagonal element [40, 49]. The MCMM method is fitted to reproduce *ab initio* energies and gradients. Thus, the MCMM potential is a proper function for use to compute nuclear quantum effects and KIE.

2.1.2 QM Potentials

It is possible to treat the entire enzyme–solvent system by semiempirical or first principles QM methods [50–56]. Although this approach has the advantage of avoiding the intermediate parameterization step and has been applied successfully to a variety of condensed-phase systems, the computational costs are still too large to be practical for free energy simulations of enzymatic reactions with appreciable barriers. A semiempirical molecular orbital approach has been used in a variety of applications to biological systems [57–60]. In most enzyme reactions, it is not clear whether there is a need to treat the entire enzyme by quantum mechanics.

2.1.3 Combined QM/MM Potentials

The most promising approaches for modeling enzymatic reactions are QM/MM methods, in which a system is divided into a QM region and an MM region

[30, 31, 46, 61–70]. The QM region typically includes atoms that are directly involved in the chemical step and they are treated explicitly by a QM electronic structure method, whereas the MM region consists of the rest of the system and is approximated by an MM force field. This way of combining QM with MM was initially developed for gas-phase calculations by Warshel and Karplus [71], and it was applied to enzyme systems by Warshel and Levitt [62]. Molecular dynamics and Monte Carlos simulations using combined QM/MM potentials began to emerge more than 10 years later [30, 66, 72]. The QM/MM potential is given by [31, 65]:

$$U_{tot} = \langle \Psi(S) | H^{o}_{qm}(S) + H_{qm/mm}(S) | \Psi(S) \rangle + U_{mm}, \tag{10}$$

where $H^{o}_{qm}(S)$ is the Hamiltonian of the QM subsystem (the substrate and key amino acid residues) in the gas phase, U_{mm} is the classical (MM) potential energy of the remainder of the system, $H_{qm/mm}(S)$ is the QM/MM interaction Hamiltonian between the two regions, and $\Psi(S)$ is the molecular wavefunction of the QM-subsystem optimized for $H^{o}_{qm}(S) + H_{qm/mm}(S)$.

We have found that it is most convenient to rewrite Eq. 10 as follows [30, 66]:

$$U_{tot} = E^{o}_{qm}(S) + \Delta E_{qm/mm}(S) + U_{mm}, \tag{11}$$

where $E^{o}_{qm}(S)$ is the energy of an isolated QM subsystem in the gas phase:

$$E^{o}_{qm}(S) = \left\langle \Psi^{o}(S) \left| H^{o}_{qm}(S) \right| \Psi^{o}(S) \right\rangle. \tag{12}$$

In Eq. 11, $\Delta E_{qm/mm}(S)$ is the interaction energy between the QM and MM regions, corresponding to the energy change of transferring the QM subsystem from the gas phase into the condensed phase, which is defined by:

$$\Delta E_{qm/mm}(S) = \langle \Psi(S) | H^{o}_{qm}(S) + H_{qm/mm}(S) | \Psi(S) \rangle - E^{o}_{qm}(S). \tag{13}$$

In Eqs. 11 to 13, we have identified the energy terms involving electronic degrees of freedom by E and those purely empirical functions by U, the combination of which is also an empirical potential.

Eq. 11 is especially useful in that the total energy of a hybrid QM and MM system is separated into two "*independent*" terms—the gas-phase energy and the interaction energy—which can now be evaluated using different QM methods. There is sometimes confusion regarding the accuracy of applications using semiempirical QM/MM potentials [42]. Eq. 11 illustrates that there are two issues. The first is the intrinsic performance of the model, e.g., the $E^{o}_{qm}(S)$ term, which is indeed not adequate using semiempirical models and which would require *extremely* high-level QM methods to achieve the desired accuracy. This is only possible by using CCSD(T), CASPT2, or well-tested density functionals along with a large basis set, and none of these methods are tractable for applications to enzymes. When semiempirical methods are used, the PES for the $E^{o}_{qm}(S)$ term is either reparameterized to fit experimental data, or replaced by high-level results. These methods are only directly used without alteration in rare occasions when a semiempirical model yields good agreement with experiment [73,74]. Importantly, Eq. 11 allows us to substitute a "high-level" (HL) theory for the semiempirical intrinsic energy:

$$E_{qm}^{HL}(S) = E_{qm}^{o}(S). \tag{14}$$

Now, we have the flexibility to choose our favorite, most accurate QM methods to achieve the desired accuracy. This substitution of QM models can be made on the fly during a dynamics simulation, or post priori.

The second issue on accuracy is in the calculation of the $\Delta E_{qm/mm}(S)$ term. It was recognized early on, when QM/MM simulations were first carried out, that combined QM/MM potential is an empirical model, which contains empirical parameters and should be optimized to describe QM/MM interactions [30, 68, 75]. By systematically optimizing the associated van der Waals parameters for the "QM-atoms," both semiempirical and *ab initio* (Hartree-Fock) QM/MM potentials can yield excellent results for hydrogen-bonding and dispersion interactions in comparison with experimental data. The use of semiempirical methods, such as the Austin Model 1 (AM1) [76] or Parameterized Model 3 (PM3) [77] in QM/MM simulations has been validated through extensive studies of a variety of properties and molecular systems, including computations of free energies of solvation and polarization energies of organic compounds [30, 78], the free energy profiles for organic reactions [23, 73, 79], and the effects of solvation on molecular structures and on electronic transitions [80, 81].

For studying enzymatic reactions, it is necessary to obtain the free energy of activation, typically obtained by computing the PMF along a reaction coordinate. This requires sufficient configurational sampling through molecular dynamics simulations, which is another critical factor contributing to accuracy, and the $\Delta E_{qm/mm}(S)$ in Eq. 11 is most relevant. It must be evaluated and repeated millions of times. Consequently, a computationally efficient method, such as a semiempirical QM model, must be used. Here, we use the term "lower-level" (LL) model to denote the use of an efficient QM/MM potential, which is used on-the-fly in molecular dynamics simulations:

$$\Delta E_{qm/mm}^{LL}(S) = \Delta E_{qm/mm}(S). \tag{15}$$

This is critical for studying enzymatic reactions because it introduces the instantaneous electronic polarization of the QM subsystem caused by the thermal fluctuations of the enzyme and solvent environment [82].

By these substitutions, we obtain a highly accurate dual-level (DL) total energy for the enzyme reaction [83]:

$$U_{tot}^{DL} = E_{qm}^{HL}(S) + \Delta E_{qm/mm}^{LL}(S) + U_{mm}. \tag{16}$$

The dual-level QM/MM approach is akin to the ONIOM model developed by Morokuma and coworkers [84], and it has been used in a number of QM/MM simulations.

The QM/MM PES combines the generality of QM methods for treating chemical processes with the computational efficiency of molecular mechanics for large molecular systems. The use of an explicit electronic structure method to describe the enzyme active site is important because understanding the changes in electronic structure along the reaction path can help to design inhibitors and novel catalysts. It is also important because the dynamic fluctuations of the enzyme and aqueous solvent system have a major impact on the polarization of the species involved in the chemical reaction, which, in turn, affects the chemical reactivity [30,82]. Combined QM/MM methods have been reviewed in several articles [31,39,61,65,85].

2.2 Classical PMF

The free energy barrier $\Delta F^{\neq}_{TST}$ (or $\Delta F^{\neq}_{qm}$ when quantum effects are considered, see Eq. 26) is the most relevant quantity to quantify enzyme catalysis, and it is typically obtained by computing the PMF along the reaction coordinate. Two methods are generally used in the calculation of PMFs for reactions in solution and in enzymes, the umbrella sampling [19] technique and the free energy perturbation theory [86,87]; these are reviewed elsewhere [88,89]. The umbrella sampling technique provides a direct estimate of the relative probability of finding the reaction system at the reactant position along the reactant coordinate and at the transition state position; this estimate includes both the structural variations of the substrate and dynamic, thermal fluctuations of the enzyme along the reaction coordinate. Thus, it provides the most accurate estimate of the change in the free energy of activation $\Delta\Delta F^{\neq} = \Delta F^{\neq}_{enz} - \Delta F^{\neq}_{aq}$ from water to the enzyme using the given PES.

Because these calculations are often carried out using classical mechanics, zero-point energy and effects of the quantization of vibrational motion are neglected. Studies have shown that inclusion of these effects can significantly lower the classical free energy barrier, particularly in systems involving hydrogen transfer [90–93], and that omission of these quantum effects can lead to significant errors in computed free energies of activation, particularly in systems involving hydrogen transfer. Thus, it is desirable to make quantal corrections to the CM-PMF or to perform simulations that directly include nuclear QM effects.

2.2.1 Quasiclassical PMF

A convenient procedure to include the effect of quantization of molecular vibrations in free energy calculations is to relate the quantum PMF to the classical PMF of Eq. 8 at temperature T by [18,93]:

$$w_{\mathrm{QC}}(z) = w_{\mathrm{CM}}(z) + \Delta w_{\mathrm{vib}}(T, z), \tag{17}$$

where $\Delta w_{\mathrm{vib}}(T, z)$ is an ensemble average of the instantaneous harmonic approximation to the quantal correction to the classical vibrational free energy for vibrational modes orthogonal to the reaction coordinate, z. The subscript QC in Eq. 17 specifies the fact that tunneling and other quantum effects on the reaction coordinate

z are excluded, giving rise to the quasiclassical PMF (QC-PMF). The constraint that the modes included are orthogonal to z can be achieved by a projection operator [18, 93]. In practice [8], for each instantaneous configuration that was saved from the classical dynamics trajectory, we separate the entire N-atom system into a primary zone, consisting of N_1 atoms (typically those in the QM-subsystem) and a secondary zone (the rest of N–N_1 atoms). We then freeze all secondary atoms and evaluate the normal modes of the primary zone system under the effects of the secondary zone. In computing the QC-PMF, $3N_1 - 7$ modes at the transition state and $3N_1 - 6$ modes at the reactant state are used. To evaluate the quasiclassical free energy of activation, $\Delta F^{\neq}_{QC}$, quantum corrections must also be made to the $F^{\rm R}_{\rm CM}(z)$ term in Eq. 7. A procedure for making these corrections has been reported [18].

The rate constant with quantized vibrations using $\Delta F^{\neq}_{QC}$ is:

$$k_{QC} = \frac{1}{\beta h} e^{-\beta \Delta F^{\neq}_{QC}}, \tag{18}$$

where $\Delta F^{\neq}_{QC}$ is the quasiclassical free energy of activation. We note here that the recrossing transmission coefficient can be estimated by coupling the reaction coordinate z to the rest of the $3N_1 - 1$ degrees of freedom using the frozen bath approximation. Most applications show that the recrossing factor is in the range of 0.2 to 1. Thus, it is not further discussed in this article.

Billeter et al. [92] mixed quantum and classical molecular dynamics (MQCMD) to obtain the quantized PMF, and then a transmission coefficient was added by molecular dynamics with quantum transitions. In the MQCMD calculation, which has been applied to hydrogen transfer reactions, the three degrees of freedom for the transferring atom are treated as a quantal subsystem, and a numerical method is used to solve the three-dimensional vibrational wave function. The quantal subsystem is embedded in the rest of the system, which is treated classically [94, 95]. The large number of grid points, on which the potentials from the environment are evaluated, are the computational bottleneck in this approach, which has limited its application to quantize only one atom [92].

Free energy simulations of model proton shifts in water as well as enzymatic hydrogen transfer reactions in a number of enzymes indicate that inclusion of quantum vibrational free energy contributions reduces the classical barrier height by 2 to 4 kcal/mol and that more degrees of freedom than those associated with the migrating atoms are needed in these calculations [91–93, 96, 97].

2.2.2 Tunneling

Nuclear tunneling and other quantum effects on the reaction coordinate that are missing in the calculation of the PMF are treated separately from effects on other degrees of freedom [8], although QM effects are not uniquely separable [7]. Nevertheless, it provides useful insights into the factors of QM contributions to the evaluation of the rate constant. In path integral (see next section) or quantum-classical molecular dynamics (QCMD) calculations, such a separation is obviously not possible, and, to a certain degree, the effect of the nuclear tunneling contributions is absorbed into the computed PMF in these methods. From a more general perspective, any reliable approach to calculating tunneling effects must be multidimensional to take account the nonseparability of the reaction coordinate [8].

In the ensemble-averaged variational TST with multidimensional tunneling (EA-VTST/MT) [18, 27], nuclear quantum effects on the reaction coordinate are estimated as a transmission coefficient. Thus, the overall quantum correction factor is:

$$\kappa = \frac{k_{qm}}{k_{TST}} = \kappa_{MT} \cdot e^{-\beta\left(\Delta F_{QC}^{\neq} - \Delta F_{TST}^{\neq}\right)}, \tag{19}$$

where κ_{MT} is the multidimensional tunneling (MT) transmission coefficient. Here, we have again omitted the classical recrossing factor. The κ_{MT} coefficient is approximated as [18]:

$$\kappa_{MT} = < \kappa_{MT}(i) >_{\neq} \cong \frac{1}{M} \sum_{i=1}^{M} \kappa_{MT}(i), \tag{20}$$

where the brackets $\langle \cdots \rangle_{\neq}$ represent an ensemble average over transition state (TS) configurations at temperature T, κ_{MT} accounts for tunneling through the effective barrier and nonclassical diffractive reflection from the barrier top, i denotes a particular member of the QC transition state ensemble, and M is the total number of configurations that have been sampled in the calculation. The ensemble of TS configurations can be generated during the PMF calculation using umbrella sampling [18] or in a separate molecular dynamics simulation with the constraint that the reaction coordinate corresponds to the TS value [91, 96].

For each configuration i in the TS ensemble [18], the "tunneling factor" $\kappa_{MT}(i)$ is evaluated by a semiclassical MT approximation, either the small-curvature tunneling (SCT) approximation [98], or the microcanonically optimized multidimensional tunneling (μ OMT) approximation [99]. The latter involves choosing, at each tunneling energy, the better of the small-curvature [98] and large-curvature [100, 101] tunneling approximations.

2.2.3 Centroid Path Integral Simulations

Feynman path integral simulations provide a convenient procedure for incorporating quantum effects on vibrations [102]. In this approach, the ensemble averages for the quantum system can be obtained by carrying out a classical simulation in which the quantized particles are represented by ring polymers of classical particles. To determine the PMF, the average position (centroid) of the quantized particles is used as a classical variable [103–107], leading to a method called path integral quantum TST (PI-QTST). Although this approach has been successful for some problems and it was applied to an enzyme reaction [108], it has been noted that difficulty exists in using the approach for asymmetric reactions at low temperature [109–112].

Hwang et al. used an approach called quantized classical path (QCP), in which quantum effects are formulated as a correction to the classical PMF [97, 107]. Thus, the classical simulations and quantum corrections are fully separated, making it particularly attractive and efficient to model enzymatic reactions [6, 97, 107, 113].

In the discrete path integral method, each quantized nucleus is represented by a ring of P quasiparticles called beads, whose coordinates are denoted as $\mathbf{r} = \{\mathbf{r}_i; i = 1, \cdots, P\}$, with a definition of $\mathbf{r}_{P+1} = \mathbf{r}_1$. Each bead is connected to its two neighbors via harmonic springs, and is subjected to a fraction, $1/P$, of the full classical potential, $U(\mathbf{r}_i, \mathbf{S})$, where $\mathbf{S}$ represents all classical protein–solvent coordinates. The following discussion can easily be extended to many-quantized particles. In a centroid path integral, the centroid position, $\bar{\mathbf{r}}$, is used as the principle variable and the canonical QM partition function of the hybrid system can be written as follows:

$$Q_P^{qm} = \int d\mathbf{S} \int d\bar{\mathbf{r}} \left(\frac{P}{2\pi\lambda^2}\right)^{3P/2} \int d\mathbf{R} e^{-\beta V^{qm}(\mathbf{r},\mathbf{S})}, \tag{21}$$

where $V^{qm}(\mathbf{r}, \mathbf{S})$ is the effective QM potential, $\int d\mathbf{R} = \int d\mathbf{r}_1 \cdots \int d\mathbf{r}_P \delta(\bar{\mathbf{r}})$; the centroid coordinate, $\bar{\mathbf{r}}$, of the quasiparticles is defined as $\bar{\mathbf{r}} = 1/P \sum_{i=1}^{P} \mathbf{r}_i$; and the de Broglie thermal wavelength, λ^2, of a particle of mass M is $\lambda^2 = \beta\bar{h}^2/M$. The key result in the hybrid classical and path integral approach or quantized classical path is that the quantum partition function can be rewritten as the double averages [97, 107, 114–116]:

$$Q_P^{qm} = Q_P^{cm} << e^{-\beta\Delta\overline{U}(\bar{\mathbf{r}},\mathbf{S})} >_{FP,\bar{\mathbf{r}}}>_{U_{tot}}, \tag{22}$$

where Q_P^{cm} is the classical partition function [102], the average $< \cdots >_{U_{tot}}$ is obtained according the potential $U_{tot}(\bar{\mathbf{r}}, \mathbf{S})$, which is of QM/MM type, but purely classical in nuclear degrees of freedom, and the inner average $< \cdots >_{FP,\bar{\mathbf{r}}}$ represents free particle sampling carried out without the external potential $U_{tot}(\bar{\mathbf{r}}, \mathbf{S})$:

$$< \cdots >_{FP,\bar{\mathbf{r}}} = \frac{\int d\mathbf{r}_P \{\cdots\} \delta(\bar{\mathbf{r}}) e^{-(P/2\lambda^2)\Sigma_i^P (\Delta\mathbf{r}_i)^2}}{\int d\mathbf{r}_P \delta(\bar{\mathbf{r}}) e^{-(P/2\lambda^2)\Sigma_i^P (\Delta\mathbf{r}_i)^2}}, \tag{23}$$

where $\Delta\mathbf{r}_i = \mathbf{r}_i - \mathbf{r}_{i+1}$. This procedure was initially used by Sprik et al. for a system consisting of one electron embedded in random hard spheres [114]. As Warshel pointed out [97, 107], the expression of Eq. 22 is particularly useful because the quantum free energy of the system can be obtained by first carrying out classical trajectories for averaging classical configurations $(\bar{\mathbf{r}}, \mathbf{S})$, and then determining the quantum contributions through free particle sampling by path integral simulations (Eq. 23).

Based on Eq. 22, the QM PMF, defined as a function of the centroid reaction coordinate, $\bar{z}$, can be readily expressed by:

$$w_{QM}(\bar{z}) = w_{CM}(\bar{z}) - k_B T \ln << e^{-\beta\Delta\overline{U}(\bar{z})} >_{\mathrm{FP},\bar{z}}>_{U_{tot}}, \tag{24}$$

where $w_{QM}(\bar{z})$ and $w_{CM}(z = \bar{z})$ are the centroid QM and CM PMF, respectively, and the average potential energy is given as follows:

$$\Delta\overline{U}(\bar{z}) = \frac{1}{P}\sum_{i}^{P}\{U(\mathbf{r}_i, \mathbf{S}) - U(\bar{\mathbf{r}}, \mathbf{S})\}. \tag{25}$$

Finally, the quantum free energy of activation (in Eq. 4) can be obtained by:

$$\Delta F_{qm}^{\neq} = \left[w_{QM}\left(\bar{z}_{qm}^{\neq}\right) - w_{QM}\left(\bar{z}_{qm}^{R}\right)\right], \tag{26}$$

where the symbol $\bar{z}_{qm}^{\neq}$ specifies the value of the centroid reaction coordinate, at which $w_{QM}(\bar{z})$ has the maximum value, and $\bar{z}_{qm}^{R}$ is the coordinate at the reactant state.

A bisection sampling scheme has been developed for centroid path integral simulations [115, 116], based on the original procedure of Ceperley [117], and this method has been implemented in the context of the quantized classical path simulation strategy (BQCP). Through a series of investigations [79, 118], it has been demonstrated that the BQCP sampling procedure can yield rapidly converging results [115, 116], which has been a major problem for application to enzymes. In general, any particle position of the cyclic quasiparticles can be expressed as:

$$\mathbf{r}_i = \lambda_M \theta_i; \quad i = 1, 2, \cdots, \mathrm{P}, \tag{27}$$

where the vector θ_i is a generalized positive vector, properly scaled, generated randomly according to the free particle distribution, and associated with earlier levels of bisection sampling. The specific details have been given in references [115, 116, 119]. Note that the beads positions are dependent on the particle mass via λ_M.

2.3 Computation of KIE

An integrated path integral–free energy perturbation and umbrella sampling (PI-FEP/UM) method has been developed to compute KIE, in which molecular dynamics simulations are first carried out to obtain the CM-PMF using umbrella sampling. Then, the nuclear coordinates of atoms associated with the chemical reaction are quantized by a path integral with the constraint that the centroid positions coincide with their corresponding classical coordinates. KIE are evaluated by free energy perturbation between heavy and light atom masses, which is related to the quantized quasiparticle positions.

Considering an atom transfer reaction in which the light atom of mass M_L is replaced by a heavier isotope of mass M_H, we use exactly the same sequence of random numbers, i.e., displacement vectors, to generate the bisection path integral distribution for both isotopes to obtain the free particle distribution. Thus, the resulting coordinates of these two bead-distributions differ only by the ratio of the corresponding masses:

$$\frac{\mathbf{r}_{i,L}}{\mathbf{r}_{i,H}} = \frac{\lambda_{M_L}\theta_i}{\lambda_{M_H}\theta_i} = \sqrt{\frac{M_H}{M_L}}; \quad i = 1, 2, \cdots, \mathrm{P}, \tag{28}$$

where $\mathbf{r}_{i,L}$ and $\mathbf{r}_{i,\mathrm{H}}$ are the coordinates for bead i of the corresponding light and heavy isotopes.

The KIE can be computed by considering the ratio between the heavy and light QM partition functions. At a given reaction coordinate value $\bar{z}$ (centroid coordinates), the ratio can be obtained exactly by free energy perturbation:

$$\frac{Q_{QM}^{H}(\bar{z})}{Q_{QM}^{L}(\bar{z})} = \frac{< \delta(z-\bar{z}) < e^{-\frac{\beta}{P}\Sigma_i \Delta U_i^{L\to H}} e^{-\beta\Delta\overline{U}_L} >_{FP,L}>_{U_{tot}}}{< \delta(z-\bar{z}) e^{-\beta[F_L(\bar{z},\mathbf{S})-F_{FP}^{\mathrm{o}}]} >_{U_{tot}}}, \tag{29}$$

where the superscripts or subscripts L and H specify computations performed using light or heavy isotopes, and $\Delta U_i^{L\to H} = U_{tot}(\mathbf{r}_{i,H}) - U_{tot}(\mathbf{r}_{i,L})$ represents the difference in potential energy at the heavy and light bead positions, $\mathbf{r}_{i,H}$ and $\mathbf{r}_{i,L}$. In Eq. 29, we obtain the free energy (inner average) difference between the heavy and light isotopes by carrying out the bisection path integral sampling with the light atom and then perturbing the heavy isotope positions according to Eq. 28. Then, the free energy difference between the light and heavy isotope ensembles is weighted by free energy (outer average).

The KIE are computed as follows:

$$KIE = \frac{k^L}{k^H} = \left[\frac{Q_{QM}^{L}\left(\bar{z}_L^{\neq}\right)}{Q_{QM}^{H}\left(\bar{z}_H^{\neq}\right)}\right]\left[\frac{Q_{QM}^{H}\left(\bar{z}_H^{R}\right)}{Q_{QM}^{L}\left(\bar{z}_L^{R}\right)}\right] e^{-\beta\left\{F_{\mathrm{CM}}^{\mathrm{R}}\left(\bar{z}_L^{R}\right)-F_{\mathrm{CM}}^{\mathrm{R}}\left(\bar{z}_H^{R}\right)\right\}}. \tag{30}$$

Note that $F_{\mathrm{CM}}^{\mathrm{R}}(\bar{z}_L^{R})$ and $F_{\mathrm{CM}}^{\mathrm{R}}(\bar{z}_H^{R})$ are the free energies of the mode in the reactant (R) state that correlates with the progress coordinate z for the light and heavy isotopes. A method for estimating their values has been described [18].

3 Illustrative Examples

3.1 The Symmetric Eckart Barrier

We first present the result for a model system of proton transfer barrier described by the symmetric Eckart potential:

$$V(z) = V_{\max}\cosh^2\left(\frac{\pi z}{l}\right), \tag{31}$$

Table 1 Computed QM correction factor for hydrogen and H/D KIE for the Eckart potential

Temperature (K)	κ		κ^H/κ^D	
	BQCP	Exact	PI-FEP/UM	Exact
500	1.43	1.53	1.19	1.23
400	1.75	1.92	1.31	1.37
350	2.09	2.23	1.43	1.51
300	2.73	3.15	1.62	1.76
250	4.46	5.31	2.08	2.28
200	11.5	15.0	3.33	3.84
100	1.30×10^5	1.80×10^5	156	106

Table 2 Computed and experimental primary $^{12}C/^{13}C$ and secondary $^{14}N/^{15}N$ KIE for the decarboxylation of N-methyl picolinate at 25°C in water

	$^{12}k/^{13}k$	$^{14}k/^{15}k$
Exp (25°C)	1.0281 ± 0.0003	1.0070 ± 0.0003
PI-FEP/UM	1.0318 ± 0.0028	1.0083 ± 0.0016

where $V_{max} = 5.7307$ kcal/mol and $l = 1.0967$ Å [120, 121]. The PI-BQCP simulation was carried out by 10,000 Monte Carlo steps using 32 beads for the quantized particle. The computed quantum correction factors at various temperatures are given in Table 1, which are in reasonable agreement with the exact values [104, 120–122]. Similar results were also obtained using 64 beads by Hwang and Warshel [107]. The deviation of the present BQCP simulation from the exact values is likely caused by a combination of factors including the centroid approximation near the top of the barrier reducing the computed tunneling contributions, and a finite number of beads used in the sampling. The results listed in Table 1 were obtained with 32 beads in BQCP sampling, but we have tested the convergence by using 256 beads at 300 K, which yielded results within 5% of that for the 32-bead system.

Table 1 also lists the computed KIE for a hypothetic proton and deuterium transfer over this barrier, which is compared with the exact results (the computed KIE included only the quantum correction terms). In this comparison, the quantum correction factors for the hydrogen and deuterium transfer reactions are computed independently along the "reaction coordinate," z (Eq. 31). It is evident that the errors in the computed KIE are reduced relative to that of the transmission coefficient (Table 2). At 300 K, the KIE is underestimated by 0.14, which translates to an error of 8% in comparison to that from the exact data.

3.2 *The Decarboxylation of N-Methyl Picolinate in Water*

The decarboxylation of N-methyl picolinate was used as a model to probe the mechanism of the uncatalyzed reaction in water for orotidine 5′-monophosphate decarboxylase (Scheme 1). The primary and secondary heavy atom KIE were determined

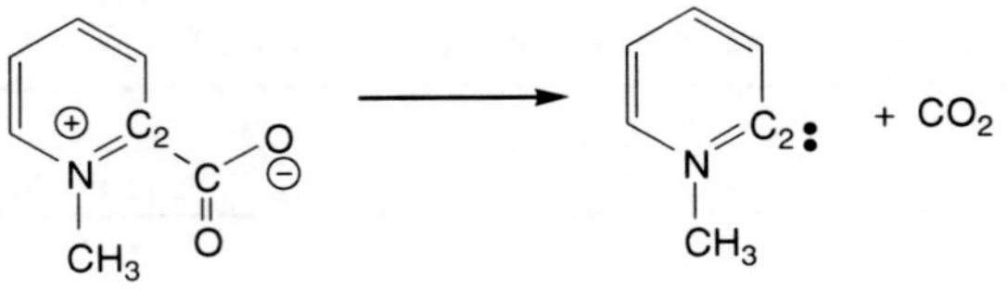

Scheme 1

by Rishavy and Cleland at an elevated temperature and then extrapolated to 25°C [123]. Classical molecular dynamics simulations and umbrella sampling were first carried out for a system containing the *N*-methyl picolinate, treated by the AM1 Hamiltonian, in a cubic box ($30 \times 30 \times 30\,\text{Å}^3$) of 888 water molecules, described by the TIP3P potential [119]. In all calculations, long-range electrostatic interactions were treated by the particle mesh–Ewald method for QM/MM potentials [124]. The CM-PMF as a function of the C_2–C_{O2} distance reaction coordinate was obtained at 25°C and 1 atm through a total of 2 ns of dynamics simulations. Then, the coordinates from the classical trajectory were used in path integral BQCP simulations, a total of 97,800 classical configurations was used for each isotope (^{12}C, ^{13}C, ^{14}N, and ^{15}N) for the decarboxylation reaction, combined with 10 path-integral steps per classical configuration. Each quantized atom was described by 32 beads.

Solvent effects are significant, increasing the free energy barrier by 15.2 kcal/mol to a value of 26.8 kcal/mol, which is accompanied by a net free energy of reaction of 24.7 kcal/mol. The large solvent effect is caused by the presence of a positive charge on the pyridine nitrogen, which is annihilated in the decarboxylation reaction. There is only approximately 2 kcal/mol of CO_2 recombination barrier, which would cause difficulty for computing KIE using only a single transition structure.

Both the ^{12}C/^{13}C primary KIE and the ^{14}N/^{15}N secondary KIE have been determined (Table 2), with the immediate adjacent atoms around the isotopic substitution site quantized as well. Figure 1 shows the absolute quantum correction factor (in kcal/mol) for both ^{12}C and ^{13}C at the carboxyl position, and Fig. 2 depicts their difference to illustrate the computational sensitivity. First, we note that the nuclear quantum effects are not negligible even for bond cleavage involving two carbon atoms, which reduce the free energy barrier by 0.45 kcal/mol. The computed intrinsic ^{13}C primary KIE, without including the $F_{\mathrm{CM}}^{\mathrm{R}}(\bar{z}^R)$ term in Eq. 30, is 1.0318 ± 0.0028 at 25°C for the decarboxylation of *N*-methyl picolinate in water (Table 1). To emphasize the sensitivity of the computational result, the computed KIE is equivalent to a free energy difference of merely 0.0187 kcal/mol (Fig. 2), which is feasible by the use of free energy perturbation/umbrella sampling techniques. For comparison, the experimental value is 1.0281 ± 0.0003 at 25°C. For the secondary ^{15}N KIE, the PI-FEP/UM simulation yields an average value of 1.0083 ± 0.0016, which may be compared with experiment (1.0070 ± 0.0003) [123]. The agreement between theory and experiment is excellent, which provides support for a unimolecular decarboxylation mechanism in this model reaction.

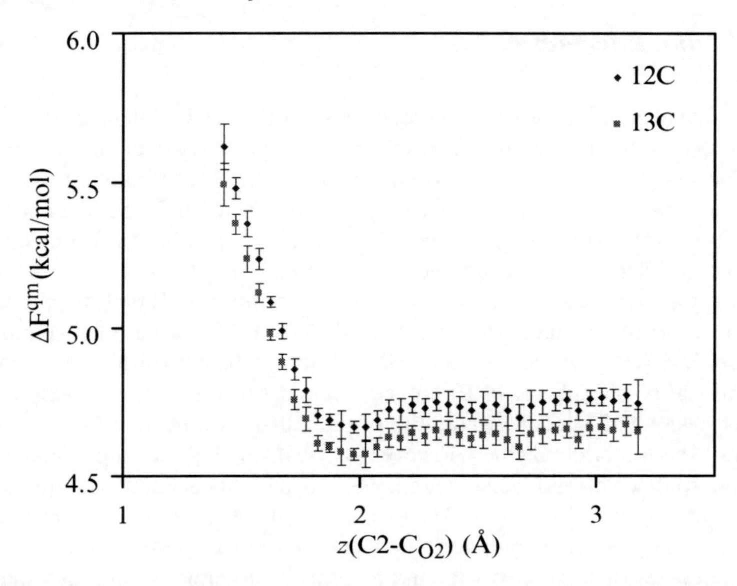

Fig. 1 Nuclear QM free energy corrections for the decarboxylation reaction of *N*-methyl picolinate in aqueous solution

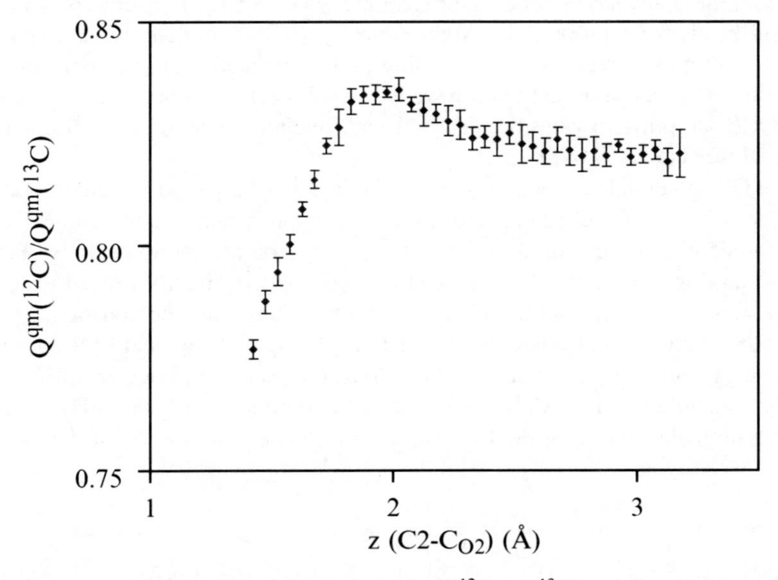

Fig. 2 The ratio of the QM partition functions between ^{12}C and ^{13}C isotopic substitutions at the carboxyl carbon position from the PI-FEP/UM method

3.3 Alanine Racemase

Alanine racemase catalyzes the interconversion of L- and D-alanine, the latter of which is an essential component in the peptidoglycan layer of the bacterial cell wall. The biosynthesis of D-Ala is unique to bacteria [125], making the enzyme alanine racemase (AlaR) an attractive target for antibacterial drug design [126]. The chemical transformation is illustrated in Scheme 2, which has been modeled by a combined QM/MM potential in molecular dynamics simulations.

Following the procedure described above, the primary KIE both for the forward and reverse processes have been determined [118, 127]. In these calculations, the semiempirical AM1 formalism was used to describe the active site, which includes the pyridoxal 5′-phosphate (PLP) cofactor-bound substrate and the acid and base residues Lys39 and Tyr265′, where the prime indicates a residue from the second subunit of the dimeric enzyme. However, the AM1 model was reparameterized to model the AlaR-catalyzed racemization, yielding a highly accurate Hamiltonian that is comparable to mPW1PW91/6-311++G(3df,2p) calculations. Stochastic boundary molecular dynamics simulations were carried out for a system of a 30 Å sphere about the center of the active site, and a series of umbrella sampling simulations were performed, cumulating a total of 24 ns statistical sampling (with a 1-fs integration step) to yield the classical PMF. Then, the transferring proton, the donor, and acceptor heavy atoms for each process are quantized by a centroid path integral with 32 beads for each particle. Approximately 15,000 configurations saved from these trajectories in regions corresponding to the Michaelis complex reactant state, transition state, and product state were extracted, each of which was subjected to 10 BQCP sampling to yield the centroid path integral–quantum corrections to the classical free energy profile.

The QM potentials of mean force are displayed in Fig. 3, which incorporate the QM corrections to the classical PMF from molecular dynamics simulations. For the L $\rightarrow$ D alanine racemization in AlaR, the first proton abstraction step by Tyr265′ is rate limiting, and, thus, the observed rate constant is directly related to this reaction step, $k_{obs} = k_1$, and the KIE is computed using this rate constant. Figure 3 shows that inclusion of QM contributions to the computed classical PMF lowers the free energy barrier by 2.60 and 1.74 kcal/mol for proton and deuteron transfer to Tyr265′ phenolate ion in AlaR. This leads to a computed intrinsic KIE of 4.21 for the α-proton abstraction in the L $\rightarrow$ D alanine conversion; the $F^{\mathrm{R}}_{\mathrm{CM}}(\bar{z}^R)$ has been

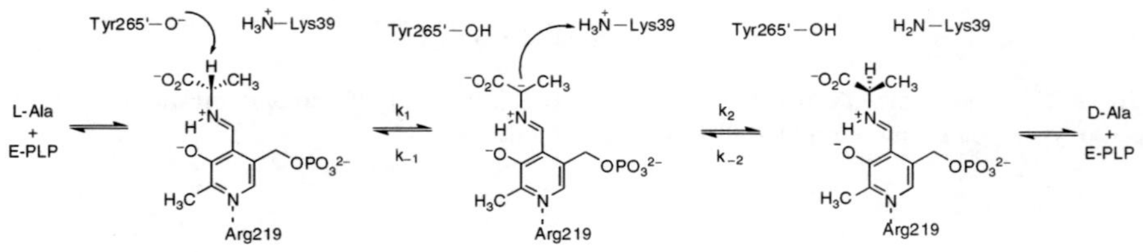

Scheme 2

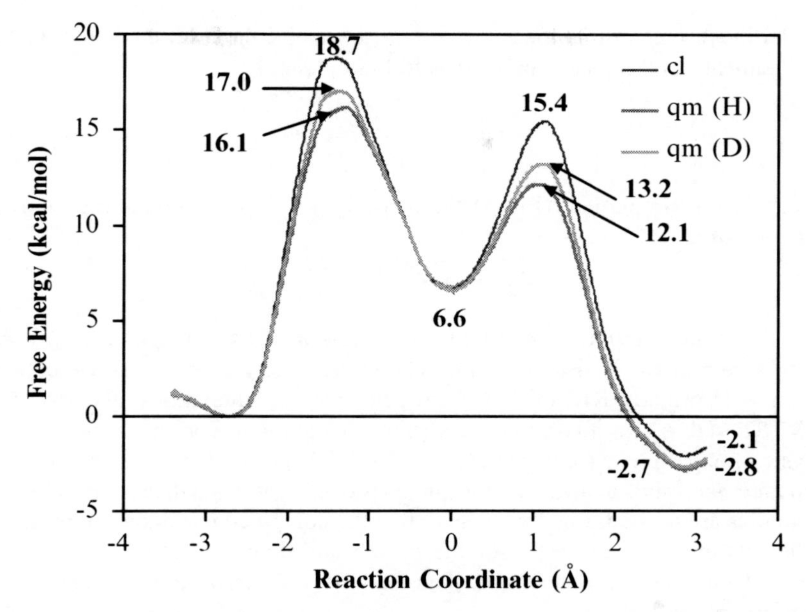

Fig. 3 Computed CM and QM potentials of mean force for the proton and deuteron abstraction of Ala-PLP by Tyr265′, and the reprotonation of Ala-PLP carbanion intermediate by Lys39 in alanine racemase

neglected in these calculations. The computational result is much greater than the experimental value of 1.9 [128], which may be a result of one factor or a combination of factors, including the complexity involved in the analysis of experimental kinetic data, the possibility that the experimental KIE was not exactly the intrinsic value, and computational uncertainty.

For comparison, the primary intrinsic KIE was also computed using the EA-VTST/MT method [8, 18], yielding a value of 3.97 for the Tyr265′ proton abstraction, in close agreement with the path integral simulation results. Interestingly, the EA-VTST/MT method allows the separation of the total nuclear quantum effects into vibrational motions and tunneling. We found that the dominant QM contribution is caused by the change in zero-point energy in going from the reactant state to the transition state, and that hydrogen tunneling is negligible, with an average transmission factor of 1.14 and 1.31 for the hydrogen and deuterium. Furthermore, the net quantum effects, from EA-VTST/MT calculations, lower the classical barrier by 2.71 and 1.89 kcal/mol for the hydrogen and deuterium transfer reactions, respectively, in accord with the BQCP calculations. The average recrossing transmission factors, Γ, were also computed using the EA-VTST/MT method, which are 0.96 and 0.94 for the hydrogen and deuterium transfers, respectively.

For the D $\rightarrow$ L conversion of alanine by AlaR, the actual proton transfer step is not rate limiting. If we neglect the complexity of reaction steps involving internal and external aldimine exchange, substrate binding, and product release, i.e., we only consider the proton abstraction and reprotonation steps, the overall rate constant for the "chemical step" can be expressed as follows [127]:

$$k_{eff} = \frac{k_{-2}k_{-1}}{k_{-1} + k_2}. \tag{32}$$

Because our simulations (Fig. 3) show that $k_2 >> k_{-1}$, we obtain the following rate expression:

$$k_{eff} = \frac{k_{-2}k_{-1}}{k_2} = \frac{k_{-1}}{K_2}, \tag{33}$$

where K_2 is the equilibrium constant for the proton transfer reaction from L-Ala-PLP to a neutral Lys39. Based on the relative free energies in Fig. 3, we obtained an estimated primary KIE of 3.06 from path integral simulations, and 2.65 from EA-VTST/MT. Again, these values are greater than that from a multicomponent analysis of experimental kinetic data (1.43 ± 0.20) [128]. The difference between computational results and values from analysis of experimental data suggest that further studies are needed. Finally, we note that the computed recrossing transmission coefficients are 0.81 and 0.87 for the proton and deuteron transfer reactions to Lys39, which are somewhat smaller than the L to D isomerization, but they are still very close to unity.

4 Concluding Remarks

We presented a method for incorporating quantum mechanics into enzyme kinetics modeling. Three aspects are emphasized: 1) the PES is represented by combined QM/MM method in which electronic structure theory is used to describe bond forming and breaking processes, 2) quasiclassical PMF is obtained by incorporating quantum vibrational free energies, and 3) quantum effects on the reaction coordinate motion are estimated by the use of MT methods. In the EA-VTST/MT, these three aspects of QM effects can be individually treated, providing useful insights into the mechanism of enzymatic reactions. Centroid path integral simulations that make QM corrections to the CM PMF are also described. In this method, the nuclear quantum vibrational and tunneling contributions are not separable. An integrated centroid path integral–free energy perturbation and umbrella sampling (PI-FEP/UM) method along with a bisection sampling procedure was summarized, which provides an accurate, easily convergent method for computing KIE for chemical reactions in solution and in enzymes. These methods are illustrated by applications to a model process in the gas phase, the decarboxylation reaction of N-methyl picolinate in water, and the proton abstraction and reprotonation process catalyzed by alanine racemase. These examples show that the incorporation of QM effects is essential for enzyme kinetics simulations. These computational approaches provided insights and helped to interpret experimental data, such as KIE.

Acknowledgements This work was supported in part by the National Institutes of Health and by the Army Research Laboratory through the Army High-Performance Computing Research Center (AHPCRC) under the auspices of Army Research Laboratory DAAD 19-01-2-0014. JG is grateful to Professor Modesto Orozco for hospitality during his sabbatical leave at the Barcelona Supercomputing Center, and acknowledges support from the Ministerio de Educación y Ciencia, Spain.

References

1. Sumner, J. B. (1926) "THE ISOLATION AND CRYSTALLIZATION OF THE ENZYME UREASE. PRELIMINARY PAPER," *J. Biol. Chem.* **69**, 435–441.
2. Pauling, L. (1946) "Molecular architecture and biological reactions," *Chem. Eng. News* **24**, 1375.
3. Schowen, R. L. (1978) in *Transition States of Biochemical Processes* (Gandour, R. D., and Schowen, R. L., Eds.) pp 77–114, Plenum Press, New York.
4. Gao, J., Ma, S., Major, D. T., Nam, K., Pu, J., and Truhlar, D. G. (2006) "Mechanisms and free energies of enzymatic reactions," *Chem. Rev.* **106**, 3188–3209.
5. Garcia-Viloca, M., Gao, J., Karplus, M., and Truhlar, D. G. (2004) "How enzymes work: Analysis by modern rate theory and computer simulations," *Science* **303**, 186–195.
6. Villa, J. and Warshel, A. (2001) "Energetics and dynamics of enzymatic reactions," *J. Phys. Chem. B* **105**, 7887–7907.
7. Gao, J. and Truhlar, D. G. (2002) "Quantum mechanical methods for enzyme kinetics," *Ann. Rev. Phys. Chem.* **53**, 467–505.
8. Pu, J., Gao, J., and Truhlar, D. G. (2006) "Multidimensional tunneling, recrossing, and the transmission coefficient for enzymatic reactions," *Chem. Rev.* **106**, 3140–3169.
9. Wolfenden, R. and Snider, M. J. (2001) "The depth of chemical time and the power of enzymes as catalysts," *Acc. Chem. Res.* **34**, 938–945.
10. Snider, M. G., Temple, B. S., and Wolfenden, R. (2004) "The path to the transition state in enzyme reactions: A survey of catalytic efficiencies," *J. Phys. Org. Chem.* **17**, 586–591.
11. Truhlar, D. G., Garrett, B. C., and Klippenstein, S. J. (1996) "Current status of transition-state theory," *J. Phys. Chem.* **100**, 12771–12800.
12. Fernandez-Ramos, A., Miller, J. A., Klippenstein, S. J., and Truhlar, D. G. (2006) "Modeling the kinetics of bimolecular reactions," *Chem. Rev.* **106**, 4518–4584.
13. Bennett, C. H. (1977) "Molecular dynamics and transition state theory: the simulation of infrequent events," *ACS Symp. Ser.* **46**, 63–97.
14. Chandler, D. (1978) "Statistical mechanics of isomerization dynamics in liquids and the transition state approximation," *J. Chem. Phys.* **68**, 2959–2970.
15. Neria, E. and Karplus, M. (1997) "Molecular dynamics of an enzyme reaction: proton transfer in TIM," *Chem. Phys. Lett.* **267**, 26–30.
16. Nam, K., Prat-Resina, X., Garcia-Viloca, M., Devi-Kesavan, L. S., and Gao, J. (2004) "Dynamics of an enzymatic substitution reaction in haloalkane dehalogenase," *J. Am. Chem. Soc.* **126**, 1369–1376.
17. Voth, G. A. and Hochstrasser, R. M. (1996) "Transition state dynamics and relaxation processes in solutions: a frontier of physical chemistry," *J. Phys. Chem.* **100**, 13034–13049.
18. Alhambra, C., Corchado, J., Sanchez, M. L., Garcia-Viloca, M., Gao, J., and Truhlar, D. G. (2001) "Canonical variational theory for enzyme kinetics with the protein mean force and multidimensional quantum mechanical tunneling dynamics. Theory and application to liver alcohol dehydrogenase," *J. Phys. Chem. B* **105**, 11326–11340.
19. Valleau, J. P. and Torrie, G. M. (1977) in *Modern Theoretical Chemistry* (Berne, B. J., Ed.) pp 169–194, Plenum, New York.
20. Jorgensen, W. L. (1989) "Free energy calculations: a breakthrough for modeling organic chemistry in solution," *Acc. Chem. Res.* **22**, 184–189.

21. Hwang, J. K., King, G., Creighton, S., and Warshel, A. (1988) "Simulation of free energy relationships and dynamics of SN2 reactions in aqueous solution," *J. Am. Chem. Soc.* **110**, 5297–5311.
22. Gertner, B. J., Bergsma, J. P., Wilson, K. R., Lee, S., and Hynes, J. T. (1987) "Nonadiabatic solvation model for SN2 reactions in polar solvents," *J. Chem. Phys.* **86**, 1377–1386.
23. Gao, J. (1996) "Hybrid quantum mechanical/molecular mechanical simulations: an alternative avenue to solvent effects in organic chemistry," *Acc. Chem. Res.* **29**, 298–305.
24. Muller, R. P. and Warshel, A. (1995) "Ab initio calculations of free energy barriers for chemical reactions in solution," *J. Phys. Chem.* **99**, 17516–17524.
25. Mo, Y. and Gao, J. (2000) "Ab initio QM/MM simulations with a molecular orbital-valence bond (MOVB) method: application to an SN2 reaction in water," *J. Comput. Chem.* **21**, 1458–1469.
26. Truhlar, D. G. and Garrett, B. C. (2000) "Multidimensional transition state theory and the validity of Grote-Hynes theory," *J. Phys. Chem. B* **104**, 1069–1072.
27. Truhlar, D. G., Gao, J., Alhambra, C., Garcia-Viloca, M., Corchado, J., Sanchez, M. L., and Villa, J. (2002) "The incorporation of quantum effects in enzyme kinetics modeling," *Acc. Chem. Res.* **35**, 341–349.
28. MacKerell, A. D., Jr., Bashford, D., Bellott, M., Dunbrack, R. L., Evanseck, J. D., Field, M. J., Fischer, S., Gao, J., Guo, H., Ha, S., Joseph-McCarthy, D., Kuchnir, L., Kuczera, K., Lau, F. T. K., Mattos, C., Michnick, S., Ngo, T., Nguyen, D. T., Prodhom, B., Reiher, W. E., III, Roux, B., Schlenkrich, M., Smith, J. C., Stote, R., Straub, J., Watanabe, M., Wiorkiewicz-Kuczera, J., Yin, D., and Karplus, M. (1998) "All-atom empirical potential for molecular modeling and dynamics studies of proteins," *J. Phys. Chem. B* **102**, 3586–3616.
29. Ponder, J. W. and Case, D. A. (2003) "Advances in Protein Chemisty", edited by V. Daggett, QM/MM and related approaches, *Adv. Protein Chem.* **66**, 27–85.
30. Gao, J. and Xia, X. (1992) "A prior evaluation of aqueous polarization effects through Monte Carlo QM-MM simulations," *Science* **258**, 631–635.
31. Gao, J. (1995) in *Rev. Comput. Chem.* (Lipkowitz, K. B., and Boyd, D. B., Eds.) Methods and applications of combined QM/MM methods. pp 119–185, VCH, New York.
32. Chandrasekhar, J., Smith, S. F., and Jorgensen, W. L. (1984) "SN2 reaction profiles in the gas phase and aqueous solution," *J. Am. Chem. Soc.* **106**, 3049–3050.
33. Gao, J. (1991) "A priori computation of a solvent-enhanced SN2 reaction profile in water: the Menshutkin reaction," *J. Am. Chem. Soc.* **113**, 7796–7797.
34. Donini, O., Darden, T., and Kollman, P. A. (2000) "QM-FE calculations of aliphatic hydrogen abstraction in citrate synthase and in solution: reproduction of the effect of enzyme catalysis and demonstration that an enolate rather than an enol is formed," *J. Am. Chem. Soc.* **122**, 12270–12280.
35. Zhang, Y. K., Liu, H. Y., and Yang, W. T. (2000) "Free energy calculation on enzyme reactions with an efficient iterative procedure to determine minimum energy paths on a combined ab initio QM/MM potential energy surface," *J. Chem. Phys.* **112**, 3483–3492.
36. Liu, H., Lu, Z., Cisneros, G. A., and Yang, W. (2004) "Parallel iterative reaction path optimization in ab initio quantum mechanical/molecular mechanical modeling of enzyme reactions," *J. Chem. Phys.* **121**, 697–706.
37. Warshel, A. and Weiss, R. M. (1980) "An empirical valence bond approach for comparing reactions in solutions and in enzymes," *J. Am. Chem. Soc.* **102**, 6218–6226.
38. Warshel, A. (1991) *Computer Modeling of Chemical Reactions in Enzymes and Solutions*, Wiley, New York.
39. Aaqvist, J., Fothergill, M., and Warshel, A. (1993) "Computer simulation of the carbon dioxide/bicarbonate interconversion step in human carbonic anhydrase I," *J. Am. Chem. Soc.* **115**, 631–635.
40. Kim, Y., Corchado, J. C., Villa, J., Xing, J., and Truhlar, D. G. (2000) "Multiconfiguration molecular mechanics algorithm for potential energy surfaces of chemical reactions," *J. Chem. Phys.* **112**, 2718–2735.

41. Olsson, M. H. M. and Warshel, A. (2004) "Solute solvent dynamics and energetics in enzyme catalysis: the SN2 reaction of dehalogenase as a general benchmark," *J. Am. Chem. Soc.* **126**, 15167–15179.
42. Warshel, A. (2003) "Computer simulations of enzyme catalysis: Methods, progress, and insights," *Ann. Rev. Biophys. Biomol. Struct.* **32**, 425–443.
43. Aqvist, J. and Warshel, A. (1993) "Simulation of enzyme reactions using valence bond force fields and other hybrid quantum/classical approaches," *Chem. Rev.* **93**, 2523–2544.
44. Warshel, A., Sharma, P. K., Kato, M., Xiang, Y., Liu, H., and Olsson, M. H. M. (2006) "Electrostatic basis for enzyme catalysis," *Chem. Rev.* **106**, 3210–3235.
45. Mo, Y., Zhang, Y., and Gao, J. (1999) "A simple electrostatic model for trisilylamine: theoretical examinations of the n→sigma. Negative hyperconjugation, p.pi.→d.pi. bonding, and stereoelectronic interaction," *J. Am. Chem. Soc.* **121**, 5737–5742.
46. Mo, Y. and Gao, J. (2000) "An ab initio molecular orbital-valence bond (MOVB) method for simulating chemical reactions in solution," *J. Phys. Chem. A* **104**, 3012–3020.
47. Gao, J. and Mo, Y. (2000) "Simulation of chemical reactions in solution using an ab initio molecular orbital-valence bond model," *Prog. Theor. Chem. Phys.* **5**, 247–268.
48. Gao, J., Garcia-Viloca, M., Poulsen, T. D., and Mo, Y. (2003) "Solvent effects, reaction coordinates, and reorganization energies on nucleophilic substitution reactions in aqueous solution," *Adv. Phys. Org. Chem.* **38**, 161–181.
49. Albu, T. V., Corchado, J. C., and Truhlar, D. G. (2001) "Molecular mechanics for chemical reactions: a standard strategy for using multiconfiguration molecular mechanics for variational transition state theory with optimized multidimensional tunneling," *J. Phys. Chem. A* **105**, 8465–8487.
50. Liu, H., Elstner, M., Kaxiras, E., Fraunheim, T., Hermans, J., and Yang, W. (2001) "Quantum mechanics simulation of protein dynamics on long timescale," *Proteins: Struct., Funct. Gen.* **44**, 484–489.
51. Monard, G. and Merz, K. M., Jr. (1999) "Combined quantum mechanical/molecular mechanical methodologies applied to biomolecular systems," *Acc. Chem. Res.* **32**, 904–911.
52. Car, R. and Parrinello, M. (1985) "Unified approach for molecular dynamics and density-functional theory," *Phys. Rev. Lett.* **55**, 2471–2474.
53. Tuckerman, M. E., Laasonen, K., Sprik, M., and Parrinello, M. (1994) "Ab initio simulations of water and water ions," *J. Phys.: Conden. Matter* **6**, A93–A100.
54. Sprik, M., Hutter, J., and Parrinello, M. (1996) "Ab initio molecular dynamics simulation of liquid water: comparison of three gradient-corrected density functionals," *J. Chem. Phys.* **105**, 1142–1152.
55. Rothlisberger, U., Carloni, P., Doclo, K., and Parrinello, M. (2000) "A comparative study of galactose oxidase and active site analogs based on QM/MM Car-Parrinello simulations," *J Biol Inorg Chem* **5**, 236–250.
56. Rohrig, U. F., Guidoni, L., and Rothlisberger, U. (2005) *Chem Phys Chem* **6**, 1836.
57. York, D. M., Lee, T.-S., and Yang, W. (1998) "Quantum mechanical treatment of biological macromolecules in solution using linear-scaling electronic structure methods," *Phys. Rev. Lett.* **80**, 5011–5014.
58. Titmuss, S. J., Cummins, P. L., Rendell, A. P., Bliznyuk, A. A., and Gready, J. E. (2002) "Comparison of linear-scaling semiempirical methods and combined quantum mechanical/molecular mechanical methods for enzymic reactions. II. An energy decomposition analysis," *J. Comput. Chem.* **23**, 1314–1322.
59. Van der Vaart, A. and Merz, K. M., Jr. (1999) "The role of polarization and charge transfer in the solvation of biomolecules," *J. Am. Chem. Soc.* **121**, 9182–9190.
60. Stewart, J. J. P. (1996) "Application of localized molecular orbitals to the solution of semiempirical self-consistent field equations," *Int. J. Quantum Chem.* **58**, 133–146.
61. Gao, J. and Thompson, M. A. (1998) *Combined Quantum Mechanical and Molecular Mechanical Methods*, Vol. 712, American Chemical Society, Washington, DC.
62. Warshel, A. and Levitt, M. (1976) "Theoretical studies of enzymic reactions: dielectric, electrostatic and steric stabilization of the carbonium ion in the reaction of lysozyme," *J. Mol. Biol.* **103**, 227–249.

63. Singh, U. C. and Kollman, P. A. (1986) "A combined ab initio quantum mechanical and molecular mechanical method for carrying out simulations on complex molecular systems: applications to the CH3Cl + Cl-exchange reaction and gas phase protonation of polyenes.," *J. Comput. Chem.* **7**, 718–730.
64. Tapia, O., Lluch, J. M., Cardenas, R., and Andres, J. (1989) "Theoretical study of solvation effects on chemical reactions. A combined quantum chemical/Monte Carlo study of the Meyer-Schuster reaction mechanism in water," *J. Am. Chem. Soc.* **111**, 829–835.
65. Field, M. J., Bash, P., A., and Karplus, M. (1990) "A combined quantum mechanical and molecular mechanical potential for molecular dynamics simulations," *J. Comput. Chem.* **11**, 700–733.
66. Gao, J. (1992) "Absolute free energy of solvation from Monte Carlo simulations using combined quantum and molecular mechanical potentials," *J. Phys. Chem.* **96**, 537–540.
67. Stanton, R. V., Hartsough, D. S., and Merz, K. M., Jr. (1993) "Calculation of solvation free energies using a density functional/molecular dynamics coupled potential," *J. Phys. Chem.* **97**, 11868–11870.
68. Freindorf, M. and Gao, J. (1996) "Optimization of the Lennard-Jones parameters for a combined ab initio quantum mechanical and molecular mechanical potential using the 3-21G basis set.," *J. Comput. Chem.* **17**, 386–395.
69. Hillier, I. H. (1999) "Chemical reactivity studied by hybrid QM/MM methods," *Theochem* **463**, 45–52.
70. Morokuma, K. (2002) *Phil. Trans. Roy. Soc. London, Ser. A* **360**, 1149.
71. Warshel, A. and Karplus, M. (1972) "Calculation of ground and excited state potential surfaces of conjugated molecules. I. Formulation and parametrization," *J. Amer. Chem. Soc.* **94**, 5612–5625.
72. Bash, P. A., Field, M. J., and Karplus, M. (1987) "Free energy perturbation method for chemical reactions in the condensed phase: a dynamic approach based on a combined quantum and molecular mechanics potential," *J. Am. Chem. Soc.* **109**, 8092–8094.
73. Gao, J. (1995) "An automated procedure for simulating chemical reactions in solution. Application to the decarboxylation of 3-carboxybenzisoxazole in water," *J. Am. Chem. Soc.* **117**, 8600–8607.
74. Wu, N., Mo, Y., Gao, J., and Pai, E. F. (2000) "Electrostatic stress in catalysis: structure and mechanism of the enzyme orotidine monophosphate decarboxylase," *Proc. Natl. Acad. Sci. U.S.A.* **97**, 2017–2022.
75. Gao, J. (1994) "Computation of intermolecular interactions with a combined quantum mechanical and classical approach," *ACS Symp. Ser.* **569**, 8–21.
76. Dewar, M. J. S., Zoebisch, E. G., Healy, E. F., and Stewart, J. J. P. (1985) "Development and use of quantum mechanical molecular models. 76. AM1: a new general purpose quantum mechanical molecular model," *J. Am. Chem. Soc.* **107**, 3902–3909.
77. Stewart, J. J. P. (1989) "Optimization of parameters for semiempirical methods I. Method," *J. Comp. Chem.* **10**, 209–220.
78. Orozco, M., Luque, F. J., Habibollahzadeh, D., and Gao, J. (1995) "The polarization contribution to the free energy of hydration. [Erratum to document cited in CA122:299891]," *J. Chem. Phys.* **103**, 9112.
79. Major, D. T., York, D. M., and Gao, J. (2005) "Solvent polarization and kinetic isotope effects in nitroethane deprotonation and implications to the nitroalkane oxidase reaction," *J. Am. Chem. Soc.* **127**, 16374–16375.
80. Gao, J. (1994) "Monte Carlo quantum mechanical-configuration interaction and molecular mechanics simulation of solvent effects on the n → pi.* blue shift of acetone," *J. Am. Chem. Soc.* **116**, 9324–9328.
81. Gao, J. and Byun, K. (1997) "Solvent effects on the n → pi* transition of pyrimidine in aqueous solution," *Theor. Chem. Acc.* **96**, 151–156.
82. Garcia-Viloca, M., Truhlar, D. G., and Gao, J. (2003) "Importance of substrate and cofactor polarization in the active site of dihydrofolate reductase," *J. Mol. Biol.* **327**, 549–560.

83. Byun, K., Mo, Y., and Gao, J. (2001) "New insight on the origin of the unusual acidity of Meldrum's acid from ab initio and combined QM/MM simulation study," *J. Am. Chem. Soc.* **123**, 3974–3979.
84. Maseras, F. and Morokuma, K. (1995) "IMOMM: a new integrated ab initio + molecular mechanics geometry optimization scheme of equilibrium structures and transition states," *J. Comput. Chem.* **16**, 1170–1179.
85. Mulholland, A. J. (2001) "The QM/MM approach to enzymatic reactions," *Theor. Comput. Chem.* **9**, 597–653.
86. Zwanzig, R. (1954) "High-temperature equation of state by a perturbation method. I. Nonpolar gases," *J. Chem. Phys.* **22**, 1420–1426.
87. Jorgensen, W. L. and Ravimohan, C. (1985) "Monte Carlo simulation of differences in free energies of hydration," *J. Chem. Phys.* **83**, 3050–3054.
88. Kollman, P. (1993) "Free energy calculations: Applications to chemical and biochemical phenomena," *Chem. Rev.* **93**, 2395–2417.
89. Simonson, T. (2001) in *Computational Biochemistry and Biophysics* (Becker, O. M., MacKerell, A. D., Jr., Roux, B., and Watanabe, M., Eds.) pp. 169–197. Dekker, New York.
90. Espinosa-Garcia, J., Corchado, J. C., and Truhlar, D. G. (1997) "importance of quantum effects for C-H bond activation reactions," *J. Am. Chem. Soc.* **119**, 9891–9896.
91. Alhambra, C., Gao, J., Corchado, J. C., Villa, J., and Truhlar, D. G. (1999) "Quantum mechanical dynamical effects in an enzyme-catalyzed proton transfer reaction," *J. Am. Chem. Soc.* **121**, 2253–2258.
92. Billeter, S. R., Webb, S. P., Iordanov, T., Agarwal, P. K., and Hammes-Schiffer, S. (2001) "Hybrid approach for including electronic and nuclear quantum effects in molecular dynamics simulations of hydrogen transfer reactions in enzymes," *J. Chem. Phys.* **114**, 6925–6936.
93. Garcia-Viloca, M., Alhambra, C., Truhlar, D. G., and Gao, J. (2001) "Inclusion of quantum-mechanical vibrational energy in reactive potentials of mean force," *J. Chem. Phys.* **114**, 9953–9958.
94. Hammes-Schiffer, S. (1998) "Mixed quantum/classical dynamics of hydrogen transfer reactions," *J. Phys. Chem. A* **102**, 10443–10454.
95. Webb, S. P. and Hammes-Schiffer, S. (2000) "Fourier grid Hamiltonian multiconfigurational self-consistent-field: A method to calculate multidimensional hydrogen vibrational wavefunctions," *J. Chem. Phys.* **113**, 5214–5227.
96. Alhambra, C., Corchado, J., Sanchez, M. L., Gao, J., and Truhlar, D. G. (2000) "Quantum dynamics of hydride transfer in enzyme catalysis," *J. Am. Chem. Soc.* **122**, 8197–8203.
97. Hwang, J.-K. and Warshel, A. (1996) "How important are quantum mechanical nuclear motions in enzyme catalysis?," *J. Am. Chem. Soc.* **118**, 11745–11751.
98. Liu, Y. P., Lynch, G. C., Truong, T. N., Lu, D. H., Truhlar, D. G., and Garrett, B. C. (1993) "Molecular modeling of the kinetic isotope effect for the [1,5]-sigmatropic rearrangement of cis-1,3-pentadiene," *J. Am. Chem. Soc.* **115**, 2408–2415.
99. Liu, Y. P., Lu, D. H., Gonzalez-Lafont, A., Truhlar, D. G., and Garrett, B. C. (1993) "Direct dynamics calculation of the kinetic isotope effect for an organic hydrogen-transfer reaction, including corner-cutting tunneling in 21 dimensions," *J. Am. Chem. Soc.* **115**, 7806–7817.
100. Garrett, B. C., Truhlar, D. G., Wagner, A. F., and Dunning, T. H., Jr. (1983) "Variational transition state theory and tunneling for a heavy-light-heavy reaction using an ab initio potential energy surface. Atomic chlorine-37 + hydrogen chloride [H(D)35Cl] .fwdarw. hydrogen chloride [H(D)37Cl] + atomic chlorine-35," *J. Chem. Phys.* **78**, 4400–4413.
101. Fernandez-Ramos, A. and Truhlar, D. G. (2001) "Improved algorithm for corner-cutting tunneling calculations," *J. Chem. Phys.* **114**, 1491–1496.
102. Feynman, R. P. and Hibbs, A. R. (1965) *Quantum Mechanics and Path Integrals*, McGraw-Hill, New York.
103. Gillan, M. J. (1988) "The quantum simulation of hydrogen in metals," *Phil. Mag. A* **58**, 257–283.
104. Voth, G. A., Chandler, D., and Miller, W. H. (1989) "Rigorous formulation of quantum transition state theory and its dynamical corrections," *J. Chem. Phys.* **91**, 7749–7760.

105. Messina, M., Schenter, G. K., and Garrett, B. C. (1993) "Centroid-density, quantum rate theory: variational optimization of the dividing surface," *J. Chem. Phys.* **98**, 8525–8536.
106. Cao, J. and Voth, G. A. (1994) "The formulation of quantum statistical mechanics based on the Feynman path centroid density. V. Quantum instantaneous normal mode theory of liquids," *J. Chem. Phys.* **101**, 6184–6192.
107. Hwang, J. K. and Warshel, A. (1993) "A quantized classical path approach for calculations of quantum mechanical rate constants," *J. Phys. Chem.* **97**, 10053–10058.
108. Thomas, A., Jourand, D., Bret, C., Amara, P., and Field, M. J. (1999) "Is there a covalent intermediate in the viral neuraminidase reaction? A hybrid potential free-energy study," *J. Am. Chem. Soc.* **121**, 9693–9702.
109. Makarov, D. E. and Topaler, M. (1995) "Quantum transition-state theory below the crossover temperature," *Phys. Rev. E: Stat. Phys., Plasmas, Fluids, Relat. Interdiscip. Top.* **52**, 178–188.
110. Messina, M., Schenter, G. K., and Garrett, B. C. (1995) "A variational centroid density procedure for the calculation of transmission coefficients for asymmetric barriers at low temperature," *J. Chem. Phys.* **103**, 3430–3435.
111. Mills, G., Schenter, G. K., Makarov, D. E., and Jonsson, H. (1997) "Generalized path integral based quantum transition state theory," *Chem. Phys. Lett.* **278**, 91–96.
112. Jang, S. and Voth, G. A. (2000) "A relationship between centroid dynamics and path integral quantum transition state theory," *J. Chem. Phys.* **112**, 8747–8757. Erratum: 2001. **114**, 1944.
113. Feierberg, I., Luzhkov, V., and Aqvist, J. (2000) "Computer simulation of primary kinetic isotope effects in the proposed rate-limiting step of the glyoxalase I catalyzed reaction," *J. Biol. Chem.* **275**, 22657–22662.
114. Sprik, M., Klein, M. L., and Chandler, D. (1985) "Staging: a sampling technique for the Monte Carlo evaluation of path integrals," *Phys. Rev. B* **31**, 4234–4244.
115. Major, D. T. and Gao, J. (2005) "Implementation of the bisection sampling method in path integral simulations," *J. Mol. Graph. Model.* **24**, 121–127.
116. Major, D. T., Garcia-Viloca, M., and Gao, J. (2006) "Path integral simulations of proton transfer reactions in aqueous solution using combined QM/MM potentials," *J. Chem. Theory Comput.* **2**, 236–245.
117. Ceperley, D. M. (1995) "Path integrals in the theory of condensed helium," *Rev. Mod. Phys.* **67**, 279–355.
118. Major, D. T., Nam, K., and Gao, J. (2006) "Transition state stabilization and a-amino carbon acidity in alanine racemase," *J. Am. Chem. Soc.* **128**, 8114–8115.
119. Major, D. T. and Gao, J. (2007) "An integrated path intergral and free-energy perturbation-umbrella sampling method for computing kinetic isotope effects of chemical reactions in solution and in enzymes," *J. Chem. Theory Comput.* **3**, 949–960.
120. Ramirez, R. (1997) "Dynamics of quantum particles of by path-integral centroid simulations: The symmetric Eckart barrier," *J. Chem. Phys.* **107**, 3550–3557.
121. Johnston, H. S. (1966) *Gas Phase Reaction Rate Theory*, Ronald Press, New York.
122. Shavitt, I. (1959) "Calculation of the rates of the ortho-para conversions and isotope exchanges in hydrogen," *J. Chem. Phys.* **31**, 1359.
123. Rishavy, M. A. and Cleland, W. W. (2000) "Determination of the mechanism of orotidine 5′-monophosphate decarboxylase by isotope effects," *Biochemistry* **39**, 4569–4574.
124. Nam, K., Gao, J., and York, D. M. (2005) "An efficient linear-scaling ewald method for long-range electrostatic interactions in combined QM/MM calculations," *J. Chem. Theory Comput.* **1**, 2–13.
125. Toney, M. D. (2005) "Reaction specificity in pyridoxal phosphate enzymes," *Arch. Biochem. Biophys.* **433**, 279–283.
126. Ondrechen, M. J., Briggs, J. M., and McCammon, J. A. (2001) "A model for enzyme-substrate interaction in alanine racemase," *J. Am. Chem. Soc.* **123**, 2830–2834.
127. Major, D. T. and Gao, J. (2006) "A combined quantum mechanical and molecular mechanical study of the reaction mechanism and a-amino acidity in alanine racemase," *J. Am. Chem. Soc.* **128**, 16345–16357.
128. Spies, M. A., Woodward, J. J., Watnik, M. R., and Toney, M. D. (2004) "Alanine racemase free energy profiles from global analyses of progress curves," *J. Am. Chem. Soc.* **126**, 7464–7475.

Chapter 4
Comparison of Protein Force Fields for Molecular Dynamics Simulations

Olgun Guvench and Alexander D. MacKerell, Jr.

Summary In the context of molecular dynamics simulations of proteins, the term "force field" refers to the combination of a mathematical formula and associated parameters that are used to describe the energy of the protein as a function of its atomic coordinates. In this review, we describe the functional forms and parameterization protocols of the widely used biomolecular force fields Amber, CHARMM, GROMOS, and OPLS-AA. We also summarize the ability of various readily available noncommercial molecular dynamics packages to perform simulations using these force fields, as well as to use modern methods for the generation of constant-temperature, constant-pressure ensembles and to treat long-range interactions. Finally, we finish with a discussion of the ability of these force fields to support the modeling of proteins in conjunction with nucleic acids, lipids, carbohydrates, and/or small molecules.

Keywords: Amber · CHARMM · GROMOS · Molecular dynamics · OPLS-AA · Protein

1 Introduction

Classical molecular dynamics (MD) simulations of proteins are founded on the idea of using a differentiable function of the atomic coordinates to represent the energy of the system. This function is an approximation of the true quantum mechanical (QM) wavefunction. The function's partial derivatives with respect to the atomic Cartesian coordinates yield forces that can then be used to propagate the system through time using classical mechanics. In addition to being dependent on the atomic coordinates, the function's value also depends on a set of parameters that describe the geometric and energetic properties of interparticle interactions. Unlike the coordinates, these parameters are invariant during the course of a simulation. The combination of the mathematical function and the parameters is commonly referred to as a "force field."

From: Methods in Molecular Biology, vol. 443, Molecular Modeling of Proteins
Edited by Andreas Kukol © Humana Press, Totowa, NJ

In this review, we will focus on the widely used biomolecular protein force fields Amber [1], CHARMM [2], GROMOS [3], and OPLS-AA [4], which account for the majority of recently published MD simulations of proteins. All four of these force fields have been developed by academic research groups and, accordingly, the associated parameters have been peer reviewed and made publicly available. A non-exhaustive list of other force fields for protein simulations includes CVFF [5], ECEPP [6–9], ENCAD [10, 11], MM4 [12–17], MMFF [18–24], and UFF [25]; further examples can be found in two recent reviews [26, 27]. Although we limit our discussion to the preceding four force fields, we do note that the ENCAD force field has been used extensively by Daggett and coworkers to study protein folding via explicit solvent MD simulations [28], and the ECEPP and UNRES force fields of Scheraga and coworkers have been used for protein structure prediction with an implicit solvent treatment [29].

We begin the review with a description of the mathematical functional forms for these force fields, first focusing on what they have in common and then pointing out how they differ. We then give an overview of parameter development and speak to the particulars of each force field. We follow this with a discussion of present-day approaches to handling long-range interactions and generating appropriate ensembles at constant temperature and pressure that allow for direct comparisons with experimental data. We follow with a list of widely available academically distributed software packages that support MD simulations using one or more of these force fields. We finish with an overview of the present-day ability of these force fields to support not only simulations of proteins but also of proteins in conjunction with nucleic acids, lipids, carbohydrates, and/or small molecules.

2 Force Field Functional Forms

The underlying functional forms of the Amber, CHARMM, GROMOS, and OPLS-AA force fields can be readily understood by the molecular properties they seek to represent. These can be classified into two groups: bonded and nonbonded. The terms representing bonded interactions seek to account for the stretching of bonds, the bending of valence angles, and the rotation of dihedrals. The terms representing nonbonded interactions aim to capture electrostatics, dispersion, and Pauli exclusion. Thus, energy terms common to these force fields are:

$$E_{\text{bonded}} = \sum_{\text{bonds}} K_b(b-b_0)^2 + \sum_{\text{angles}} K_\theta(\theta-\theta_0)^2 + \sum_{\text{dihedrals}} K_\chi[1+\cos(n\chi-\sigma)] \quad (1)$$

and

$$E_{\text{nonbonded}} = \sum_{\substack{\text{nonbonded}\\ \text{pairs } ij}} \left(\varepsilon_{ij} \left[\left(\frac{R_{\min,ij}}{r_{ij}} \right)^{12} - 2 * \left(\frac{R_{\min,ij}}{r_{ij}} \right)^{6} \right] + \frac{q_i q_j}{r_{ij}} \right), \quad (2)$$

where E_{bonded} is the contribution to the total energy from bonded interactions and $E_{\text{nonbonded}}$ is the contribution from nonbonded interactions. The total energy is then:

$$E_{\text{total}} = E_{\text{bonded}} + E_{\text{nonbonded}} + E_{\text{other}}, \quad (3)$$

where E_{other} includes any force field-specific terms, as described subsequently.

The first term in Eq. 1 is a sum over all bonded pairs of atoms and describes the stretching of bonds; b is the interatom distance (i.e., bond length); and K_b and b_0 are parameters describing the stiffness and the equilibrium length of the bond, respectively. The term has the same quadratic form as that of Hooke's law for the potential energy of a spring. The second term involves triplets of atoms, e.g., A, B, and C, where A is bonded to B and B is bonded to C, and describes the bending of angles. θ is the angle formed by the two bond vectors, K_θ and θ_0 are the parameters describing the stiffness and equilibrium geometry of the angle, and, similar to the term for bond stretching, the term is quadratic. The third and final term in Eq. 1 is a sum over quadruplets of atoms A, B, C, and D, where A is bonded to B, B to C, and C to D, and describes the energetics associated with rotation of the dihedral angle defined by those four atoms. Because such rotation is necessarily periodic in nature, a cosine function is used. χ is the value of the dihedral, K_χ is the energetic parameter that determines barrier heights, n is the periodicity or multiplicity, and σ is the phase. The addition of 1 in this term is used so that the energy is equal to or greater than zero. In addition, the equation may be extended in a Fourier series where the term is applied more than once to a given dihedral in which the different terms are associated with different periodicities (although the other parameters may also differ). It should be noted that the bonded terms are also referred to as internal or intramolecular interactions.

Equation 2 describes the nonbonded interactions; the terms external or intermolecular interactions are also used to designate these interactions. In all of the force fields discussed herein, nonbonded interactions between atoms are defined as occurring either between atoms in separate molecules or between atoms separated by three or more bonds in the same molecule. Equation 2 is composed of two parts. The first, known as the Lennard-Jones (LJ) equation, is the portion in square brackets along with the prefactor ε_{ij}, and models attractive dispersion and repulsive Pauli-exclusion interactions and is commonly referred to as the van der Waals term. As two atoms are brought together from infinite separation, the negative term in the brackets, which goes as the inverse of the interatomic separation $\boldsymbol{r}_{ij}$ to the sixth power, dominates the interaction and the atoms feel an increasing attraction with decreasing distance as the energy becomes progressively more negative. This part of the LJ equation models dispersion, and its $(1/r)^6$ form derives from the interaction energy of an instantaneous dipole with an induced dipole, according to the definition of London's dispersion. As the atoms get progressively closer, an energy minimum is reached and, at closer distances, the $(1/r)^{12}$ term, which is positive, starts to dominate and leads to increasing energy and, hence, repulsion. Its form was originally chosen based on its computational expedience because it is simply the square of $(1/r)^6$. Nonetheless, it serves as an adequate representation of the very steep repulsive energy wall that arises from Pauli exclusion as two atoms get closer than the sum of their van der Waals radii. The prefactor, ε_{ij}, is a parameter based on the types of the two interacting atoms i and j. As its value increases, the interaction minimum becomes deeper and the repulsive wall steeper. $R_{\mathrm{min},ij}$ is a parameter that also depends on the types of the two interacting atoms and defines the distance at which the LJ energy is a minimum. The LJ equation is sometimes written as $\sum_{\substack{\text{nonbonded}\\ \text{pairs } ij}} \left(\frac{A_{ij}}{r_{ij}^{12}} - \frac{B_{ij}}{r_{ij}^{6}}\right)$, and can be equated to the form in Eq. 2 using the relations $A_{ij} = \varepsilon_{ij} R_{\mathrm{min},ij}{}^{12}$ and $B_{ij} = 2\varepsilon_{ij} R_{\mathrm{min},ij}{}^{6}$.

The second part of Eq. 2 is Coulomb's law and is used to model the electrostatic interaction between nonbonded pairs of atoms. As with the LJ equation, r_{ij} is the interatomic distance. q_i and q_j are the parameters that describe the effective charges on atoms i and j. It is important to note that the effective charge parameters are not simply unit charges located on formally charged atoms. Rather they are partial atomic charges with noninteger values that are selected to represent the overall charge distribution of a molecule. Thus, even the hydrogen atoms on aliphatic carbon atoms can have charges of approximately 0.05 to 0.1 electrons in biomolecular force fields. Naturally, the sum of the partial charges in a molecule must equal the molecule's net formal charge. In addition, in the case of metal ions, the charge is typically assigned the formal charge (e.g., +1 for the sodium ion).

It is important to emphasize that nonbonded interactions involve only pairs of atoms. Early simulations of noble gases, which used only the LJ equation, showed that fitting ε_{ij} and R_{min} to reproduce the adiabatic potential energy surface of a dimer of the noble gas led to inaccuracies in the thermodynamic properties of the system as calculated from simulations of a large number of such particles using these parameters [30]. This deficit was also seen in liquid water simulation that included both LJ and Coulomb terms and used parameters derived from the QM interaction potential of two water molecules [31, 32]. In reality, multibody terms involving three or more atoms simultaneously contribute to the total energy in multiparticle systems, whereas Eq. 2 is limited to a sum over pairs of atoms for computational tractability. Thus, current biomolecular force fields, whose foremost goal it is to capture the energetics of biomolecules in their physiologically relevant condensed-phase milieu, use *effective* pairwise-nonbonded potentials, also known as additive or nonpolarizable models. That is, the parameters in Eq. 2 are developed with the constraint of accurately modeling condensed-phase properties, although this may come at the expense of deviating from gas-phase QM dimerization energies.

As a final word regarding the nonbonded contribution to the energy, one should note that there is no explicit term for hydrogen bonding. In all of the force fields discussed herein, biologically important hydrogen bonds are handled by the combination of the LJ and Coulomb terms. Amber, CHARMM, GROMOS, and OPLS-AA only include interaction sites for bonded and nonbonded interactions at the location of the atomic nuclei. This model is generally very good with respect to being able to reproduce hydrogen bond energies and geometries [33], although it can lead to deviations from QM results with respect to the angular dependence of hydrogen bonding energy in certain cases [34]. The angular dependence of hydrogen bonding can be improved by the inclusion of additional interaction sites, for example, at the position of lone pairs, as was used for modeling sulfur atoms in an earlier Amber force field [35]. Indeed, the TIP5P water model, which is a five-point water model with interaction sites at lone pair positions as well as atomic positions, exhibits excellent thermodynamic, dynamic, and structural properties in liquid water simulations [36–38]. Up to 90% of the particles in a simulation of a solvated protein belong to the water molecules thus, going from a three-particle water model with interaction sites at the location of the oxygen and hydrogen atoms to a five-point water model with additional interaction sites at the lone pair positions nearly doubles the number of particles in the system. Because nonbonded interactions are pairwise, this near doubling would lead to a threefold to fourfold increase in the number of interactions and a likewise increase in computational cost. Therefore, current simulations generally use three-particle water models lacking lone pair positions, such as TIP3P [39] and SPC/E [40], and the representation of the proteins in these force fields likewise does not include lone pair positions.

Having concluded the discussion of the similarities in the functional forms of the Amber, CHARMM, GROMOS, and OPLS-AA, we now move on to the differences. Two primary differences exist in the bonded portion of the force fields. The first is the variable use of "improper" dihedrals, which can be used to maintain chirality or planarity at an atom center with bonds to three other atoms. For example, in the case of the NH group in the protein backbone amide bond, the improper dihedral angle would be defined by the atoms H-C-C_α-N, with N having bonds to H, C, and C_α. In the case of Amber and OPLS-AA, improper dihedral angles contribute to the energy via the dihedral term in Eq. 1, and are applied to planar groups and use a periodicity $n = 2$. The CHARMM and GROMOS force fields add a separate term for improper dihedral energy that has a quadratic dependence on the value of the improper dihedral, similar to the terms for bonds and angles. This is particularly important in the case of the GROMOS force field, which does not include particle positions for hydrogen atoms bonded to aliphatic carbons; improper dihedral terms serve to preserve chirality at these carbon centers. The second difference is that the CHARMM force field adds a Urey-Bradly angle term, which treats the two terminal atoms in an angle (i.e., 1,3 atoms) with a quadratic term that depends on the atom–atom distance. The improper dihedral and Urey-Bradly angle terms provide additional degrees of freedom for the accurate reproduction of vibrational spectra during parametrization, as discussed in the next section.

Similar to the bonded terms, two primary differences exist with respect to the treatment of nonbonded interactions. The first is in the combining rules used for the determination of the LJ parameters ε_{ij} and $R_{\mathrm{min},ij}$. The subscript "*ij*" associated with these parameters exists to make explicit their dependence on the atom type of both atom i and atom j. The concept of "atom type" allows for the assignment of different parameters to atoms of identical atomic number depending on the chemical context. For example, a hydrogen atom bonded to an oxygen atom has a different atom type and associated bonded and nonbonded parameters than a hydrogen atom bonded to an aromatic carbon atom for all four of the force fields discussed.

The OPLS-AA and GROMOS force fields use geometric combining rules for both ε_{ij} and $R_{\mathrm{min},ij}$, that is, $\varepsilon_{ij} = (\varepsilon_i \varepsilon_j)^{1/2}$ and $R_{\mathrm{min},ij} = (R_{\mathrm{min},i} R_{\mathrm{min},j})^{1/2}$, where the parameters with single subscripts refer to the parameters for that particular atom's atom type. CHARMM and Amber also use the geometric mean for calculating ε_{ij}, but use the arithmetic mean, $\frac{1}{2}(R_{\mathrm{min},i} + R_{\mathrm{min},j})$, for $R_{\mathrm{min},ij}$.

The second difference in the nonbonded part of the force fields is the handling of 1,4-nonbonded interactions, that is, those between atoms A and D in the dihedral A-B-C-D. The four force fields apply various scaling constants to the LJ and Coulomb interactions between these atoms pairs. For example, Amber scales 1,4-LJ interactions by 1/2 and Coulomb interactions by 1/1.2. OPLS-AA applies a scale factor of 1/2 to both interactions. GROMOS takes a case-by-case approach that includes having different LJ parameters for a particular atom type if it is in a polar or ionic interaction versus a nonpolar interaction, scaling by 0 1,4-nonbonded interactions in aromatic rings, allowing for a different set of LJ parameters for atoms i and j if they are 1,4 relative to each other, and having particular atom type pairings for which the LJ parameters are directly defined instead of being derived by combining rules. The CHARMM force field takes a simple approach by not scaling 1,4-nonbonded interactions, although, for a few atom type pairs, special 1,4-LJ parameters are applied.

The special treatment of 1,4-nonbonded interactions was largely motivated by the desire to more readily reproduce dihedral rotation energetics while facilitating the use of the same nonbonded parameters for intermolecular and intramolecular interactions [41]. The pairwise-additive nature of the nonbonded interactions prevents the lowering of rotational barriers caused by electronic polarization, hence, the application of empirical scaling values. The lack of consensus for how best to treat such interactions suggests that no single best solution exists to this problem. Indeed, the deficits of trying to capture the conformational energetics of a molecule caused by changes in dihedral angles using a combination of LJ, Coulomb, and cosine terms becomes particularly apparent in a molecule with more than one dihedral degree of freedom. The archetypical example is the alanine dipeptide whose energy depends simultaneously on the ϕ and ψ dihedral angles (Fig. 1). The different approaches to treating 1,4-nonbonded dihedral interactions combined with differences in the partial charge and LJ parameters lead to different energy surfaces (i.e., ϕ/ψ surface or Ramachandran surface) for the alanine dipeptide, depending on which force field is used. This is reflected in the solvated energetics of the alanine dipeptide, for which

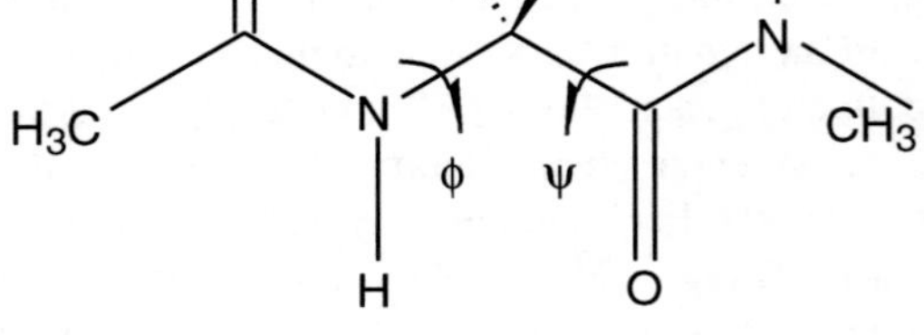

Fig. 1 The alanine dipeptide

both the depth and shape of the free energy contours of the alanine dipeptide in water are substantially different for the Amber, CHARMM, and OPLS-AA force fields [26]. Additionally, the force fields in their original versions do not reproduce the distribution of ϕ and ψ dihedral angles in comparison with either protein crystallographic data [42] or quantum chemical calculations using explicit solvent [43]. Both the Amber and OPLS-AA force fields have undergone revisions of their ϕ and ψ dihedral parameters in an effort to improve the conformational energetics of the polypeptide backbone [44–46]. In the case of Amber, this has led to a number of descendents to the original Cornell et al. Amber force field. According to the Amber 9 MD software documentation, this latter force field is no longer considered as the Amber default force field, unlike in earlier versions of the software, although it is the force field we will focus on here because of its wide past and present use and the lack of a single default force field in the Amber 9 software package; we refer interested readers to the Amber 9 documentation for a full listing and description of the available Amber force fields.

Similar to other protein force fields, the CHARMM force field has been found to have inaccuracies in the backbone conformational energetics of peptides. This originally manifested as an unrealistically large proportion of π-helical conformations for helical peptides [47], and was later found to be a result of deficiencies in the original ϕ and ψ dihedral parameters, a problem that was shown to be present in a number of other empirical force fields to varying degrees [48]. Instead of refitting these parameters such as in the case of Amber and OPLS-AA, a different approach was taken by adding a new "correction map" (CMAP) term to the potential energy equation. This additional term uses a grid-based interpolation scheme with values of points on the grid having a dependency on two dihedrals simultaneously [42, 49]. Thus, any ϕ/ψ energy surface can be exactly reproduced by the CHARMM force field using the CMAP term.

3 Parameter Optimization

The optimization of force field parameters involves adjusting parameter values until the force field is able to reproduce a set of target data to within a prescribed threshold. The target data include some subset of experimental spectroscopic, thermodynamic, and crystallographic data as well as data computed using QM methods. Typical examples of experimental target data include vibrational spectra; heats of vaporization; densities; solvation free energies; microwave, electron, or X-ray diffraction structures; and relative conformational energies and barrier heights. Typical examples of computed QM target data include vibrational spectra; minimum energy geometries; dipole moments; conformational energies and barrier heights; electrostatic potentials; and dimerization energies.

The Amber, CHARMM, GROMOS, and OPLS-AA force fields for proteins each target a different subset of the possible experimental and QM data, although there is substantial overlap between the subsets. The most commonality exists in the development of the bond stretching and angle bending parameters, which are uniformly aimed at reproducing either experimental or computed infrared spectra, and experimental or computed geometries. In fact, the OPLS-AA bond and angle parameters are largely those of the Amber force field, along with some from the CHARMM force field [4]. All of these force fields similarly use experimental and computed QM conformational energies as target data for the development of dihedral parameters. Thus, there is consensus with respect to the development of the bonded parameters, which is likely a reflection of the influence of the pioneering efforts made in the development of earlier force fields designed to reproduce conformations and energetics of small molecules [6, 50–52].

Similar to the development of the bonded parameters, the nonbonded LJ parameters have been developed in a similar manner across all four force fields, and follow closely the protocol that was developed for the original OPLS force field [53], which lacked interaction sites for nonpolar hydrogens, in contrast to the more recent all-atom OPLS force field (OPLS-AA) [4]. This approach involves condensed-phase simulations at constant temperature and pressure for a variety of pure liquids, such as alkanes, alcohols, amides, aromatics, etc., that have moieties chemically similar to those found in proteins. The LJ parameters are then adjusted to reproduce experimental heats of vaporization and densities for these liquids, with the underlying philosophy that the parameters will be adjusted so that the pairwise-additive form of the force field nonbonded terms will effectively reproduce condensed-phase properties and, thus, be suitable for, e.g., solvated protein simulations. The GROMOS LJ parameterization protocol also incorporates experimental atomic polarizabilities [3].

Naturally, the Coulomb partial atomic charge parameters also contribute to the heats of vaporization and densities. In the case of OPLS-AA, the partial charges are empirically adjusted along with the LJ parameters during the fitting to the experimental heats of vaporization and densities. In contrast, the Amber force field uses computed QM electrostatic potential surfaces as the target data for partial charge determination [41, 54–56]. Thus, Amber partial charges aim to reproduce molecules' gas-phase electrostatic potentials. CHARMM takes yet another approach. Similar to Amber, computed QM data are used. However, these data are the dimerization energies and minimum-energy interaction distances for small molecule–water dimers [33, 57]. This approach is aimed at balancing water–protein, water–water, and protein–protein interaction energies in the condensed phase. Finally, the GROMOS force field, similar to OPLS-AA, targets thermodynamic data in the refinement of partial atomic charge parameters. However, the solvation free energies of model compounds, both in water and in cyclohexane, are also included as target data [3,58]. The principle aim of this approach is to properly capture the partitioning of protein moieties between aqueous and nonaqueous environments, for example, between the solvent-exposed surface of a globular protein and its hydrophobic core.

The similar approach among the force fields with regard to the derivation of LJ parameters puts constraints on the possible values of the partial charges. Thus, all four force fields tend toward similar LJ and partial charge parameters. However, because small parameter differences can affect a force field's ability to reproduce experimental thermodynamic and computed QM data, each of the force fields most likely does best at reproducing the particular data for which its partial charges were parameterized. For example, the Amber, CHARMM, and OPLS-AA models, which were optimized based on pure solvent properties, all have a tendency to overestimate the free energies of solvation of model compounds representative of protein functional groups [59, 60]. Because the various nonbonded parameterization schemes all are reasonable choices, it becomes a matter of opinion regarding which, if any, particular choice of target data for partial-charge parameterization is better.

Of note is the fact that conformational energetics, and to a lesser extent bond and angle geometries and simulated vibrational spectra, depend on not just the dihedral, bond, and angle parameters but also the nonbonded parameters. For example, in the rotation of ethane about its CC bond, not only the dihedral parameters but also the LJ and charge parameters on the hydrogens affect the energy surface. Thus, force field development is necessarily an iterative self-consistent process. Any time the nonbonded parameters are changed, the ability of the force field to reproduce all of the target data must be checked and bonded parameters updated accordingly. Likewise, when bonded parameters are changed, all of the target data must be checked and nonbonded parameters updated as needed. Changes in nonbonded parameters change not only intermolecular interactions and, hence, condensed-phase properties, but also intramolecular interactions and, therefore, conformational energetics, thereby requiring adjustment of dihedral parameters. Similarly, dihedral parameters affect intramolecular conformational energetics, which contribute to the heat of vaporization. The intramolecular conformational energetics also determine preferred intramolecular geometries and molecular volumes, and, hence, the density of the liquid. As a result, nonbonded parameters require validation and may need adjustment after dihedral parameters have been altered. Thus, because of the need for self-consistent iterative refinement of the bonded and nonbonded parameters, force field parameter development can be a computationally demanding and labor-intensive process.

Protein force field development has evolved beyond the paradigm of solely using small molecule compounds to derive force field parameters, in which the parameters from model compounds are combined to produce the protein force field. The ever-increasing speed of computers has enabled recent efforts to include larger molecules directly in the parameterization process. Amber ϕ and ψ dihedral parameters have been adjusted to better reproduce the experimental conformational properties of structured peptides [45]. Data from MD simulations on myoglobin were used in the final stages of optimization of the CHARMM force field [2], whereas the parameters for the grid-based CHARMM CMAP term for the backbone ϕ/ψ energy have been developed using a large collection of protein x-ray crystallographic data [49]. Finally, the OPLS-AA ϕ and ψ dihedral parameters have been updated based not only on the alanine dipeptide but also on the alanine tetrapeptide [46].

We end this section by noting that the properties of a condensed-phase system simulated using MD depend not only on the force field but also on the simulation methodology. In particular, properties depend on methods for generating the desired thermodynamic ensemble, typically constant temperature and constant pressure, and for handling long-range nonbonded interactions. Accordingly, the next section is dedicated to a discussion of these points.

4 Modern Methods for Constant Temperature and Pressure and the Treatment of Long-Range Interactions

Current simulations of proteins in an explicit solvent environment typically involve systems on the order of 25,000 to several hundred thousand atoms and 75 to 150 Å in dimension. The application of periodic boundary conditions is the most well established and widely used method for preventing boundary artifacts (i.e., edge effects) that can arise because of the finite size of the protein–solvent system [30]. Under periodic boundary conditions, the system interacts with images of itself in every dimension. Thus, when a molecule drifts outside the span of the primary system it drifts back in from the opposite side.

MD simulations are based on integration of Newton's equations of motion. According to these equations, energy is a constant. Additionally, the definition of periodic boundary conditions implies that the system has a volume, which is also constant. Thus, the simplest MD simulation of a system under periodic boundary conditions is under constant energy and constant volume conditions. However, laboratory conditions are constant temperature and constant volume, corresponding to the Gibbs free energy. Thus, a real system is free to exchange energy with its surroundings both through heat and through pressure–volume work.

Unlike Monte Carlo simulations, which readily generate constant temperature–constant pressure ensembles by the application of temperature as a parameter in the definition of the method and changes in volume as Monte Carlo trial moves [61–63], MD methodology requires nontrivial changes to the equations of motion so as to generate a proper constant temperature–constant pressure ensemble under the constraint of continuous dynamics. Although it is possible to apply velocity reassignment and Monte Carlo volume changes during the course of a MD simulation, these disrupt the continuity of the dynamics. MD shows superior ability relative to Monte Carlo to equilibrate heterogeneous systems because continuous dynamics allows the system to retain a "memory" of where it has been and gives it momentum through phase space [64–66], thus, the importance of preserving continuous dynamics.

Newton's equations of motion have been extended to allow for the continuous transfer of heat and pressure–volume work between the system and additional degrees of freedom [67–72]. Owing to these developments, continuous dynamics at constant temperature and constant pressure are now possible. These methods are rigorous from a theoretical perspective in that not only do they assure that the average temperature and pressure of the system remain constant, but they also assure that fluctuations in energy and volume are consistent with those in a real system at constant temperature and pressure (i.e., reproduce a Boltzmann distribution). This is in contrast to the "weak coupling" or "Berendsen" algorithm, which also produces constant average temperature and pressure with continuous dynamics, but does not reproduce proper fluctuations in energy and volume and, therefore, does not correspond to the Gibbs thermodynamic ensemble [71, 73]. Although weak coupling was a pioneering effort aimed at producing constant temperature, constant pressure simulations with continuous dynamics and benefits from being a straightforward extension to a constant energy–constant volume code, it has been superseded by the described methods and should be avoided except in the heating and equilibration stages of a simulation, although it can be very useful in these stages because it will prevent large oscillations in temperature and/or volume for an MD trajectory started from a conformation that is far from equilibrium.

The simplest treatment of the long-range nonbonded interactions in a periodic system is the minimum image condition, in which a particle interacts with the closest images of the other particles in the system [30]. However, this can become very computationally costly as the system size increases, because the number of pairwise interactions goes as the number of particles squared. To lessen the computational burden, various truncation schemes have been developed whereby nonbonded interactions beyond a cutoff distance are ignored, and a smoothing function is typically applied to ensure continuity in the forces [74]. Such procedures can be a particularly severe approximation in the case of electrostatic interactions, whose energy goes as $1/r_{ij}$. As a result, not only do the sums of these interactions continue to make a substantial contribution to the total system energy beyond the typical cutoff length of approximately 10 Å, but pairwise forces are also nonnegligible beyond the cutoff.

The introduction of Ewald sums [75] into biomolecular simulations under periodic boundary conditions has essentially solved the problem of long-range electrostatic truncation. The system is treated as being infinitely periodic, and interaction energies and forces beyond the cutoff length are calculated using the Ewald algorithm, which works in reciprocal space, instead of being calculated directly using Coulomb's law. Although the original Ewald formalism is computationally expensive, recent developments based on grid-based treatments of reciprocal space, including particle-mesh Ewald (PME) methods, allow for the rigorous treatment of long-range electrostatics in a computationally efficient manner [76–79]. Ewald methods impose the constraint of charge neutrality on the system. Thus, neutralizing counterions are typically added to the system to cancel out any excess net formal charge on the protein, although the use of a uniform neutralizing plasma does allow for the application of Ewald methods to systems with nonzero net charge.

The fast $(1/r_{ij})^6$ decay of the attractive portion of the LJ potential means that the neglect of these interactions beyond the cutoff is significantly less of an approximation than doing the same for Coulomb interactions. Pairwise forces arising from LJ interactions approach 0 by 10 Å, making the direct truncation of these interactions less problematic than for electrostatic interactions. However, because all of the LJ interactions are favorable (attractive) at long distances, the sum total of these pairwise interactions does result in a nonnegligible contribution to the total energy and system pressure. This is particularly important in the case of constant pressure MD calculations, because neglecting this contribution will have an effect on the average volume of the system. This can be corrected by the application of a long-range correction to the truncated LJ potential, thereby recovering the contribution of LJ interactions beyond the cutoff to the pressure and, hence, the average volume [30, 80].

In summary, it is recommended that simulations under periodic boundary conditions be performed with theoretically rigorous continuous dynamics methods for maintaining constant temperature and pressure. Additionally, the use of PME is recommended for the handling of long-range electrostatic interactions beyond the cutoff at approximately 10 Å and the use of the long-range correction for the treatment of LJ interactions beyond the cutoff. These methods properly address the original limitations of MD simulations with periodic boundary conditions and are becoming the *de facto* standard.

Although the above methods may be considered state-of-the art for simulations of solvated proteins as well as other biological systems, alternative methods that may be used are worth noting. These, in particular, include stochastic boundary conditions and implicit solvent models. Both methods are particularly useful in situations in which doing a fully solvated simulation under periodic boundary conditions becomes computationally infeasible because of the large number of solvent molecules. In the case of stochastic boundary conditions, only the part of the protein that is of interest is solvated in a droplet of water. A reaction field is applied beyond the radius of the water droplet to prevent edge artifacts, and the portion of the protein outside of the water droplet is constrained so that it does not denature [81–83]. A recent study has shown that this approach yields excellent results for the solvation free energies of amino acid side chain analogs in comparison with periodic boundary conditions with PME electrostatics [60]. Continuum solvent models are the limiting case of this approach, in that no explicit water molecules are included at all. Rather, the water environment is included in a mean-field fashion, with additional energy terms introduced to account for the solvation of polar and charged groups, charge screening by solvent, and the hydrophobic effect, such as in the widely used generalized Born model and its variants [84–91]. As a further approximation to reduce computation time, it is possible to apply a continuum solvent model to the full protein while only a small portion of the protein is allowed to move, which can be useful in the study of protein–ligand interactions [92]. Current work in the field of implicit solvent models aims to include lipid bilayers in an implicit fashion [93, 94] as well as to account for the missing solute–solvent LJ interactions [95, 96].

We conclude this section with a cautionary word. Because the described MD simulation methods have only recently become standardized, none of the force fields were developed using a complete combination of proper thermostating and barostating, Ewald sums for long-range Coulomb interactions, and the long-range correction for LJ interactions. All force field development work in our laboratory, which is the primary developer of the CHARMM force field, now uses these modern methods for constant temperature and pressure and the treatment of long-range interactions. We anticipate that other groups will adopt this as standard practice for force field parameter development, if they have not already done so, because of the wide availability of these methods in various MD software packages, the fact that most recent MD simulations using these force fields use these methods, and, most importantly, the theoretical soundness of these methods. In the next section, we turn our attention to MD software packages and their support for these force fields and MD methods.

5 MD Packages

The Amber, CHARMM, and GROMOS force fields have been primarily developed in the context of the Amber [97–99], CHARMM [100], and GROMOS [101, 102] molecular mechanics and dynamics software packages, whereas OPLS-AA has been developed in the context of the BOSS [103] and MCPro software packages [104]. However, these force fields are accessible through a variety of different MD packages. We limit our discussion to a number of academically developed software packages that can be obtained either free or for a minor cost for academic research, and whose full documentation sets can be freely obtained on the World Wide Web: Amber (http://amber.scripps.edu), CHARMM (http://www.charmm.org), GROMACS (http://www.gromacs.org) [105–107], NAMD (http://www.ks.uiuc.edu/Research/namd) [108–111], and Tinker (http://dasher.wustl.edu/tinker). According to their respective web sites, the most recent versions of these programs at the time of writing were: Amber v. 9, CHARMM v. c32b2, GROMACS v. 3.3.2, NAMD v. 2.6, and Tinker v. 4.2. We have not included GROMOS in this list because its full documentation is not available via the World Wide Web. BOSS and MCPro are not included because they are primarily Monte Carlo software packages and this review is focused on MD. Excellent overviews of the GROMOS, BOSS, and MCPro programs, as well as of the IMPACT software that supports MD simulations using OPLS-AA, have recently been authored by their developers in a special issue of the Journal of Computational Chemistry, Volume 26, Number 16, "Special emphasis issue on biomolecular simulations," along with overviews of Amber, GROMACS, and NAMD, and those articles provide descriptions that are much more detailed than is possible within the scope of this review [99,102,104,107,111,112]. A similar review of the CHARMM software package is anticipated in 2008.

Table 1 summarizes the force fields, methods, and academic costs associated with each software package. This table has been designed in the context of this particular review. It is important to emphasize that the various packages have a host of features that are not represented in the table. Amber, CHARMM, and Tinker include very large feature sets with regards to molecular modeling and dynamics methodology. GROMACS has been designed to maximize the speed of computation. NAMD has been designed to take advantage of hardware platforms running large numbers of processors in parallel. We note that all of these software packages support PME for long-range electrostatics, although there is variable support for the recommended thermostating and barostating methods and the long-range LJ correction. We also note that all of these packages are under active development and are continually upgraded with respect to feature sets, performance, and compatibility with different hardware platforms, and, therefore, we recommend that users refer directly to the respective web sites for the most up-to-date information.

Table 1 Supported force fields, methods, and academic cost of the Amber, CHARMM, GROMACS, NAMD, and Tinker MD software packages

	Amber v. 9	CHARMM v. c32b2	GROMACS v. 3.3.2	NAMD v. 2.6	Tinker v. 4.2
Amber	X	X	X	X	X
CHARMM		X		X	X[a]
GROMOS			X		
OPLS-AA		X	X	X	X
Thermostat[b]		X	X	[c]	X
Barostat[b]		X	X	X	
PME	X	X	X	X	X
LRC	X	X	X		
Parallel[d]	X	X	X	X	
Academic cost	$400	$600	$0	$0	$0

All data in the table is from the software packages' respective web sites and the documentation available therein.

[a]No support for the additional grid-based CHARMM force field energy term.

[b]All packages provide means for thermostating and barostating. "X" indicates support for the recommended thermostating and barostating methods.

[c]Lacks recommended thermostat, but such thermostating can be achieved in constant temperature–constant pressure simulations through use of the Langevin piston barostat.

[d]Ability to run MD on multiple CPUs in parallel.

6 Incorporating Nucleic Acids, Lipids, Carbohydrates, and Small Molecules into Simulations with Proteins

The ability to simulate proteins in an aqueous environment provides many opportunities for studying protein folding, structure, function, dynamics, and thermodynamics [113]. It is nevertheless of great interest and importance to be able to expand MD simulations of proteins to include not only polypeptides in water, but also solvated proteins in conjunction with nucleic acids, lipids, carbohydrates, and small molecules. Such simulations are important in the study of protein–nucleic acid interactions, membrane-spanning proteins, glycosylated proteins, and protein substrates, inhibitors, and cofactors. Because parameter development is both a CPU- and labor-intensive process, the extension of force fields to include these other nonprotein entities is no small undertaking. Nevertheless, significant progress has been made with respect to the development of force field parameters for all of these classes of compounds. In this final section, we give a brief overview of the state of the Amber, CHARMM, GROMOS, and OPLS-AA force fields with respect to the modeling of these compounds.

6.1 Nucleic Acids

Both the Amber and CHARMM force fields include nucleic acid parameter sets that have seen wide use in simulations of DNA and RNA. These parameter sets were specifically developed for nucleic acids simulations and are, thus, highly optimized [1, 114]. Recent examples of simulations with the Amber force field include studies of the motions of ions around DNA [115], the TATA/TBP DNA/protein interaction [116], and ribosomal RNA [117]. The CHARMM force field has recently been used to study ribosomal peptidyl transfer [118], the reorganization of solvent in DNA conformational changes [119], and protein-facilitated DNA base flipping [120]. Of note is that these examples include heterogeneous protein–nucleic acid simulations for both force fields.

Very recently, a new GROMOS nucleic acid force field has been released [121]. It updates the previous parameter set through the addition of new backbone dihedral energy terms and the refinement of existing ones, as well as the addition of explicit aromatic hydrogen atoms to the nucleotide bases to better account for aromatic–aromatic interactions. Unlike the most recent GROMOS protein force field, this nucleic acid force field did not use solvation free energies as parameterization target data. Whether this has an impact on the simulation of protein–nucleic acid systems will likely become apparent in the near future as simulations using this force field emerge. The OPLS-AA force field does include atom types appropriate for modeling nucleic acid polymers, although no simulations of nucleic acids have been reported. If calculations on nucleic acids were undertaken with OPLS-AA, care should be taken to validate the force field with regard to the phosphodiester backbone and sugar moieties whose conformational energetics are particularly sensitive to the accuracy of the dihedral parameters.

Although it may be tempting, we recommend against mixing force fields to perform simulation of heterogeneous systems, for example simulating a protein–DNA complex using OPLS-AA protein parameters and CHARMM DNA parameters. Reliable results in such systems depend critically on the proper balance of the nonbonded parameters because these determine the intermolecular interactions. Because the nonbonded parameters are developed differently for each of the force fields, and, furthermore, because of differences in the LJ parameter combining rules, achieving the proper balance of solute–solute, solute–solvent, and solvent–solvent interactions in a simulation with multiple parameter sets is probably unlikely and certainly untested.

6.2 Lipids

The CHARMM lipid force field has seen extensive effort put into development, testing, and application. Since the original version for modeling saturated [122] and unsaturated phospholipids [123], there have been improvements in the treatment of the dihedral and LJ parameters and extension to the original lipid parameter set to include polyunsaturated chains, such that there is now support for accurate modeling of a wide variety of lipid systems [124–126]. In addition to studies of lipid–water systems, carried out during the development process and corroborating the ability of the force field to reproduce experimental observables, there have been recent simulations of protein–lipid–water systems that include G-protein-coupled receptors [127], viral fusion peptides in bilayers and micelles [128], and the conduction of water through a transmembrane water channel [129]. It is important to note that the published improvements and extensions to this force field are reflected in the latest distribution of the CHARMM parameters (freely available for download from http://www.pharmacy.umaryland.edu/faculty/amackere/force_fields.htm), and future applications of the force field should use the newest available version, which is always considered the "official" CHARMM force field.

Studies of peptides and proteins interacting with bilayers using the latest version of GROMOS protein force field are eagerly awaited because solvation free energies of the amino acid analogs in this force field were tuned for both cyclohexane and water, and this should help ensure proper protein behavior in both low-dielectric membrane and high-dielectric water environments. Previous versions of the GROMOS force field have been widely used for simulations in protein–lipid systems in water, and often use the dipalmitoylphosphatidylcholine (DPPC) lipid parameterization of Berger et al. [130] although a DPPC model based on GROMOS alkane parameters has recently been proposed [131]; the results of these simulations will serve as a useful benchmark for comparison with the newest parameter set. Recent studies have included simulations of peptide–bilayer systems [132, 133] and ion channels [134].

Although MD simulations of protein–lipid–water systems are dominated by the CHARMM and GROMOS force fields, lipid–water simulations are also possible with the Amber and OPLS-AA force fields. With Amber, only a limited number of lipid simulations have been reported [135], including heterogeneous simulations of peptides [136] and membrane channels [137]. Studies of lipid–water systems using OPLS-AA would be of interest because of the force field's strong emphasis on reproducing condensed-phase properties of model compounds, and would serve to validate its application to protein–lipid–water systems.

6.3 Carbohydrates

Carbohydrates provide a particular challenge with regard to force field development because of the large number of intramolecular hydrogen bonds that can be formed. Thus, the nonbonded and dihedral parameters must be carefully parameterized to correctly capture intramolecular energies and geometries. Furthermore, many biologically interesting carbohydrates are geometric isomers of each other. For example, glucose, galactose, and mannose differ only in the chirality of various hydroxyl- and hydrogen-bearing carbon centers. Therefore, without resorting to special atoms types for each monosaccharide, the force field parameters, including dihedral terms, must be able to simultaneously account for the energetics of the different geometric isomers.

The Amber carbohydrate force field is named GLYCAM [138–140]. Unlike the protein and nucleic acid force fields that were developed by Peter Kollman and coworkers, GLYCAM was developed and continues to undergo development primarily in the laboratory of Robert Woods. Nonetheless, GLYCAM is part of the Amber force field insofar as it is documented in the Amber software distribution and discussed in detail in a recent review of the Amber software [99]. Of the four force fields, protein–carbohydrate simulations with Amber/GLYCAM predominate, with recent simulations including enzyme–carbohydrate-analog complexes [141], protein–carbohydrate solutions [142], glycosylated peptides [143], and protein–oligosaccharide complexes [144]. However, to quantitatively reproduce the energetics of the exocyclic hydroxymethyl rotation, it was shown that 1,4-nonbonded interactions should not be scaled [145], in contrast with the protein and nucleic acid Amber force fields, and this can potentially complicate the simulation of protein–carbohydrate systems.

The OPLS-AA carbohydrate force field has done a notably good job of targeting the intramolecular energetics of a large set of pyranose monosaccharide geometric isomers and their conformations, as well as the glycosidic linkage [146]. The use of standard OPLS-AA atoms types and their associated partial charges and LJ parameters keeps it consistent with the OPLS-AA protein force field and, therefore, enables protein–carbohydrate simulations using OPLS-AA, such as a recent one of protein–disaccharide complexes [147]. A revision to include 1,5 and 1,6 nonbonded scale factors was proposed and shown to increase the accuracy of the carbohydrate intramolecular energetics [148]. However, these nonstandard scale factors make it incompatible with rest of the OPLS-AA force field, which limits scaling to only 1,4-nonbonded parameters.

The CHARMM force field has recently been expanded to include parameters that reproduce well the hydroxymethyl rotation in solvated glucose and galactose [149]. Because of its recent development, it has not yet seen use in protein–carbohydrate simulations. To accommodate the simulation of more diverse carbohydrate systems with the CHARMM force field, our laboratory has recently begun a large effort to develop a CHARMM carbohydrate force field. Its development is following the same procedures as used for the CHARMM protein, nucleic acid, and lipid force fields, with the aim of facilitating the accurate simulation of heterogenous biomolecular systems. In addition, very recently, a new GROMOS force field for hexopyranose-based carbohydrates has been developed and validated with respect to experimental data via solvated MD simulations of monosaccharides and disaccharides [150]. Similar to the recent GROMOS nucleic acid parameters, but unlike the recent GROMOS protein parameters, the force field development process did not target solvation free energies. Thus, similar to the GROMOS nucleic acid force field, it would be useful to compare simulation results of heterogeneous systems using the most recent GROMOS protein force field as well as the previous protein force field version [151, 152], which was parameterized in a fashion similar to the new carbohydrate force field.

6.4 Small Molecules

Amber, CHARMM, GROMOS, and OPLS-AA all provide a reasonably large palette of atoms types such that many organic small molecules can be represented by assigning atom types based on chemical similarity. The Amber force field includes the general Amber Force Field (GAFF) [153], which is not simply a parameter set but rather a software package designed to generate an Amber force field model for an input molecule. This makes it particularly attractive from an end-user standpoint by facilitating automated construction of a force field model of an arbitrary small molecule and having that model be consistent with the Amber protein force field. OPLS-AA, with its emphasis on condensed-phase simulations of small molecules, provides a diverse set of compounds and may be a good choice, although atom type assignment must be done by hand. In addition to highly optimized parameters for proteins, nucleic acids, and lipids, the CHARMM force field includes a large collection of similarly parameterized small molecules whose parameters can be used for simulation of substrates, cofactors, inhibitors, etc., with the limitation that, similar to the case of OPLS-AA, atom type assignment must be done by hand; work on a more general CHARMM force field is in progress in our laboratory. Finally, the GROMOS force field atom type palette, which derives from parameters for biopolymers, also provides a reasonable amount of diversity for the construction of force field models of small molecules.

One key issue for all of the above-mentioned force fields is that although sufficient, if not good to excellent, parameters already exist for the bond, angle, LJ, and Coulomb terms, the dihedral term can pose a difficulty in the modeling of small molecules. Conformational energetics can be context dependent, especially in cases in which there are multiple and/or strong intramolecular nonbonded interactions. This, therefore, can impede the transferability of dihedral parameters developed in the context of biopolymers or simple small molecules to larger, conformationally flexible small molecules, and validation by comparing with QM conformational energies is recommended. To the benefit of the uninitiated, there are many commercially available QM programs with intuitive graphical user interfaces that greatly facilitate molecule construction and energy calculations. Modern desktop computers are capable of QM calculations on small molecules at the HF/6-31G(d) model chemistry level, which is often satisfactory for conformational energetics, although MP2/6-31G(d) or higher is preferable. Comparison of force field and QM results for conformational surfaces of small molecules is recommended to serious deficiencies in the force field representation of the conformational energetics, and adjustment of dihedral parameters is warranted when there are large differences between the QM and force field results.

7 Conclusion

The outlook for force field-based simulations of proteins and heterogeneous protein-containing systems is bright. The field of simulation benefits enormously from the vibrant commodity computer hardware industry, which continues to push the limits of computing power and does so at very affordable prices. Although all of the force fields detailed in this review—Amber, CHARMM, GROMOS, and OPLS-AA—treat proteins at an often satisfactory level of accuracy, it will not be surprising if they all will require further revision as MD simulations probe larger systems at increasingly longer time scales and, in doing so, expose currently unknown deficiencies. Finally, a number of laboratories are currently undertaking the development of force fields that include explicit treatment of electronic polarizability [154–157], laying the groundwork for the next generation of empirical force fields for proteins and other biological molecules.

Acknowledgements Financial support from the National Institutes of Health (NIH; R01GM051501 and R01GM070855 to ADM, and F32CA119771 to OG) and from the University of Maryland Computer-Aided Drug Design Center is acknowledged.

References

1. Cornell, W. D., Cieplak, P., Bayly, C. I., Gould, I. R., Merz Jr., K. M., Ferguson, D. M., Spellmeyer, D. C., Fox, T., Caldwell, J. W., and Kollman, P. A. (1995) A second generation force field for the simulation of proteins, nucleic acids, and organic molecules. *J. Am. Chem. Soc.* **117**, 5179–5197.
2. MacKerell, A. D., Jr., Bashford, D., Bellott, M., Dunbrack, R. L., Evanseck, J. D., Field, M. J., Fischer, S., Gao, J., Guo, H., Ha, S., Joseph-McCarthy, D., Kuchnir, L., Kuczera, K., Lau, F. T. K., Mattos, C., Michnick, S., Ngo, T., Nguyen, D. T., Prodhom, B., Reiher, W. E., Roux, B., Schlenkrich, M., Smith, J. C., Stote, R., Straub, J., Watanabe, M., Wiórkiewicz-Kuczera, J., Yin, D., and Karplus, M. (1998) All-atom empirical potential for molecular modeling and dynamics studies of proteins. *J. Phys. Chem. B* **102**, 3586–3616.
3. Oostenbrink, C., Villa, A., Mark, A. E., and Van Gunsteren, W. F. (2004) A biomolecular force field based on the free enthalpy of hydration and solvation: The GROMOS force-field parameter sets 53A5 and 53A6. *J. Comput. Chem.* **25**, 1656–1676.
4. Jorgensen, W. L., Maxwell, D. S., and Tirado-Rives, J. (1996) Development and testing of the OPLS all-atom force field on conformational energetics and properties of organic liquids. *J. Am. Chem. Soc.* **118**, 11225–11236.
5. Kitson, D. H. and Hagler, A. T. (1988) Theoretical studies of the structure and molecular dynamics of a peptide crystal. *Biochemistry* **27**, 5246–5257.
6. Momany, F. A., Carruthers, L. M., McGuire, R. F., and Scheraga, H. A. (1974) Intermolecular potentials from crystal data. III. Determination of empirical potentials and application to the packing configurations and lattice energies in crystals of hydrocarbons, carboxylic acids, amines, and amides. *J. Phys. Chem.* **78**, 1595–1620.
7. Momany, F. A., McGuire, R. F., Burgess, A. W., and Scheraga, H. A. (1975) Energy parameters in polypeptides. VII. Geometric parameters, partial atomic charges, nonbonded interactions, hydrogen bond interactions, and intrinsic torsional potentials for the naturally occurring amino acids. *J. Phys. Chem* **79**, 2361–2381.
8. Némethy, G., Pottle, M. S., and Scheraga, H. A. (1983) Energy parameters in polypeptides. 9. Updating of geometrical parameters, nonbonded interactions, and hydrogen bond interactions for the naturally occurring amino acids. *J. Phys. Chem.* **87**, 1883–1887.
9. Némethy, G., Gibson, K. D., Palmer, K. A., Yoon, C. N., Paterlini, G., Zagari, A., Rumsey, S., and Scheraga, H. A. (1992) Energy parameters in polypeptides. 10. Improved geometrical

parameters and nonbonded interactions for use in the ECEPP/3 algorithm, with application to proline-containing peptides. *J. Phys. Chem.* **96**, 6472–6484.
10. Levitt, M. (1983) Molecular dynamics of native protein: I. Computer simulation of trajectories. *J. Mol. Biol.* **168**, 595–620.
11. Levitt, M., Hirshberg, M., Sharon, R., and Daggett, V. (1995) Potential energy function and parameters for simulations of the molecular dynamics of proteins and nucleic acids in solution. *Comput. Phys. Commun.* **91**, 215–231.
12. Allinger, N. L., Chen, K. S., and Lii, J. H. (1996) An improved force field (MM4) for saturated hydrocarbons. *J. Comput. Chem.* **17**, 642–668.
13. Nevins, N., Chen, K. S., and Allinger, N. L. (1996) Molecular mechanics (MM4) calculations on alkenes. *J. Comput. Chem.* **17**, 669–694.
14. Nevins, N., Lii, J. H., and Allinger, N. L. (1996) Molecular mechanics (MM4) calculations on conjugated hydrocarbons. *J. Comput. Chem.* **17**, 695–729.
15. Nevins, N. and Allinger, N. L. (1996) Molecular mechanics (MM4) vibrational frequency calculations for alkenes and conjugated hydrocarbons. *J. Comput. Chem.* **17**, 730–746.
16. Allinger, N. L., Chen, K. S., Katzenellenbogen, J. A., Wilson, S. R., and Anstead, G. M. (1996) Hyperconjugative effects on carbon-carbon bond lengths in molecular mechanics (MM4). *J. Comput. Chem.* **17**, 747–755.
17. Langley, C. H. and Allinger, N. L. (2002) Molecular mechanics (MM4) calculations on amides. *J. Phys. Chem. A* **106**, 5638–5652.
18. Halgren, T. A. (1996) Merck molecular force field. 1. Basis, form, scope, parameterization, and performance of MMFF94. *J. Comput. Chem.* **17**, 490–519.
19. Halgren, T. A. (1996) Merck molecular force field. 2. MMFF94 van der Waals and electrostatic parameters for intermolecular interactions. *J. Comput. Chem.* **17**, 520–552.
20. Halgren, T. A. (1996) Merck molecular force field. 3. Molecular geometries and vibrational frequencies for MMFF94. *J. Comput. Chem.* **17**, 553–586.
21. Halgren, T. A. and Nachbar, R. B. (1996) Merck molecular force field. 4. Conformational energies and geometries for MMFF94. *J. Comput. Chem.* **17**, 587–615.
22. Halgren, T. A. (1996) Merck molecular force field. 5. Extension of MMFF94 using experimental data, additional computational data, and empirical rules. *J. Comput. Chem.* **17**, 616–641.
23. Halgren, T. A. (1999) MMFF VI. MMFF94s option for energy minimization studies. *J. Comput. Chem.* **20**, 720–729.
24. Halgren, T. A. (1999) MMFF VII. Characterization of MMFF94, MMFF94s, and other widely available force fields for conformational energies and for intermolecular-interaction energies and geometries. *J. Comput. Chem.* **20**, 730–748.
25. Rappé, A. K., Casewit, C. J., Colwell, K. S., Goddard, W. A., and Skiff, W. M. (1992) UFF, a full periodic table force field for molecular mechanics and molecular dynamics simulations. *J. Am. Chem. Soc.* **114**, 10024–10035.
26. Ponder, J. W. and Case, D. A. (2003) Force fields for protein simulations, in *Protein Simulations*, Vol. 66, Academic Press, San Diego, pp. 27–85.
27. MacKerell, A. D., Jr. (2004) Empirical force fields for biological macromolecules: Overview and issues. *J. Comput. Chem.* **25**, 1584–1604.
28. Daggett, V. (2006) Protein folding-simulation. *Chem. Rev.* **106**, 1898–1916.
29. Oldziej, S., Czaplewski, C., Liwo, A., Chinchio, M., Nanias, M., Vila, J. A., Khalili, M., Arnautova, Y. A., Jagielska, A., Makowski, M., Schafroth, H. D., Kazmierkiewicz, R., Ripoll, D. R., Pillardy, J., Saunders, J. A., Kang, Y. K., Gibson, K. D., and Scheraga, H. A. (2005) Physics-based protein-structure prediction using a hierarchical protocol based on the UNRES force field: Assessment in two blind tests. *Proc. Natl. Acad. Sci. U. S. A.* **102**, 7547–7552.
30. Allen, M. P. and Tildesley, D. J. (1987) Computer simulation of liquids, Oxford University Press, Oxford.
31. Matsuoka, O., Clementi, E., and Yoshimine, M. (1976) CI study of the water dimer potential surface. *J. Chem. Phys.* **64**, 1351–1361.
32. Owicki, J. C. and Scheraga, H. A. (1977) Monte Carlo calculations on the isothermal-isobaric ensemble. 1. Liquid water. *J. Am. Chem. Soc.* **99**, 7403–7412.

33. Reiher, W. E. (1985), Theoretical studies of hydrogen bonding, Ph.D. thesis, Harvard University.
34. Morozov, A. V., Kortemme, T., Tsemekhman, K., and Baker, D. (2004) Close agreement between the orientation dependence of hydrogen bonds observed in protein structures and quantum mechanical calculations. *Proc. Natl. Acad. Sci. U. S. A.* **101**, 6946–6951.
35. Weiner, S. J., Kollman, P. A., Case, D. A., Singh, U. C., Ghio, C., Alagona, G., Profeta, S., and Weiner, P. A. (1984) A new force field for molecular mechanical simulation of nucleic acids and proteins. *J. Am. Chem. Soc.* **106**, 765–784.
36. Mahoney, M. W. and Jorgensen, W. L. (2000) A five-site model for liquid water and the reproduction of the density anomaly by rigid, nonpolarizable potential functions. *J. Chem. Phys.* **112**, 8910–8922.
37. Mahoney, M. W. and Jorgensen, W. L. (2001) Diffusion constant of the TIP5P model of liquid water. *J. Chem. Phys.* **114**, 363–366.
38. Jorgensen, W. L. and Tirado-Rives, J. (2005) Potential energy functions for atomic-level simulations of water and organic and biomolecular systems. *Proc. Natl. Acad. Sci. U. S. A.* **102**, 6665–6670.
39. Jorgensen, W. L., Chandrasekhar, J., Madura, J. D., Impey, R. W., and Klein, M. L. (1983) Comparison of simple potential functions for simulating liquid water. *J. Chem. Phys.* **79**, 926–935.
40. Berendsen, H. J. C., Grigera, J. R., and Straatsma, T. P. (1987) The missing term in effective pair potentials. *J. Phys. Chem.* **91**, 6269–6271.
41. Cornell, W. D., Cieplak, P., Bayly, C. E., and Kollman, P. A. (1993) Application of RESP charges to calculate conformational energies, hydrogen bond energies, and free energies of solvation. *J. Amer.Chem.Soc.* **115**, 9620–9631.
42. MacKerell, A. D., Jr., Feig, M., and Brooks, C. L., III, (2004) Improved treatment of the protein backbone in empirical force fields. *J. Am. Chem. Soc.* **126**, 698–699.
43. Hu, H., Elstner, M., and Hermans, J. (2003) Comparison of a QM/MM force field and molecular mechanics force fields in simulations of alanine and glycine "dipeptides" (Ace-Ala-Nme and Ace-Gly-Nme) in water in relation to the problem of modeling the unfolded peptide backbone in solution. *Proteins* **50**, 451–463.
44. Wang, J. M., Cieplak, P., and Kollman, P. A. (2000) How well does a restrained electrostatic potential (RESP) model perform in calculating conformational energies of organic and biological molecules? *J. Comput. Chem.* **21**, 1049–1074.
45. Okur, A., Strockbine, B., Hornak, V., and Simmerling, C. (2003) Using PC clusters to evaluate the transferability of molecular mechanics force fields for proteins. *J. Comput. Chem.* **24**, 21–31.
46. Kaminski, G. A., Friesner, R. A., Tirado-Rives, J., and Jorgensen, W. L. (2001) Evaluation and reparametrization of the OPLS-AA force field for proteins via comparison with accurate quantum chemical calculations on peptides. *J. Phys. Chem. B* **105**, 6474–6487.
47. Shirley, W. A. and Brooks, C. L., III (1997) Curious structure in "canonical" alanine based peptides. *Proteins* **28**, 59–71.
48. Feig, M., MacKerell, A. D., Jr., and Brooks, C. L., III (2003) Force field influence on the observation of pi-helical protein structures in molecular dynamics simulations. *J. Phys. Chem. B* **107**, 2831–2836.
49. MacKerell, A. D., Jr., Feig, M., and Brooks, C. L., III (2004) Extending the treatment of backbone energetics in protein force fields: Limitations of gas-phase quantum mechanics in reproducing protein conformational distributions in molecular dynamics simulations. *J. Comput. Chem.* **25**, 1400–1415.
50. Jacob, E. J., Thompson, H. B., and Bartell, L. S. (1967) Influence of nonbonded interactions on molecular geometry and energy: Calculations for hydrocarbons based on Urey-Bradley field. *J. Chem. Phys.* **47**, 3736–3753.
51. Lifson, S. and Warshel, A. (1968) Consistent force field for calculations of conformations, vibrational spectra, and enthalpies of cycloalkane and *n*-alkane molecules. *J. Chem. Phys.* **49**, 5116–5129.

52. Allinger, N. L., Tribble, M. T., Miller, M. A., and Wertz, D. H. (1971) Conformational analysis. LXIX. An improved force field for the calculation of the structures and energies of hydrocarbons. *J. Am. Chem. Soc.* **93**, 1637–1648.
53. Jorgensen, W. L. and Tirado-Rives, J. (1988) The OPLS potential function for proteins. Energy minimizations for crystals of cyclic peptides and crambin. *J. Amer. Chem. Soc.* **110**, 1657–1666.
54. Bayly, C. I., Cieplak, P., Cornell, W. D., and Kollman, P. A. (1993) A well-behaved electrostatic potential based method using charge restraints for deriving atomic charges: The RESP model. *J. Phys. Chem.* **97**, 10269–10280.
55. Fox, T. and Kollman, P. A. (1998) Application of the RESP methodology in the parametrization of organic solvents. *J. Phys. Chem. B* **102**, 8070–8079.
56. Cieplak, P., Cornell, W. D., Bayly, C. I., and Kollman, P. K. (1995) Application of the multimolecule and multiconformational RESP methodology to biopolymers: Charge derivation for DNA, RNA, and proteins. *J. Comput. Chem.* **16**, 1357–1377.
57. Neria, E., Fischer, S., and Karplus, M. (1996) Simulation of activation free energies in molecular systems. *J. Chem. Phys.* **105**, 1902–1919.
58. Oostenbrink, C., Soares, T. A., van der Vegt, N. F. A., and van Gunsteren, W. F. (2005) Validation of the 53A6 GROMOS force field. *Eur. Biophys. J. Biophys. Lett.* **34**, 273–284.
59. Shirts, M. R., Pitera, J. W., Swope, W. C., and Pande, V. S. (2003) Extremely precise free energy calculations of amino acid side chain analogs: Comparison of common molecular mechanics force fields for proteins. **119**, 5740–5761.
60. Deng, Y. and Roux, B. (2004) Hydration of amino acid side chains: Non-polar and electrostatic contributions calculated from staged molecular dynamics free energy simulations with explicit water molecules. *J. Phys. Chem. B* **108**, 16567–16576.
61. Metropolis, N., Rosenbluth, A. W., Rosenbluth, M. N., Teller, A. H., and Teller, E. (1953) Equation of state calculations by fast computing machines. *J. Chem. Phys.* **21**, 1087–1092.
62. Wood, W. W. (1968) Monte Carlo calculations for hard disks in the isothermal-isobaric ensemble. *J. Chem. Phys.* **48**, 415–434.
63. McDonald, I. R. (1969) Monte Carlo calculations for one- and two-component fluids in the isothermal-isobaric ensemble. *Chem. Phys. Lett.* **3**, 241–243.
64. Maurits, N. M., Zvelindovsky, A. V., Sevink, G. J. A., van Vlimmeren, B. A. C., and Fraaije, J. (1998) Hydrodynamic effects in three-dimensional microphase separation of block copolymers: Dynamic mean-field density functional approach. *J. Chem. Phys.* **108**, 9150–9154.
65. Groot, R. D., Madden, T. J., and Tildesley, D. J. (1999) On the role of hydrodynamic interactions in block copolymer microphase separation. *J. Chem. Phys.* **110**, 9739–9749.
66. Shelley, J. C., Shelley, M. Y., Reeder, R. C., Bandyopadhyay, S., Moore, P. B., and Klein, M. L. (2001) Simulations of phospholipids using a coarse grain model. *J. Phys. Chem. B* **105**, 9785–9792.
67. Andersen, H. C. (1980) Molecular dynamics simulations at constant pressure and/or temperature. *J. Chem. Phys.* **72**, 2384–2393.
68. Parrinello, M. and Rahman, A. (1981) Polymorphic transitions in single crystals: A new molecular dynamics method. *J. Appl. Phys.* **52**, 7182–7190.
69. Nosé, S. (1984) A molecular dynamics method for simulations in the canonical ensemble. *Mol. Phys.* **52**, 255–268.
70. Hoover, W. G. (1985) Canonical dynamics: Equilibrium phase-space distributions. *Phys. Rev. A* **31**, 1695–1697.
71. Feller, S. E., Zhang, Y. H., Pastor, R. W., and Brooks, B. R. (1995) Constant pressure molecular dynamics simulation: The Langevin piston method. *J. Chem. Phys.* **103**, 4613–4621.
72. Tuckerman, M. E. and Martyna, G. J. (2000) Understanding modern molecular dynamics: Techniques and applications. *J. Phys. Chem. B* **104**, 159–178.
73. Berendsen, H. J. C., Postma, J. P. M., van Gunsteren, W. F., Di Nola, A., and Haak, J. R. (1984) Molecular dynamics with coupling to an external bath. *J. Chem. Phys.* **81**, 3684–3690.
74. Steinbach, P. J. and Brooks, B. R. (1994) New spherical-cutoff methods for long-range forces in macromolecular simulation. *J. Comput. Chem.* **15**, 667–683.

75. Ewald, P. P. (1921) Die Berechnung optischer und elektrostatischer Gitterpotentiale. *Ann. Phys.* **64**, 253–287.
76. Hockney, R. W. and Eastwood, J. W. (1981) Computer simulations using particles, McGraw Hill, New York.
77. Darden, T., York, D., and Pedersen, L. (1993) Particle mesh Ewald: An N•log(N) method for Ewald sums in large systems. *J. Chem. Phys.* **98**, 10089–10092.
78. Essmann, U., Perera, L., Berkowitz, M. L., Darden, T. A., Lee, H., and Pedersen, L. G. (1995) A smooth particle mesh Ewald method. *J. Chem. Phys.* **103**, 8577–8593.
79. Darden, T., Toukmaji, A., and Pedersen, L. (1997) Long-range electrostatic effects in biomolecular systems. *J. Chem. Phys.* **94**, 1346–1364.
80. Lagüe, P., Pastor, R. W., and Brooks, B. R. (2004) Pressure-based long-range correction for Lennard-Jones interactions in molecular dynamics simulations: Application to alkanes and interfaces. *J. Phys. Chem. B* **108**, 363–368.
81. Brooks, C. L., III, and Karplus, M. (1983) Deformable stochastic boundaries in molecular dynamics. *J. Chem. Phys.* **79**, 6312–6325.
82. Brünger, A. T., Brooks, C. L., III, and Karplus, M. (1984) Stochastic boundary conditions for molecular dynamics simulations of ST2 water. *Chem. Phys. Lett.* **105**, 495.
83. Im, W., Bernéche, S., and Roux, B. (2001) Generalized solvent boundary potential for computer simulations. *J. Chem. Phys.* **114**, 2924–2937.
84. Still, W. C., Tempczyk, A., Hawley, R. C., and Hendrickson, T. (1990) Semianalytical treatment of solvation for molecular mechanics and dynamics. *J. Am. Chem. Soc.* **112**, 6127–6129.
85. Hawkins, G. D., Cramer, C. J., and Truhlar, D. G. (1995) Pairwise solute descreening of solute charges from a dielectric medium. *Chem. Phys. Lett.* **246**, 122–129.
86. Qui, D., Shenkin, P. S., Hollinger, F. P., and Still, W. C. (1997) The GB/SA continuum model for solvation. A fast analytical method for the calculation of approximate Born radii. *J. Phys. Chem. A* **101**, 3005–3014.
87. Ghosh, A., Rapp, C. S., and Friesner, R. A. (1998) Generalized born model based on a surface integral formulation. *J. Phys. Chem. B* **102**, 10983–10990.
88. Dominy, B. N. and Brooks, C. L., III (1999) Development of a generalized born model parametrization for proteins and nucleic acids. *J. Phys. Chem. B* **103**, 3765–3773.
89. Onufriev, A., Bashford, D., and Case, D. A. (2000) Modification of the generalized Born model suitable for macromolecules. *J. Phys. Chem. B* **104**, 3712–3720.
90. Lee, M. S., Salsbury, F. R., and Brooks, C. L., III (2002) Novel generalized Born methods. *J. Chem. Phys.* **116**, 10606–10614.
91. Im, W. P., Lee, M. S., and Brooks, C. L., III (2003) Generalized born model with a simple smoothing function. *J. Comput. Chem.* **24**, 1691–1702.
92. Guvench, O., Shenkin, P., Kolossvary, I., and Still, W. C. (2002) Application of the frozen atom approximation to the GB/SA continuum model for solvation free energy. *J. Comput. Chem.* **23**, 214–221.
93. Im, W., Feig, M., and Brooks, C. L., III (2003) An implicit membrane generalized born theory for the study of structure, stability, and interactions of membrane proteins. *Biophys. J.* **85**, 2900–2918.
94. Tanizaki, S. and Feig, M. (2005) A generalized Born formalism for heterogeneous dielectric environments: Application to the implicit modeling of biological membranes. *J. Chem. Phys.* **122**.
95. Gallicchio, E., Zhang, L. Y., and Levy, R. M. (2002) The SGB/NP hydration free energy model based on the surface generalized born solvent reaction field and novel nonpolar hydration free energy estimators. *J. Comput. Chem.* **23**, 517–529.
96. Gallicchio, E. and Levy, R. M. (2004) AGBNP: An analytic implicit solvent model suitable for molecular dynamics simulations and high-resolution modeling. *J. Comput. Chem.* **25**, 479–499.
97. Weiner, P. K. and Kollman, P. A. (1981) AMBER: Assisted model building with energy refinement. A general program for modeling molecules and their interactions. *J. Comput. Chem.* **2**, 287–303.

98. Pearlman, D. A., Case, D. A., Caldwell, J. W., Ross, W. S., Cheatham, T. E., Debolt, S., Ferguson, D., Seibel, G., and Kollman, P. (1995) AMBER, a package of computer programs for applying molecular mechanics, normal mode analysis, molecular dynamics and free energy calculations to simulate the structural and energetic properties of molecules. *Comput. Phys. Commun.* **91**, 1–41.
99. Case, D. A., Cheatham, T. E., Darden, T., Gohlke, H., Luo, R., Merz, K. M., Onufriev, A., Simmerling, C., Wang, B., and Woods, R. J. (2005) The Amber biomolecular simulation programs. *J. Comput. Chem.* **26**, 1668–1688.
100. Brooks, B. R., Bruccoleri, R. E., Olafson, B. D., States, D. J., Swaminathan, S., and Karplus, M. (1983) CHARMM: A program for macromolecular energy, minimization, and dynamics calculations. *J. Comput. Chem.* **4**, 187–217.
101. Scott, W. R. P., Hünenberger, P. H., Tironi, I. G., Mark, A. E., Billeter, S. R., Fennen, J., Torda, A. E., Huber, T., Krüger, P., and van Gunsteren, W. F. (1999) The GROMOS biomolecular simulation program package. *J. Phys. Chem. A* **103**, 3596–3607.
102. Christen, T., Hünenberger, P. H., Bakowies, D., Baron, R., Burgi, R., Geerke, D. P., Heinz, T. N., Kastenholz, M. A., Krautler, V., Oostenbrink, C., Peter, C., Trzesniak, D., and Van Gunsteren, W. F. (2005) The GROMOS software for biomolecular simulation: GROMOS05. *J. Comput. Chem.* **26**, 1719–1751.
103. Jorgensen, W. L. (1998) BOSS – Biochemical and organic simulation system, in *The Encyclopedia of Computational Chemistry* (Schleyer, P. v. R., ed.), Vol. 5, John Wiley and Sons, Ltd., Athens, GA.
104. Jorgensen, W. L. and Tirado-Rives, J. (2005) Molecular modeling of organic and biomolecular systems using BOSS and MCPRO. *J. Comput. Chem.* **26**, 1689–1700.
105. Berendsen, H. J. C., van der Spoel, D., and van Drunen, R. (1995) GROMACS—A message-passing parallel molecular-dynamics implementation. *Comput. Phys. Commun.* **91**, 43–56.
106. Lindahl, E., Hess, B., and van der Spoel, D. (2001) GROMACS 3.0: A package for molecular simulation and trajectory analysis. *J. Mol. Model.* **7**, 306–317.
107. van der Spoel, D., Lindahl, E., Hess, B., Groenhof, G., Mark, A. E., and Berendsen, H. J. C. (2005) GROMACS: Fast, flexible, and free. *J. Comput. Chem.* **26**, 1701–1718.
108. Nelson, M., Humphrey, W., Kufrin, R., Gursoy, A., Dalke, A., Kale, L., Skeel, R., and Schulten, K. (1995) MDScope—A visual computing environment for structural biology. *Comput. Phys. Commun.* **91**, 111–133.
109. Nelson, M. T., Humphrey, W., Gursoy, A., Dalke, A., Kale, L. V., Skeel, R. D., and Schulten, K. (1996) NAMD: A parallel, object oriented molecular dynamics program. *Int. J. Supercomput. Appl. High Perform. Comput.* **10**, 251–268.
110. Kale, L., Skeel, R., Bhandarkar, M., Brunner, R., Gursoy, A., Krawetz, N., Phillips, J., Shinozaki, A., Varadarajan, K., and Schulten, K. (1999) NAMD2: Greater scalability for parallel molecular dynamics. *J. Comput. Phys.* **151**, 283–312.
111. Phillips, J. C., Braun, R., Wang, W., Gumbart, J., Tajkhorshid, E., Villa, E., Chipot, C., Skeel, R. D., Kale, L., and Schulten, K. (2005) Scalable molecular dynamics with NAMD. *J. Comput. Chem.* **26**, 1781–1802.
112. Banks, J. L., Beard, H. S., Cao, Y. X., Cho, A. E., Damm, W., Farid, R., Felts, A. K., Halgren, T. A., Mainz, D. T., Maple, J. R., Murphy, R., Philipp, D. M., Repasky, M. P., Zhang, L. Y., Berne, B. J., Friesner, R. A., Gallicchio, E., and Levy, R. M. (2005) Integrated modeling program, applied chemical theory (IMPACT). *J. Comput. Chem.* **26**, 1752–1780.
113. Karplus, M., and McCammon, J. A. (2002) Molecular dynamics simulations of biomolecules. *Nat. Struct. Biol.* **9**, 646–652.
114. MacKerell, A. D., Jr., Wiórkiewicz-Kuczera, J., and Karplus, M. (1995) An all-atom empirical energy function for the simulation of nucleic acids. *J. Am. Chem. Soc.* **117**, 11946–11975.
115. Ponomarev, S. Y., Thayer, K. M., and Beveridge, D. L. (2004) Ion motions in molecular dynamics simulations on DNA. *Proc. Natl. Acad. Sci. U. S. A.* **101**, 14771–14775.
116. Zhang, Q. and Schlick, T. (2006) Stereochemistry and position-dependent effects of carcinogens on TATA/TBP binding. *Biophys. J.* **90**, 1865–1877.
117. Spackova, N. and Sponer, J. (2006) Molecular dynamics simulations of sarcin-ricin rRNA motif. *Nucleic Acids Res.* **34**, 697–708.

118. Trobro, S. and Aqvist, J. (2006) Analysis of predictions for the catalytic mechanism of ribosomal peptidyl transfer. *Biochemistry* **45**, 7049–7056.
119. Pastor, N. (2005) The B- to A-DNA transition and the reorganization of solvent at the DNA surface. *Biophys. J.* **88**, 3262–3275.
120. Huang, N., Banavali, N. K., and MacKerell, A. D., Jr. (2003) Protein-facilitated base flipping in DNA by cytosine-5-methyltransferase. *Proc. Natl. Acad. Sci. U. S. A.* **100**, 68–73.
121. Soares, T. A., Hünenberger, P. H., Kastenholz, M. A., Krautler, V., Lenz, T., Lins, R. D., Oostenbrink, C., and Van Gunsteren, W. F. (2005) An improved nucleic acid parameter set for the GROMOS force field. *J. Comput. Chem.* **26**, 725–737.
122. Schlenkrich, M., Brinkman, J., MacKerell, A. D., Jr., and Karplus, M. (1996) An empirical potential energy function for phospholipids: Criteria for parameter optimization and applications, in *Membrane Structure and Dynamics* (Merz, K. M., and Roux, B., eds.), Birkhauser, Boston, pp. 31–81.
123. Feller, S. E., Yin, D. X., Pastor, R. W., and MacKerell, A. D., Jr. (1997) Molecular dynamics simulation of unsaturated lipid bilayers at low hydration: Parameterization and comparison with diffraction studies. *Biophys. J.* **73**, 2269–2279.
124. Yin, D. X. and MacKerell, A. D., Jr. (1998) Combined ab initio empirical approach for optimization of Lennard-Jones parameters. *J. Comput. Chem.* **19**, 334–348.
125. Feller, S. E. and MacKerell, A. D., Jr. (2000) An improved empirical potential energy function for molecular simulations of phospholipids. *J. Phys. Chem. B* **104**, 7510–7515.
126. Feller, S. E., Gawrisch, K., and MacKerell, A. D., Jr. (2002) Polyunsaturated fatty acids in lipid bilayers: Intrinsic and environmental contributions to their unique physical properties. *J. Am. Chem. Soc.* **124**, 318–326.
127. Spijker, P., Vaidehi, N., Freddolino, P. L., Hilbers, P. A. J., and Goddard, W. A. (2006) Dynamic behavior of fully solvated beta 2-adrenergic receptor, embedded in the membrane with bound agonist or antagonist. *Proc. Natl. Acad. Sci. U. S. A.* **103**, 4882–4887.
128. Lagüe, P., Roux, B., and Pastor, R. W. (2005) Molecular dynamics simulations of the influenza hemagglutinin fusion peptide in micelles and bilayers: Conformational analysis of peptide and lipids. *J. Mol. Biol.* **354**, 1129–1141.
129. Tajkhorshid, E., Nollert, P., Jensen, M. O., Miercke, L. J. W., O'Connell, J., Stroud, R. M., and Schulten, K. (2002) Control of the selectivity of the aquaporin water channel family by global orientational tuning. *Science* **296**, 525–530.
130. Berger, O., Edholm, O., and Jahnig, F. (1997) Molecular dynamics simulations of a fluid bilayer of dipalmitoylphosphatidylcholine at full hydration, constant pressure, and constant temperature. *Biophys. J.* **72**, 2002–2013.
131. Chandrasekhar, I., Kastenholz, M., Lins, R. D., Oostenbrink, C., Schuler, L. D., Tieleman, D. P., and van Gunsteren, W. F. (2003) A consistent potential energy parameter set for lipids: Dipalmitoylphosphatidylcholine as a benchmark of the GROMOS96 45A3 force field. *Eur. Biophys. J. Biophys. Lett.* **32**, 67–77.
132. Glattli, A., Chandrasekhar, I., and van Gunsteren, W. F. (2006) A molecular dynamics study of the bee venom melittin in aqueous solution, in methanol, and inserted in a phospholipid bilayer. *Eur. Biophys. J. Biophys. Lett.* **35**, 255–267.
133. Leontiadou, H., Mark, A. E., and Marrink, S. J. (2006) Antimicrobial peptides in action. *J. Am. Chem. Soc.* **128**, 12156–12161.
134. Grottesi, A., Domene, C., Hall, B., and Sansom, M. S. P. (2005) Conformational dynamics of M2 helices in KirBac channels: Helix flexibility in relation to gating via molecular dynamics simulations. *Biochemistry* **44**, 14586–14594.
135. Moore, P. B., Lopez, C. F., and Klein, M. L. (2001) Dynamical properties of a hydrated lipid bilayer from a multinanosecond molecular dynamics simulation. *Biophys. J.* **81**, 2484–2494.
136. Liepina, I., Czaplewski, C., Janmey, P., and Liwo, A. (2003) Molecular dynamics study of a gelsolin-derived peptide binding to a lipid bilayer containing phosphatidylinositol 4,5-bisphosphate. *Biopolymers* **71**, 49–70.
137. Bastug, T. and Kuyucak, S. (2005) Test of molecular dynamics force fields in gramicidin A. *Eur. Biophys. J. Biophys. Lett.* **34**, 377–382.

138. Woods, R. J., Dwek, R. A., Edge, C. J., and Fraserreid, B. (1995) Molecular mechanical and molecular dynamical simulations of glycoproteins and oligosaccharides. 1. GLYCAM_93 parameter development. *J. Phys. Chem.* **99**, 3832–3846.
139. Woods, R. J. and Chappelle, R. (2000) Restrained electrostatic potential atomic partial charges for condensed-phase simulations of carbohydrates. *Theochem–J. Mol. Struct.* **527**, 149–156.
140. Basma, M., Sundara, S., Calgan, D., Vernali, T., and Woods, R. J. (2001) Solvated ensemble averaging in the calculation of partial atomic charges. *J. Comput. Chem.* **22**, 1125–1137.
141. Kawatkar, S. P., Kuntz, D. A., Woods, R. J., Rose, D. R., and Boons, G. J. (2006) Structural basis of the inhibition of Golgi alpha-mannosidase II by mannostatin A and the role of the thiomethyl moiety in ligand-protein interactions. *J. Am. Chem. Soc.* **128**, 8310–8319.
142. Lins, R. D., Pereira, C. S., and Hünenberger, P. H. (2004) Trehalose-protein interaction in aqueous solution. *Proteins* **55**, 177–186.
143. Bosques, C. J., Tschampel, S. M., Woods, R. J., and Imperiali, B. (2004) Effects of glycosylation on peptide conformation: A synergistic experimental and computational study. *J. Am. Chem. Soc.* **126**, 8421–8425.
144. Ford, M. G., Weimar, T., Kohli, T., and Woods, R. J. (2003) Molecular dynamics simulations of galectin-1-oligosaccharide complexes reveal the molecular basis for ligand diversity. *Proteins* **53**, 229–240.
145. Kirschner, K. N. and Woods, R. J. (2001) Solvent interactions determine carbohydrate conformation. *Proc. Natl. Acad. Sci. U. S. A.* **98**, 10541–10545.
146. Damm, W., Frontera, A., TiradoRives, J., and Jorgensen, W. L. (1997) OPLS all-atom force field for carbohydrates. *J. Comput. Chem.* **18**, 1955–1970.
147. Margulis, C. J. (2005) Computational study of the dynamics of mannose disaccharides free in solution and bound to the potent anti-HIV virucidal protein cyanovirin. *J. Phys. Chem. B* **109**, 3639–3647.
148. Kony, D., Damm, W., Stoll, S., and van Gunsteren, W. F. (2002) An improved OPLS-AA force field for carbohydrates. *J. Comput. Chem.* **23**, 1416–1429.
149. Kuttel, M., Brady, J. W., and Naidoo, K. J. (2002) Carbohydrate solution simulations: Producing a force field with experimentally consistent primary alcohol rotational frequencies and populations. *J. Comput. Chem.* **23**, 1236–1243.
150. Lins, R. D. and Hünenberger, P. H. (2005) A new GROMOS force field for hexopyranose-based carbohydrates. *J. Comput. Chem.* **26**, 1400–1412.
151. Daura, X., Mark, A. E., and van Gunsteren, W. F. (1998) Parametrization of aliphatic CHn united atoms of GROMOS96 force field. *J. Comput. Chem.* **19**, 535–547.
152. Schuler, L. D., Daura, X., and Van Gunsteren, W. F. (2001) An improved GROMOS96 force field for aliphatic hydrocarbons in the condensed phase. *J. Comput. Chem.* **22**, 1205–1218.
153. Wang, J. M., Wolf, R. M., Caldwell, J. W., Kollman, P. A., and Case, D. A. (2004) Development and testing of a general amber force field. *J. Comput. Chem.* **25**, 1157–1174.
154. Ren, P. Y. and Ponder, J. W. (2002) Consistent treatment of inter- and intramolecular polarization in molecular mechanics calculations. *J. Comput. Chem.* **23**, 1497–1506.
155. Patel, S., MacKerell, A. D., Jr., and Brooks, C. L. III (2004) CHARMM fluctuating charge force field for proteins: II—Protein/solvent properties from molecular dynamics simulations using a nonadditive electrostatic model. *J. Comput. Chem.* **25**, 1504–1514.
156. Kim, B. C., Young, T., Harder, E., Friesner, R. A., and Berne, B. J. (2005) Structure and dynamics of the solvation of bovine pancreatic trypsin inhibitor in explicit water: A comparative study of the effects of solvent and protein polarizability. *J. Phys. Chem. B* **109**, 16529–16538.
157. Anisimov, V. M., Lamoureux, G., Vorobyov, I. V., Huang, N., Roux, B., and MacKerell, A. D., Jr. (2005) Determination of electrostatic parameters for a polarizable force field based on the classical Drude oscillator. *J. Chem. Theory Comput.* **1**, 153–168.

Chapter 5
Normal Modes and Essential Dynamics

Steven Hayward and Bert L. de Groot

Summary Normal mode analysis and essential dynamics analysis are powerful methods used for the analysis of collective motions in biomolecules. Their application has led to an appreciation of the importance of protein dynamics in function and the relationship between structure and dynamical behavior. In this chapter, the methods and their implementation are introduced and recent developments such as elastic networks and advanced sampling techniques are described.

Keywords: Collective protein dynamics · Conformational flooding · Conformational sampling · Elastic network · Principal component analysis

1 Introduction

1.1 Standard Normal Mode Analysis

Normal mode analysis (NMA) is one of the major simulation techniques used to probe the large-scale, shape-changing motions in biological molecules [1–3]. Although it has connection to the experimental techniques of infrared and Raman spectroscopy, its recent application has been to predict functional motions in proteins or other biological molecules. Functional motions are those that relate to function and are often the consequence of binding other molecules. In NMA studies, it is always assumed that the normal modes with the largest fluctuation (lowest frequency modes) are the ones that are functionally relevant, because, like function, they exist by evolutionary design rather than by chance. The ultimate justification for this assumption must come from comparisons with experimental data and indeed studies that compare predictions of an NMA with transitions derived from multiple x-ray conformers do suggest that the low-frequency normal modes are often functionally relevant.

From: Methods in Molecular Biology, vol. 443, Molecular Modeling of Proteins
Edited by Andreas Kukol © Humana Press, Totowa, NJ

NMA is a harmonic analysis. In its purest form, it uses exactly the same force fields as used in molecular dynamics simulations. In that sense, it is accurate. However, the underlying assumption that the conformational energy surface at an energy minimum can be approximated by a parabola over the range of thermal fluctuations is known not to be correct at physiological temperatures. There exists abundant evidence, both experimental [4] and computational [5], that the harmonic approximation breaks down spectacularly for proteins at physiological temperatures, where, far from performing harmonic motion in a single energy minimum, the state point visits multiple minima crossing energy barriers of various heights. Thus, when performing NMA, one has to be aware of this assumption and its limitations at functioning temperatures.

A standard NMA requires a set of coordinates, a force field describing the interactions between constituent atoms, and software to perform the required calculations. The performance of an NMA in Cartesian coordinate space requires three main calculation steps: 1) minimization of the conformational potential energy as a function of the atomic Cartesian coordinates; 2) the calculation of the so-called "Hessian" matrix, which is the matrix of second derivatives of the potential energy with respect to the mass-weighted atomic coordinates; and 3) the diagonalization of the Hessian matrix. This final step yields eigenvalues and eigenvectors (the "normal modes"). Each of these three steps can be computationally demanding, depending on the size of the molecule. Usually, the first and final steps are the bottlenecks. Normally, energy minimization is demanding of CPU time and diagonalization is demanding of CPU time and memory because it involves the diagonalization of a $3N \times 3N$ matrix, where N is the number of atoms in the molecule. We have called this NMA "standard" NMA to distinguish it from the elastic network model NMA.

1.2 Elastic Network Models

Because of the computational difficulties of standard NMA, the current popularity of the elastic network models is not surprising. This is still an NMA, but the protein model is dramatically simplified. Tirion first introduced it into protein research [6]. As the name suggests, the atoms are connected by a network of elastic connections. The method has two main advantages over the standard NMA. The first is that there is no need for energy minimization because the distances of all of the elastic connections are taken to be at their minimum energy length. Second, the diagonalization task is greatly reduced compared with the standard NMA method because the number of atoms is reduced from the total number of atoms to the number of residues, if one uses only C^{α} atoms, as is common practice. This leads to a tenfold reduction in the number of atoms. Unlike standard NMA, elastic network models have two parameters to be set. One is the force or spring constant, normally denoted as γ or C, and the other is a cut-off distance, denoted R_c.

A pertinent question is whether the method is any less accurate than the standard NMA. Tirion showed that there is a respectable degree of correspondence between the two methods [6]. Given the drastic assumptions that are inherent in the standard NMA, the small difference between the results from these two methods is probably unimportant relative to differences between standard NMA and reality. Comparisons between movements in the low-frequency modes derived from elastic network models of 20 proteins with movements derived from pairs of x-ray structures [7] suggest the same level of moderate correspondence seen in similar studies using standard NMA. This, together with the relatively low computational cost of elastic network models, explains their current popularity in comparison with standard NMA.

1.3 Essential Dynamics and Principal Components Analysis

Because of the complexity of biomolecular systems, molecular dynamics simulations can be notoriously hard to analyze, rendering it difficult to grasp the motions of interest, or to uncover functional mechanisms. A principal components analysis (PCA) [8–10] often alleviates this problem. Similar to NMA, PCA rests on the assumption that the major collective modes of fluctuation dominate the functional dynamics. Interestingly, it has been found that the vast majority of protein dynamics can be described by a surprisingly low number of collective degrees of freedom [9]. For the analysis of protein molecular dynamics simulations, this approach has the advantage that the dynamics along the individual modes can be inspected and visualized separately, thereby allowing one to filter the main modes of collective motion from more local fluctuations. Because these principal modes of motion could, in many cases, be linked to protein function, the dynamics in the low-dimensional subspace spanned by these modes was termed "essential dynamics" [9], to reflect the notion that these are the modes essential for function. The subspace spanned by the major modes of collective fluctuations is accordingly often referred to as "essential subspace." The fact that only a small subset of the total number of degrees of freedom dominates the molecular dynamics of biomolecules not only aids the analysis and interpretation of molecular dynamics trajectories, but also opens the way to enhanced sampling algorithms that search the essential subspace in either a systematic or exploratory fashion [11–14].

In contrast to NMA, PCA of a molecular dynamics simulation trajectory does not rest on the assumption of a harmonic potential. In fact, PCA can be used to study the degree of anharmonicity in the molecular dynamics of a simulated system. For proteins, it was shown that, at physiological temperatures, especially the major modes of collective fluctuation are dominated by anharmonic fluctuations [9, 15]. Overall, protein dynamics at physiological temperatures has been described as diffusion among multiple minima [16–18]; on short timescales, the dynamics are dominated by fluctuations within a local minimum (that can be approximated well by a system's local normal modes), whereas, on longer timescales, the large fluctuations are dominated by a largely anharmonic diffusion between multiple wells.

In NMA the modes of greatest fluctuation are those with the lowest frequencies. As in PCA, no assumptions are implied regarding the harmonicity of the motion, modes are usually sorted according to variance rather than frequency. Nevertheless, the largest-amplitude modes of a PCA usually also represent the slowest dynamical transitions.

2 Theory

2.1 Standard NMA

NMA is usually performed in a vacuum, where the potential energy of a biomolecule is a complex function of its 3N coordinates, N being the number of atoms. This function is normally written in terms of its bonded and nonbonded energy terms. It is usual to use Cartesian coordinates [3], although dihedral angles have been used [1, 19, 20]. The basic idea is that, at a minimum, the potential energy function V can be expanded in a Taylor series in terms of the mass-weighted coordinates $q_i = \sqrt{m_i}\Delta x_i$, where Δx_i is the displacement of the ith coordinate from the energy minimum and m_i is that mass of the corresponding atom. If the expansion is terminated at the quadratic level, then because the linear term is zero at an energy minimum:

$$V = \frac{1}{2}\sum_{i,j=1}^{3N} \left.\frac{\partial^2 V}{\partial q_i \partial q_j}\right|_0 q_i q_j. \tag{1}$$

Thus, the energy surface is approximated by a parabola characterized by the second derivatives evaluated at the energy at the minimum. The basic, but false, assumption of NMA of biomolecules at physiological temperatures is that fluctuations still occur within this parabolic energy surface. It is known, however, that, at these temperatures, the state point moves on a complex energy surface with multiple minima, crossing energy barriers of various heights [4]. The second derivatives in Eq. 1 can be written in a matrix, which is often called the "Hessian," **F**. Determination of its eigenvalues and eigenvectors (equivalent to diagonalization) implies:

$$\mathbf{F}\mathbf{w}_j = \omega_j^2 \mathbf{w}_j, \tag{2}$$

where $\mathbf{w}_j$ is the jth eigenvector and ω_j^2 is the jth eigenvalue. There are 3N such eigenvector equations. Each eigenvector specifies a normal mode coordinate through:

$$Q_j = \sum_{i=1}^{3N} w_{ij}\, q_i. \tag{3}$$

The sum is over the elements of $\mathbf{w}_j$. Note that $|\mathbf{w}_j| = 1$. It can be shown that these normal mode coordinates oscillate harmonically and independently of each other each with the angular frequency, ω_j:

$$Q_j = A_j \cos(\omega_j t + \varepsilon_j). \tag{4}$$

Here, A_j is the amplitude and ε_j is the phase. These normal mode coordinates are collective variables because they are linear combinations of the atom-based Cartesian coordinates, as shown in Eq. 3. If a single normal mode j is activated, then:

$$\Delta x_{ij} = \frac{w_{ij}}{\sqrt{m_i}} A_j \cos(\omega_j t + \varepsilon_j), \tag{5}$$

which means that, in the jth mode, the relative displacements of the Cartesian coordinates are specified by the elements of $\mathbf{w}_j$. Each normal mode then specifies a pattern of atomic displacement. For example, in a multidomain protein, this pattern of displacement could indicate the relative movement of two domains. Figure 1b shows an example. A more thorough introduction to the theory and its application to biomolecules can be found elsewhere [21].

It can be shown that the lower the frequency, the larger the fluctuation of the corresponding normal mode coordinate [22]. It is common to compare the lowest frequency modes with functional modes derived from, e.g., a pair of x-ray structures, one bound to a functional ligand and the other unbound. The overlap with the jth mode can be defined as [23]:

$$O_j = \frac{\sum_{i=1}^{3N} \Delta x_{ij} \Delta x_i^{exp}}{\sqrt{\sum_{i=1}^{3N} (\Delta x_{ij})^2} \sqrt{\sum_{i=1}^{3N} (\Delta x_i^{exp})^2}}. \tag{6}$$

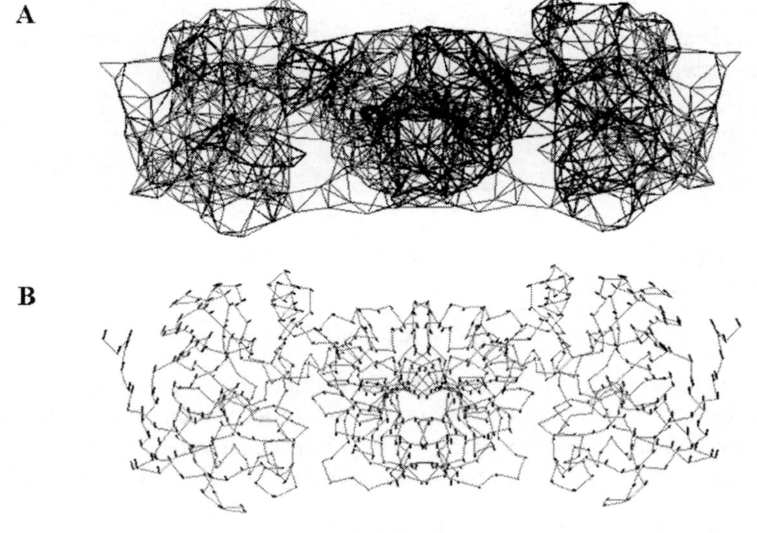

Fig. 1 (**a**) Elastic network model of the homodimeric molecule liver alcohol dehydrogenase. A cut-off distance, R_c, of 7 Å was used. (**b**) C^α trace of liver alcohol dehydrogenase, with each short line showing the displacement of the C^α in the first normal mode derived from the elastic network model shown in (**a**)

2.2 Elastic Network Models

There is, in fact, no essential difference between the elastic network NMA and the standard NMA other than the force field. In the case of the elastic network, the Hessian would be derived from the following potential energy function [6]:

$$V = \frac{\gamma}{2} \sum_{\left|r_{ij}^{0}\right| < R_C} \left(r_{ij} - r_{ij}^{0}\right)^2, \tag{7}$$

where r_{ij} is the distance between atoms i and j and r_{ij}^0 is the distance between the atoms in the reference structure, e.g., the crystallographic structure. This summation is only performed over atoms less than a cut-off distance R_c, and γ is the spring, or force constant for the elastic bond between the atoms and is the same for all atoms pairs (see Fig. 1a). The energy function in Eq. 7 seems to be the most popular, although other types of functions can be used. The network corresponding to the energy function of Eq. 7 is sometimes referred to as the anharmonic network model [24]. A Gaussian network model has a different energy function, which results in modes without any directional information [24,25] and will not be considered here. Once the function of Eq. 7 has been calculated, the procedure is exactly the same as for the standard NMA, namely, the Hessian is calculated and its eigenvalues and eigenvectors are determined. Whereas the standard NMA must be performed on all atoms as required by the force field, the elastic network model can be carried out on a subset of atoms. Often, for a protein, this would be the C^α atoms. Compared with the standard NMA, this would result in a Hessian approximately tenfold lower in order, thus, yielding considerable computational savings in the calculation of the eigenvalues and eigenvectors because these routines are normally of the order of N^3 operations, where N is the order of the Hessian matrix.

2.3 Essential Dynamics and PCA

After superposition to a common reference structure, a variance–covariance matrix of positional fluctuations is constructed:

$$\mathbf{C} = < (\mathbf{x}(t) - < \mathbf{x} >)(\mathbf{x}(t) - < \mathbf{x} >)^T > \tag{8}$$

where $<>$ denotes an ensemble average. The coordinates **x** are denoted as a function of time for clarity, but may be provided in any order and can be, for example, a molecular dynamics trajectory or a set of experimental structures. **C** is a symmetric matrix that can be diagonalized by an orthogonal coordinate transformation **T**:

$$\mathbf{C} = \mathbf{T}\Lambda\mathbf{T}^T \tag{9}$$

with Λ the diagonal (eigenvalue) matrix and **T** containing, as columns, the eigenvectors of **C**. The eigenvalues λ correspond to the mean square eigenvector coordinate fluctuation, and, therefore, contain the contribution of each principal component to the total fluctuation. The eigenvectors are usually sorted such that their eigenvalues are in decreasing order. For a system of N atoms, **C** is a 3N × 3N matrix. If at least 3N configurations are used to construct **C**, then 3N − 6 eigenvectors with nonzero eigenvalues will be obtained. Six eigenvalues should be exactly zero, of which the corresponding eigenvectors describe the overall rotation and translation (that is eliminated by the superposition). If only M configurations are available (with M < 3N), then at most M − 1 nonzero eigenvalues with corresponding eigenvectors will result. If μ_i is the ith eigenvector of **C** (the ith column of **T**), then the original configurations can be projected onto each of the principal components to yield the principal coordinates $p_i(t)$ as follows:

$$p_i(t) = \mu_i \cdot (\mathbf{x}(t) - \langle \mathbf{x} \rangle) \tag{10}$$

Note that the variance $\langle p_i^2 \rangle$ equals the eigenvalue λ_i. These projections can be easily transformed back to Cartesian coordinates for visualization purposes as follows:

$$\mathbf{x}_i'(t) = p_i(t) \cdot \mu_i + \langle \mathbf{x} \rangle. \tag{11}$$

Two sets of eigenvectors μ and ν can be compared with each other by taking inner products:

$$I_{ij} = \mu_i \cdot \nu_j. \tag{12}$$

Subspace overlaps are often calculated as summed squared inner products:

$$O_n^m = \sum_{i=1}^{n} \sum_{j=1}^{m} (\mu_i \cdot \nu_j)^2, \tag{13}$$

expressing how much of the n-dimensional subspace of set μ is contained within the m-dimensional subspace of set ν. Note that m should be larger than n to achieve full overlap (O = 1).

3 Methods

3.1 Standard NMA

For standard NMA, one needs a set of coordinates, a force field, and software to perform the calculations. Often NMA is performed using molecular mechanics software packages that are also able to perform molecular dynamics simulations, etc. For a protein, the structural information is normally held in a PDB file. The software will normally be able to interpret the file to determine the correct energy function using the selected force field. Any missing atoms should be added. Missing hydrogen atoms also need to be added but most software packages have routines to do this.

It is usual, but not a requirement of the methodology, to remove water and ligands. Once the system is prepared, the first major calculation is energy minimization.

3.1.1 Energy Minimization

The two main energy minimization routines are steepest descent and conjugate gradient. The former can be used in the initial stages, for the first 100 steps, for example, followed by the latter. Sometimes, when approaching the energy minimum, the actual minimum cannot be found because of overstepping. This can present a problem for NMA, where very precise location of the minimum is required. However, many minimizers are able to adjust the step size to avoid overstepping. Normally, minimization can be stopped when the root mean square force is approximately 10^{-4} to $10^{-12}\,\text{kcal} \cdot \text{mol}^{-1} \cdot \text{Å}^{-1}$.

3.1.2 Hessian Calculation

This step creates the Hessian matrix, which is the matrix of second derivatives of the potential energy function with respect to the mass-weighted Cartesian coordinates. It is a symmetric matrix and, therefore, it is not required to store the whole matrix.

3.1.3 Diagonalization of Hessian Matrix

This stage determines the eigenvalues and eigenvectors. Because of the large size of this $3N \times 3N$ matrix, where N is the number of atoms in the molecule, this stage often presents memory problems for large molecules (see **Note 1**). The process results in a set of 3N eigenvalues and a set of 3N eigenvectors each with 3N components. The eigenvalues are sorted in ascending order and the eigenvectors are sorted accordingly. The first six eigenvalues should have values close to zero because these correspond to the three translational and three rotational degrees of freedom for the whole molecule (see **Note 2**). The seventh eigenvector is the lowest frequency mode, and it is often predicted to be a functionally relevant mode.

3.1.4 Comparison with Experimental Results

Eq. 6 shows how to measure the overlap with a functional mode derived from, e.g., two x-ray structures. To perform this calculation, one needs to calculate the experimental displacements, Δx_i^{exp}. These displacements need to be calculated from the experimental structures oriented in the same way as the minimized structure used for the NMA. To do this, one can use a least-squares best fit routine to superpose the two experimental structures on the minimized structure.

3.2 *Elastic Network Models*

One major advantage of these models is that energy minimization is not required because the structure used is already assumed to be in an energy minimum (see **Note 3**). The steps are as follows:

- Prepare the structure, e.g., remove ligands and water molecules.
- Decide which atoms will build the network, e.g., just C^{α} atoms.
- Choose a cut-off length, R_c, which typically is 7 to 10 Å when using just C^{α} atoms (see **Note 4**).
- Choose the spring constant, γ (see **Note 5**).
- Calculate the eigenvalues and eigenvectors (see **Note 6**).

From the last step onward, there is no essential difference to the standard NMA. However, if calculations are performed on the C^{α} atoms only, then, naturally, one can only compare with the movements of the C^{α} atoms in the experimentally determined functional mode, i.e., movements of side chains cannot be compared. Tama and Sanejouand have made a comparison between the results from an elastic network model and modes derived from a pair of x-ray structures for 20 proteins [7].

3.3 *Essential Dynamics and PCA*

3.3.1 PCA of Structural Ensembles

A principal component or essential dynamics analysis may be carried out on a molecular dynamics trajectory or any other structural ensemble. It typically consists of three steps. First, the configurations from the ensemble must be superposed, to enable the filtering of internal motions from overall rotation and translation. This is usually accomplished by a least-squares fit of each of the configurations onto a reference structure (see **Note 7**). Second, this "fitted" trajectory is used to construct a variance–covariance matrix that is subsequently diagonalized. The variance–covariance matrix is a symmetric matrix containing, as elements, the covariances of the atomic displacements relative to their respective averages for each pair of atoms for the off-diagonal elements and the variances of each atom displacements along the diagonal. Atoms that move concertedly give rise to positive covariances, whereas anticorrelated motions give rise to negative entries. Noncorrelated displacements result in near-zero covariances (see also **Note 8**). Diagonalization of this covariance matrix yields a set of eigenvectors and eigenvalues, which are usually sorted such that the eigenvalues are in decreasing order. The eigenvalues represent the variance along each of the corresponding collective modes (eigenvectors) and usually a small number of modes suffice to describe the majority of the total fluctuation. As a third step, the original trajectory may be analyzed in terms of the principal components. To this end, the trajectory is projected onto each of the principal modes to yield the time behavior and distribution of each of the principal coordinates (see also **Note 9**). Often, two- or three-dimensional projections along the major principal components are used to allow a representation of the sampled distribution in configuration space or to compare multiple ensembles along the principal modes of collective fluctuation. These projections onto single or multiple principal coordinates can also be readily translated back into Cartesian space to yield an ensemble or animation of the motion along a selection of principal coordinates.

In contrast to standard NMA, a PCA can be carried out on any subset of atoms, and, for proteins, usually only C_α or backbone atoms are taken into account (see also **Note 9**). This has the advantage that the storage and diagonalization of the covariance matrix is less demanding, whereas the main collective modes are very similar to an all-atom analysis [9, 26]. An additional advantage of a backbone-only analysis is that artificial apparent correlations between slow side-chain fluctuations and backbone motions are not picked up by the analysis. A PCA may be compared with results from a standard NMA. However, to this end, one must perform an all-atom PCA and the fluctuations must be calculated from mass-weighted displacements [21]. This form of PCA is often referred to as "quasiharmonic analysis." If an all-atom analysis is required, an approximation may be used to retrieve only the principal modes of fluctuation, that alleviates the need to store and diagonalize the full matrix [26]. As mentioned above, the PCA technique is not limited to the analysis of molecular dynamics trajectories but can be carried out on any ensemble of structures. It can, e.g., be carried out to derive the principal modes from sets of x-ray structures [27], to compare simulation data with experimental conformations [28–30] (see also Fig. 2), or to derive search directions from multiple homologous structures to aid homology modeling [31].

3.3.2 Convergence of PCA Results Derived from Molecular Dynamics Simulations

Principal components derived from different simulations or simulation parts allow us to compare the major directions of configurational space and sampled regions and to judge similarity and convergence. It has been observed that sub-nanosecond protein molecular dynamics simulations suffer from a significant sampling problem, resulting in an apparently poor overlap between the principal components extracted from multiple parts of these trajectories [32, 33]. Nevertheless, it was observed that despite the fact that individual principal components may be different, the subspaces that are spanned by the major principal components converge remarkable rapidly and show a favorable agreement not only between different simulation results, but also between simulation and experiment [28, 30, 34, 35], see also Fig. 2.

The anharmonic dynamics along the principal modes of collective fluctuation that corresponds to the jumping between multiple local energy minima results in a diffusive dynamics of the principal coordinates [16, 17]. The analogy of this diffusive dynamics to a multidimensional random walk allows one to assess the convergence of the dynamics along the principal (and usually slowest) modes by comparison of the time evolution of the principal coordinates with cosines that would result from

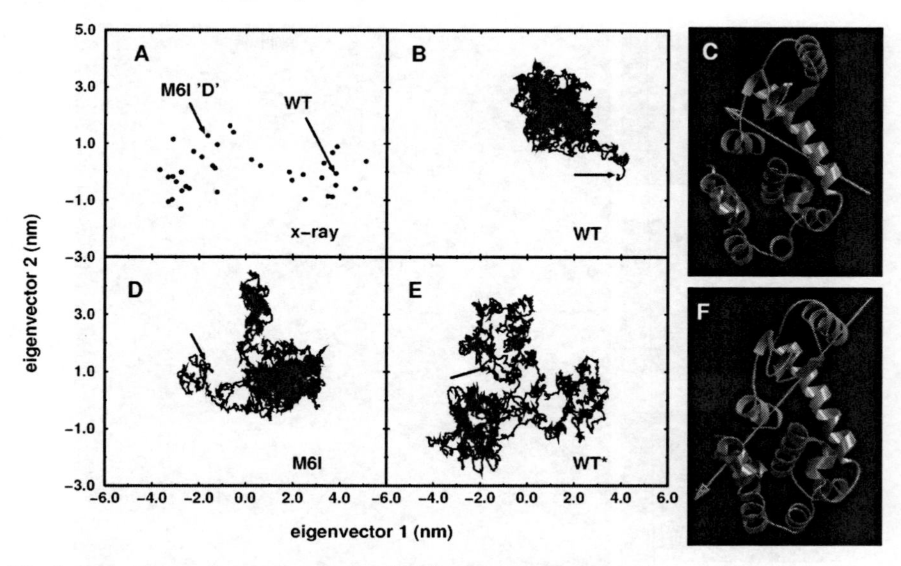

Fig. 2 PCA of a set of x-ray structures of T4-lysozyme compared with ensembles obtained from molecular dynamics simulations. Each structural ensemble is projected onto the two major principal components extracted from the x-ray ensemble (**a**, **b**, **d**, **e**). Shown are the x-ray ensemble (**a**) and three independent molecular dynamics simulations of 1 ns each (**b**, **d**, **e**). The *black arrows* depict the starting structures of the simulations (WT for wild-type; M61 'D' for the fourth conformer of the M61 mutant). The color-coded structures (**c**, **f**) depict the domain character of the motions, with the *arrow* illustrating the screw axis that describes the motion of the *red* domain with respect to the *blue* domain. The first eigenvector describes a closure motion (**c**), whereas the second eigenvector describes a twisting motion (**f**)

random diffusion [36,37]. A high cosine content typically indicates a nonconverged trajectory. Note, however, that a lack of convergence of the dynamics *along* a set of modes does not necessarily also imply that the *directions* of such modes or the subspace they span are not converged or poorly defined. Provided that a sufficiently converged trajectory is available, thermodynamic properties may be derived as ensemble averages and can be readily mapped onto the principal coordinates to yield, e.g., free energy landscapes (see Fig. 3).

3.3.3 Comparison of PCA Results from Different Sources

It is often useful to compare structural ensembles from different simulations or from experiment with each other in terms of their major principal coordinates. It is instructive to discuss three possibilities that are often used to carry out such a comparison. First, separate principal component analyses may be carried out over each individual ensemble. Subsequently, the resulting eigenvectors are compared with each other, either individually or as, e.g., a subset of major directions. Such a comparison usually involves inner products between sets of eigenvectors as a measure

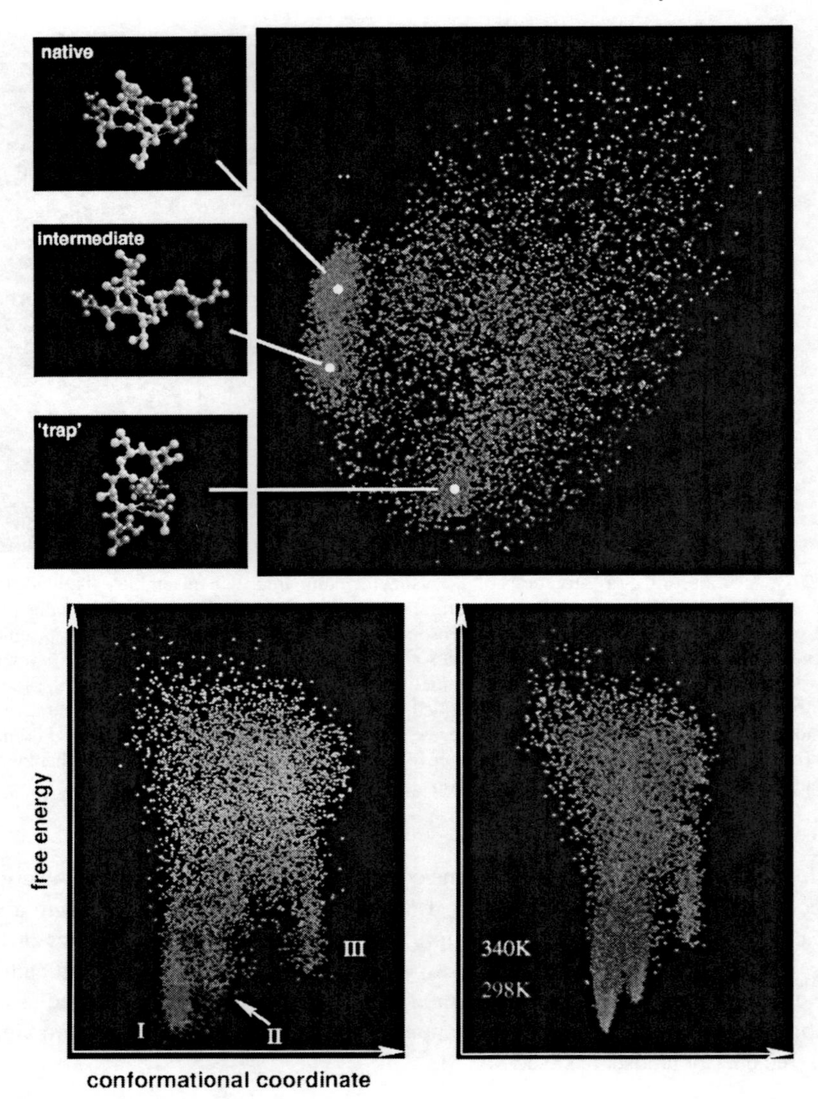

Fig. 3 PCA of a peptide trajectory that covers reversible folding and unfolding events. The *upper panel* depicts the structural ensemble projected onto the major principal modes, color coded to configurational density (*upper right panel*), together with three representative structures from the simulation (*upper left panels*). The *lower panels* depict the folding free energy landscape, revealing three low-energy configurations (see also *upper panel*). The difference of the entropic contribution at different temperatures is clearly visible (*lower right panel*)

for similarity. For sets of eigenvectors, the summed (or cumulative) squared inner product is a useful measure of similarity that is zero for orthogonal, non-overlapping subspaces and one for identical subspaces. Values from 0.3 to 0.4 already indicate significant overlap, because of the usually large dimensionality of the configuration space as compared with the analyzed subspace. Alternatively, full inner product matrices can also be used [26, 34]. This method focuses on the directions of the principal modes rather than the sampled region along the modes. Therefore, a second, complementary, method can be used to include this ensemble information. In this case, the structures from one ensemble are projected onto the eigenvectors extracted from another ensemble (usually together with the structures from that ensemble), allowing a direct comparison of the sampled regions in each of the projected ensembles. This approach has proven particularly useful for cases in which one set of eigenvectors can be regarded as a reference set, for example, those that were extracted from a set of experimental structures [28, 29]; see also Fig. 2. For cases in which there is not one natural reference set of directions, a third approach may be used. In such cases, multiple sub-ensembles may be concatenated into one meta-ensemble on which the PCA is carried out. The individual sub-ensembles can be separately projected onto this combined set of modes, allowing a direct comparison of sub-ensembles. This method has the advantage that differences between the different sub-ensembles are frequently visible along one of the combined principal modes, even for subtle effects such as the difference between an apo- or holo ensemble, or the effect of a point mutation [38].

3.3.4 Enhanced Sampling Techniques

Knowledge of the major coordinates of collective fluctuations opens the way to develop specialized simulation techniques tailored toward an efficient or even systematic sampling along these coordinates, thereby alleviating the sampling problem inherent to virtually all common computer simulations of biomolecular systems today. The first attempts in this direction were aimed at a simulation scheme in which the equations of motion were solely integrated along a selection of primary principal modes, thereby drastically reducing the number of degrees of freedom [9]. However, these attempts proved problematic because of nontrivial couplings between high- and low-amplitude modes, even though, after diagonalization, the modes are linearly independent (orthogonal). Therefore, instead, a series of other techniques has prevailed that takes into account the full-dimensional simulation system and enhance the motion along a selection of principal modes. The most common of these techniques are conformational flooding [11] and essential dynamics sampling [12–14]. In conformational flooding, an additional potential energy term that stimulates the simulated system to explore new regions of phase space is introduced on a selection of principal modes (Fig. 4), whereas, in essential dynamics, sampling a similar goal is achieved by geometrical constraints along a selection of principal modes. More recently, the concept of conformational flooding was reformulated in the context of metadynamics [39]. These techniques have in common that a sampling efficiency enhancement of up to an order of magnitude can be achieved, provided that a reasonable approximation of the principal modes has been obtained from a conventional simulation.

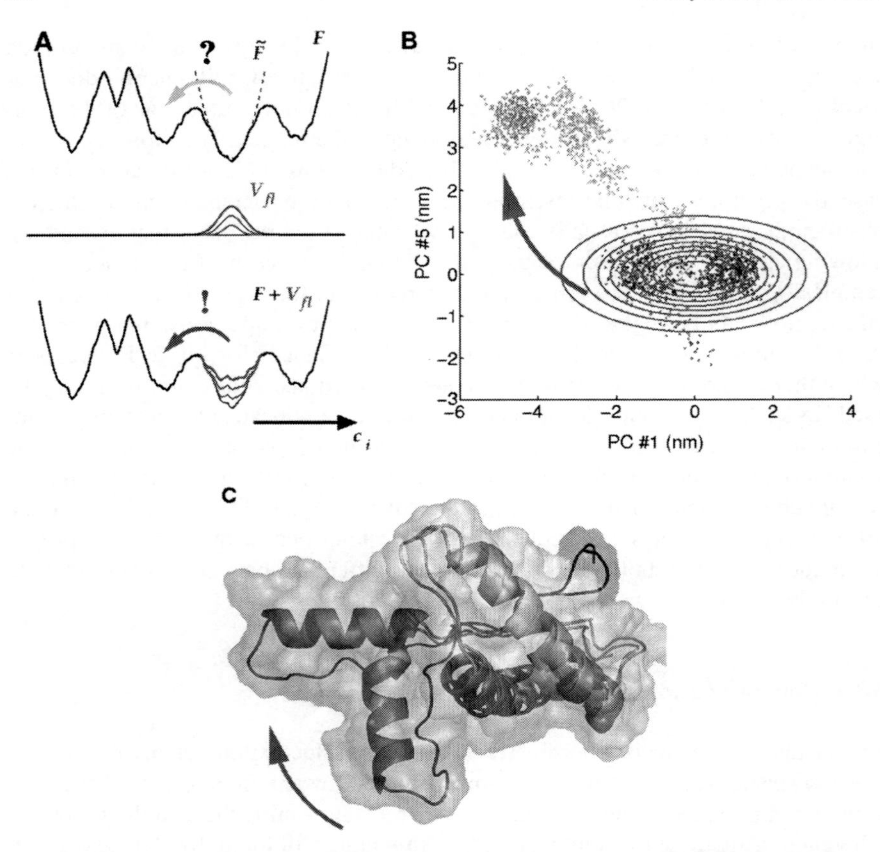

Fig. 4 Conformational flooding. (**a**) The principle of conformational flooding: configurations along principal coordinates (PC's) sampled during an molecular dynamics simulation are destabilized in a subsequent set of simulations by an additional potential energy term, V_{fl}, to enhance the probability of visiting previously unsampled minima. To this end, the original energy landscape F is locally approximated by a harmonic potential $\tilde{F}$. **b** and **c**: Application to the prion protein. The *red arrow* depicts the motion induced by the flooding potential in configuration space (**b**) and mapped onto the structure (**c**)

4 Notes

1. Diagonalization routines exert great demands on memory. For example, the routine in AMBER [40] requires $8 \times 9N(3N - 1)/2$ bytes of memory. A 400-atom system requires 1.7 Mbytes, but a 4,000-atom system requires 1.7 Gbytes [41]. A number of methodologies have been devised to overcome this memory problem. These methods are usually used to calculate only the lowest frequency eigenvectors. Another alternative is to perform dihedral angle space NMA. This reduces the number of variables by a factor of approximately 8 for proteins and approximately 11 for nucleic acids. These methods have been reviewed elsewhere [21].

2. The first six eigenvalues should be close to zero. No eigenvalues should be negative. Negative eigenvalues indicate negative curvature on the energy surface and suggest insufficient minimization.
3. A major advantage of this method over the standard NMA is that energy minimization is not required. Because energy minimization does not normally bring about large changes in conformation, it is to be expected that there would be little difference between the results from the starting structure and an energy-minimized structure.
4. It seems that choosing a value for R_c is often problematic. It relates to the radius of the first coordination shell around the selected atoms. If one uses C^{α} atoms, then its value (7–10 Å) should be longer than when using all atoms, where a value of 3 Å would be more appropriate. Some reports suggest that results do not vary dramatically with small variations in the cutoff distance [42]. Obviously, the shorter the cut-off, the greater the savings there would be in the calculation of the energy.
5. The value of γ has no effect on the eigenvectors and, thus, if one is only interested in the character of the motions, then its value is not important. However, its appropriate value is sometimes determined for x-ray structures by calculating atomic mean square fluctuations and matching them to experimentally determined B-factors. A value of 1.0 kcal/mol Å^2 might be a reasonable starting value, if no appropriate value is known.
6. Depending on the structure, some regions may be only loosely connected to the rest of the molecule, e.g., a terminal region in a protein. In such a case, the movement of this region could appear as a low-frequency mode. This may be undesirable if one is interested in global motions. Some programs (private communication from Dr. Atsushi Matsumoto) allow one to provide extra connections to these regions, thus, effectively integrating them more with the rest of the structure.
7. Before a PCA, all structures should be superimposed onto a common reference structure. This can be problematic for very flexible systems such as peptides, where the fit may be ambiguous, leading to artificial structural transitions. In certain cases, such problems may be alleviated by using a progressive fit, where each structure is superimposed onto the previous one. It is also important to note that when results of different PCAs are to be compared with each other, then each individual PCA should be based on the same reference structure used for superposition.
8. PCA is a linear analysis, i.e., only linear correlations between atomic displacements enter the covariance matrix. This means that nonlinear correlations between atom movements may be overlooked because they get spread out across multiple collective coordinates. In practice, this is usually not a big problem, except for systems that undergo large-scale rotations.
9. Similar to NMA, PCA can also be carried out in dihedral angle space [26, 43]. Although it has the advantage that it does not require superposition to a reference structure (because it is based on internal coordinates), PCA in dihedral space has two main disadvantages. First, major collective dihedral transitions do not

usually correspond to major transitions in Cartesian space. For example, a small change of one backbone dihedral in a central residue in a two-domain protein can result in a large-scale motion of the two domains with respect to each other. Although such a motion would likely be relevant, it would easily be overlooked. Second, the metric of the configuration space cannot be retained in a straightforward way. This may lead to artificial correlations between the dihedral coordinates and complicates the translation back to Cartesian space for, e.g., visualization purposes.

Acknowledgements The authors are grateful to Dr. Akio Kitao and Dr. Atsushi Matsumoto for helpful discussions, and to Oliver Lange and Helmut Grubmüller for kindly providing Fig. 4.

References

1. Go, N., Noguti, T. and Nishikawa, T. (1983). Dynamics of a small globular protein in terms of low-frequency vibrational modes. *Proc. Natl. Acad. Sci. USA* **80**, 3696–3700.
2. Levitt, M., Sander, C. and Stern, P. S. (1983). The normal modes of a protein: Native bovine pancreatic trypsin inhibitor. *Int. J. Quant. Chem.* **10**, 181–199.
3. Brooks, B. and Karplus, M. (1983). Harmonic dynamics of proteins: Normal modes and fluctuations in bovine pancreatic trypsin inhibitor. *Proc. Natl. Acad. Sci. USA* **80**, 6571–6575.
4. Austin, R. H., Beeson, K. W., Eisenstein, L., Frauenfelder, H. and Gunsalus, I. C. (1975). Dynamics of ligand binding to myoglobin. *Biochemistry* **14**, 5355–5373.
5. Elber, R. and Karplus, M. (1987). Multiple conformational states of proteins: A molecular dynamics analysis of myoglobin. *Science* **235**, 318–321.
6. Tirion, M. M. (1996). Large amplitude elastic motions in proteins from a single-parameter, atomic analysis. *Physical Review Letters* **77**, 1905–1908.
7. Tama, F. and Sanejouand, Y. H. (2001). Conformational change of proteins arising from normal mode calculations. *Protein Engineering* **14**, 1–6.
8. Garcia, A. E. (1992). Large-amplitude nonlinear motions in proteins. *Phys. Rev. Lett.* **68**, 2696–2699.
9. Amadei, A., Linssen, A. B. M. and Berendsen, H. J. C. (1993). Essential dynamics of proteins. *Proteins: Struct. Funct. Genet.* **17**, 412–425.
10. Kitao, A., Hirata, F. and Go, N. (1991). The effects of solvent on the conformation and the collective motions of protein: normal mode analysis and molecular dynamics simulations of melittin in water and in vacuum. *J. Chem. Phys.* **158**, 447–472.
11. Grubmüller, H. (1995). Predicting slow structural transitions in macromolecular systems: Conformational flooding. *Phys. Rev. E.* **52**, 2893–2906.
12. Amadei, A., Linssen, A. B. M., de Groot, B. L., van Aalten, D. M. F. and Berendsen, H. J. C. (1996). An efficient method for sampling the essential subspace of proteins. *J. Biom. Str. Dyn.* **13**, 615–626.
13. de Groot, B. L., Amadei, A., van Aalten, D. M. F. and Berendsen, H. J. C. (1996). Towards an exhaustive sampling of the configurational spaces of the two forms of the peptide hormone guanylin. *J. Biomol. Str. Dyn.* **13**, 741–751.
14. de Groot, B. L., Amadei, A., Scheek, R. M., van Nuland, N. A. J. and Berendsen, H. J. C. (1996). An extended sampling of the configurational space of HPr from *E. coli*. *Proteins: Struct. Funct. Genet.* **26**, 314–322.
15. Hayward, S., Kitao, A. and Go, N. (1995). Harmonicity and anharmonicity in protein dynamics: a normal modes and principal component analysis. *Proteins: Struct. Funct. Genet.* **23**, 177–186.

16. Kitao, A., Hayward, S. and Go, N. (1998). Energy landscape of a native protein: jumping-among-minima model. *Proteins: Struct. Funct. Genet.* **33**, 496–517.
17. Amadei, A., de Groot, B. L., Ceruso, M. A., Paci, M., Nola, A. D. and Berendsen, H. J. C. (1999). A kinetic model for the internal motions of proteins: Diffusion between multiple harmonic wells. *Proteins: Struct. Funct. Genet.* **35**, 283–292.
18. Kitao, A. and Go, N. (1999). Investigating protein dynamics in collective coordinate space. *Curr. Opin. Struct. Biol.* **9**, 143–281.
19. Kitao, A., Hayward, S. and Go, N. (1994). Comparison of normal mode analyses on a small globular protein in dihedral angle space and Cartesian coordinate space. *Biophysical Chemistry* **52**, 107–114.
20. Tirion, M. M. and ben-Avraham, D. (1993). Normal mode analysis of G-actin. *Journal of Molecular Biology* **230**, 186–195.
21. Hayward, S. (2001). Normal mode analysis of biological molecules. In *Computational Biochemistry and Biophysics* (Becker, O. M., Mackerell Jr, A. D., Roux, B. & Watanabe, M., eds.), pp. 153–168. Marcel Dekker Inc, New York.
22. Go, N. (1990). A theorem on amplitudes of thermal atomic fluctuations in large molecules assuming specific conformations calculated by normal mode analysis. *Biophysical Chemistry* **35**, 105–112.
23. Marques, O. and Sanejouand, Y.-H. (1995). Hinge-bending motion in citrate synthase arising from normal mode calculations. *Proteins* **23**, 557–560.
24. Chennubhotla, C., Rader, A. J., Yang, L. W. and Bahar, I. (2005). Elastic network models for understanding biomolecular machinery: From enzymes to supramolecular assemblies. *Physical Biology* **2**, S173–S180.
25. Bahar, I. and Rader, A. J. (2005). Coarse-grained normal mode analysis in structural biology. *Current Opinion in Structural Biology* **15**, 586–592.
26. van Aalten, D. M. F., de Groot, B. L., Berendsen, H. J. C., Findlay, J. B. C. and Amadei, A. (1997). A comparison of techniques for calculating protein essential dynamics. *J. Comp. Chem.* **18**, 169–181.
27. van Aalten, D. M. F., Conn, D. A., de Groot, B. L., Findlay, J. B. C., Berendsen, H. J. C. and Amadei, A. (1997). Protein dynamics derived from clusters of crystal structures. *Biophys. J.* **73**, 2891–2896.
28. de Groot, B. L., Hayward, S., Aalten, D. M. F. v., Amadei, A. and Berendsen, H. J. C. (1998). Domain motions in bacteriophage T4 lysozyme; a comparison between molecular dynamics and crystallographic data. *Proteins: Struct. Funct. Genet.* **31**, 116–127.
29. de Groot, B. L., Vriend, G. and Berendsen, H. J. C. (1999). Conformational changes in the chaperonin GroEL: New insights into the allosteric mechanism. *J. Mol. Biol.* **286**, 1241–1249.
30. Abseher, R., Horstink, L., Hilbers, C. W. and Nilges, M. (1998). Essential spaces defined by NMR structure ensembles and molecular dynamics simulation show significant overlap. *Proteins: Struct. Funct. Genet.* **31**, 370–382.
31. Qian, B., Ortiz, A. R. and Baker, D. (2004). Improvement of comparative model accuracy by free-energy optimization along principal components of natural structural variation. *Proc. Natl. Acad. Sci. USA* **101**, 15346–15351.
32. Balsera, M. A., Wriggers, W., Oono, Y. and Schulten, K. (1996). Principal component analysis and long time protein dynamics. *J. Phys. Chem.* **100**, 2567–2572.
33. Clarage, J. B., Romo, T., Andrews, B. K., Pettitt, B. M. and Jr., G. N. P. (1995). A sampling problem in molecular dynamics simulations of macromolecules. *Proc. Natl. Acad. Sci. USA* **92**, 3288–3292.
34. de Groot, B. L., van Aalten, D. M. F., Amadei, A. and Berendsen, H. J. C. (1996). The consistency of large concerted motions in proteins in Molecular Dynamics simulations. *Biophys. J.* **71**, 1707–1713.
35. Amadei, A., Ceruso, M. A. and Nola, A. D. (1999). On the convergence of the conformational coordinates basis set obtained by the essential dynamics analysis of proteins' molecular dynamics simulations. *Proteins: Struct. Funct. Genet.* **36**, 419–424.
36. Hess, B. (2000). Similarities between principal components of protein dynamics and random diffusion. *Phys. Rev. E* **62**, 8438–8448.

37. Hess, B. (2002). Convergence of sampling in protein simulations. *Phys. Rev. E* **65**, 031910.
38. van Aalten, D. M. F., Findlay, J. B. C., Amadei, A. and Berendsen, H. J. C. (1995). Essential dynamics of the cellular retinol binding protein—evidence for ligand induced conformational changes. *Prot. Eng.* **8**, 1129–1136.
39. Laio, A. and Parrinello, M. (2002). Escaping free-energy minima. *Proc. Natl. Acad. Sci. USA* **99**, 12562–12566.
40. Pearlman, D. A., Case, D. A., Caldwell, J. W., Ross, W. S., Cheatham, T. E., Debolt, S., Ferguson, D., Seibel, G. and Kollman, P. (1995). Amber, a package of computer-programs for applying molecular mechanics, normal-mode analysis, molecular-dynamics and free-energy calculations to simulate the structural and energetic properties of molecules. *Computer Physics Communications* **91**, 1–41.
41. Amberteam. (2004). Amber 8 users' manual.
42. Tama, F., Valle, M., Frank, J. and Brooks, C. L. (2003). Dynamic reorganization of the functionally active ribosome explored by normal mode analysis and cryo-electron microscopy. *Proc. Natl. Acad. Sci. USA* **100**, 9319–9323.
43. Mu, Y., Nguyen, P. H. and Stock, G. (2005). Energy landscape of a small peptide revealed by dihedral angle principal component analysis. *Proteins: Structure, Function, and Bioinformatics* **58**, 45–52.

Part II
Free Energy Calculations

Chapter 6
Calculation of Absolute Protein–Ligand Binding Constants with the Molecular Dynamics Free Energy Perturbation Method

Hyung-June Woo

Summary Reliable first-principles calculations of protein–ligand binding constants can play important roles in the study and characterization of biological recognition processes and applications to drug discovery. A detailed procedure for such a calculation is outlined in this chapter. The methodology is computationally implemented using the molecular dynamics sampling of relevant configurational spaces and free energy perturbation techniques. The procedure is illustrated with the model system of the phosphotyrosine peptide binding to the Src SH2 domain.

Keywords: Binding · Drug discovery · Free energy · Free energy perturbation · Ligand · Molecular dynamics · Umbrella sampling

1 Introduction

The binding of a ligand to protein receptors underlies a wide variety of recognition processes in biological systems. The understanding of such systems can be enhanced greatly by the development of reliable computational methods to calculate protein–ligand binding constants. In addition, such methodologies are of interest to pharmaceutical industries for their potential applications in drug discovery. Although the physical basis of the phenomena is, in principle, straightforward with applications of equilibrium statistical mechanics, practical implementations suitable for direct calculations using molecular simulation techniques are often challenging. Many computational methods using a range of approximations have been developed to estimate both the relative binding affinities of closely related ligands and the absolute binding constants. Widely used simplified methodologies range from docking [1], which often ignores solvation contributions and/or the flexibility of proteins and ligands, to continuum electrostatic methods, such as the molecular mechanics/Poisson–Boltzmann surface area (MM-PBSA) techniques [2, 3]. In this chapter, we will concentrate instead on an implementation [4] of the computational

From: Methods in Molecular Biology, vol. 443, Molecular Modeling of Proteins
Edited by Andreas Kukol © Humana Press, Totowa, NJ

scheme designed to calculate the absolute protein–ligand binding constant from first principles using molecular dynamics (MD) free energy perturbation techniques [5,6]. Although such *ab initio* free energy calculations [7–9] are more computationally demanding than simpler approximate treatments, their successful implementations and applications to model systems of biological significance can greatly benefit computational studies of binding in general, providing benchmarks and insights into molecular aspects of the process. The overall scheme and the model system used in this chapter are those adopted in Ref. [4], with the primary emphasis here centered on practical details and procedures of the computation. For more details of the theoretical formulations, the reader is referred to Ref. [4].

2 Theory

We adopt the scheme of free energy calculation illustrated schematically in Fig. 1. The sequence of steps is ordered such that the overall process corresponds to the reversible binding of a ligand from the bulk to the binding site of the receptor. The reverse process could have been chosen, for which each free energy term would have the opposite sign. The process can be divided into the following steps:

1. The conformation of the ligand initially in the isotropic bulk is constrained to the particular form of the bound conformation.
2. The isotropy of the space is then lifted by imposing a set of orientational constraints of the ligand orientation and center of mass axes with respect to a chosen origin at a distance r^* between the ligand and receptor.
3. The ligand center of mass is reversibly moved into the contact distance corresponding to the bound state.
4. The orientational and axial constraints on the ligand are switched off in the binding site.
5. The conformational constraint on the ligand is switched off in the binding site.

The use of Step 3 corresponding to a reversible physical separation of the ligand center of mass from the receptor (also used in Ref. [9]) is in contrast to the more conventional alchemical switching method [5–8]. In the alchemical schemes, the receptor–ligand nonbonding interaction energy terms are first reversibly turned off, after which, the ligand is reintroduced into the bulk solution. Such methods work best when the individual free energy terms calculated are relatively small in their magnitudes, as for rigid nonpolar ligands. For highly charged ligands, the absolute electrostatic solvation free energy is typically orders of magnitude larger than the net free energy of binding, making the alchemical method highly error-prone. Imposing the conformational and orientational constraints on the ligand before and after the separation [7], on the other hand, serves to restrict the configurational spaces the ligand molecule explores while detached from its native binding pocket, enhancing the efficiency of sampling.

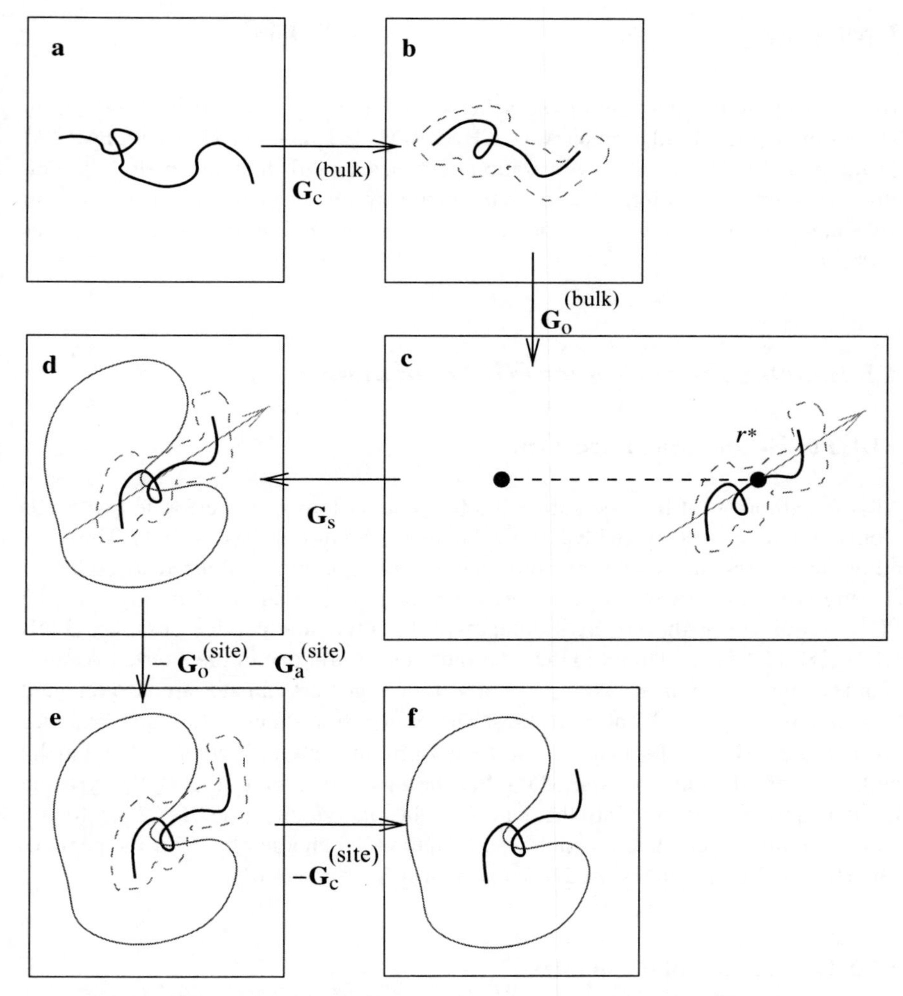

Fig. 1 Schematic illustration of the binding free energy calculation divided into steps

The equilibrium binding constant is expressed in terms of the free energy terms of the computational scheme shown in Fig. 1 as [4]:

$$K_{eq} = S^* I^* \exp\left[-\beta\left(G_c^{(bulk)} + G_o^{(bulk)} - G_a^{(site)} - G_o^{(site)} - G_c^{(site)}\right)\right], \quad (1)$$

where $G_c^{(bulk)}$ and $G_o^{(bulk)}$ are the free energy differences of imposing the conformational and orientational constraints on the ligand in the bulk, and $G_a^{(site)}$, $G_o^{(site)}$, and $G_c^{(site)}$ are the free energy differences of imposing the axial, orientational, and conformational constraints in the binding site. The two prefactors S^* and I^* are associated with the reversible radial separation of the ligand represented by the symbol G_s in Fig. 1 (Step 3).

3 Methods

The computational methods of the scheme in Fig. 1 are outlined in this section. The biomolecular simulation package CHARMM [10] version c31b1 with PARAM 27 force field [11] is used for illustration purposes. Details may differ slightly with other programs. Conditions specific to the model system chosen are indicated within parentheses in the following to facilitate applications and generalizations to other systems.

3.1 Building the System for MD Simulations

3.1.1 Protein and ligand structures

All simulations used for calculating the free energy terms are performed with all-atom explicit water setups with periodic boundary conditions (see **Note 1**). The crystallographic structure of the receptor–ligand bound complex is used to build the starting structure as well as the reference bound conformation of the ligand. The PDB coordinate of the Src SH2 domain–phosphotyrosine peptide complex (PDB ID 1LKK) [12] is used to build the reference state. Only one of the alternative side-chain rotamer positions is taken. Crystal water oxygen coordinates are also retained to form a part of the solvent molecules. The protonation states of titratable residues are assumed to be the respective most stable forms at neutral pH when isolated in the bulk solution (default in CHARMM). For the ligand peptide (pYEEI), the tyrosine is patched to be converted into the dianionic phosphotyrosine (the "TP2" patch) and its N terminus is acetylated, which results in the total charge of -5 for the peptide. The HBUILD command is used to build hydrogen atom coordinates.

3.1.2 Ligand–Receptor Complex

The ligand–receptor complex is solvated with a pre-equilibrated bulk water (TIP3P model) [13] box in orthorhombic geometry. The complex, along with the crystallographic water molecules, is placed in the box with its intermolecular axis connecting the receptor and ligand centers of mass aligned with the longest axis (x-axis) of the water box (Fig. 2). The position of the receptor center of mass is displaced by a suitable distance ($x = -10$ Å in this case) along the x-axis (see **Note 2**). A script is used to delete water molecules whose oxygen coordinates lie within 2.8 Å of any existing heavy atoms. Potassium and chloride ions are added to neutralize the overall system, with their coordinates assigned randomly away from the receptor–ligand complex. Extra ion pairs are added to simulate the condition of 150 mM ionic strength approximately using the volume of the water box. With known crystallographic coordinates fixed, the solvated complex is energy minimized (500 steps)

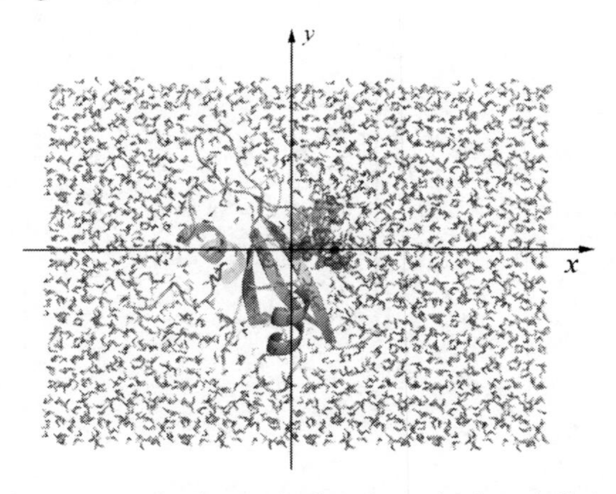

Fig. 2 The simulation geometry for the protein–ligand complex in periodic boundary conditions and the choice of axes. The protein and the ligand are shown with cartoons and spheres, respectively. The centers of mass of the protein and ligand lie on the x-axis, and the ligand is displaced toward the positive x-direction in successive windows in the radial PMF calculation

to relieve any unfavorable local configurations, followed by another minimization (500 steps) with all atoms free. In any minimization or subsequent dynamics runs, the center of mass of the protein is constrained to the initial position lying on the x-axis via a harmonic potential of moderate force constant ($\sim 10\,\text{kcal/mol}\,\text{Å}^2$) to prevent the overall drifting of the complex (see **Note 3**).

3.1.3 Ligand

The ligand is solvated in a water box of cubic geometry (with the size of the box 30 Å) for calculations of free energy terms of the peptide in bulk solution. The solvated system is energy minimized as for the complex. The center of mass of the ligand is also constrained at the box center to prevent drifting.

3.2 Imposition of Orientational Constraints

3.2.1 Coordinate Systems

Three groups of atoms each in the protein and the ligand are chosen to define the coordinate systems used for the orientational and axial constraints. The axial orientation of the ligand center of mass relative to the protein is specified by its spherical polar coordinate (r_1, θ_1, ϕ_1), defined with respect to the axes formed by the three groups P_1 (center of mass of Ile183, Leu202, Leu205, and Leu165), P_2 (center of mass of Asp171), and P_3 (center of mass of His208) of the protein (Fig. 3). The

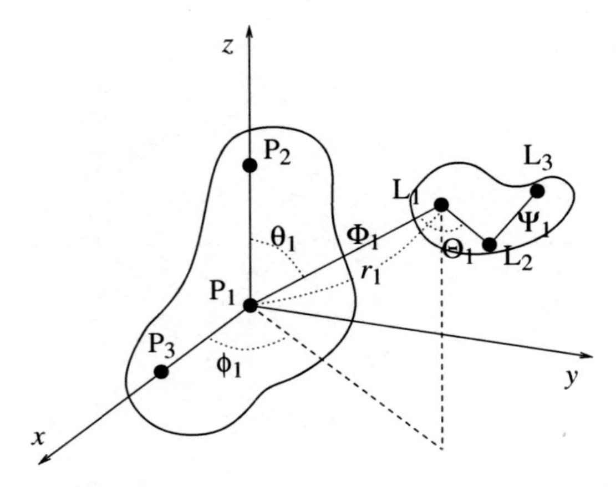

Fig. 3 Illustration of the definition of atom groups and coordinate systems for the orientational and axial constraints. P_1, P_2, and P_3 and L_1, L_2, and L_3 belong to the protein and ligand, respectively

three groups in the ligand are also chosen to define the Euler angles associated with the relative orientation of the ligand internal coordinates with respect to the protein: L_1 (center of mass of ligand), L_2 (center of mass of pTyr), and L_3 (center of mass of Ile4). The radial coordinate r_1 is the P_1 to L_1 distance, and the two spherical angles, θ_1, ϕ_1, are the angle P_2-P_1-L_1, and the dihedral P_3-P_1-P_2-L_1, respectively. The three Euler angles (Θ_1, Φ_1, Ψ_1) are taken as the angle P_1-L_1-L_2, and the dihedrals P_2-P_1-L_1-L_2 and L_2-L_1-P_1-L_3, respectively (see **Note 4**).

3.2.2 Constraints

The orientational constraint for the Euler angles is taken as:

$$u_o\left(\Theta_1, \Phi_1, \Psi_1\right) = k_o\left[\left(\Theta_1 - \Theta_1^{ref}\right)^2 + \left(\Phi_1 - \Phi_1^{ref}\right)^2 + \left(\Psi_1 - \Psi_1^{ref}\right)^2\right], \quad (2)$$

where k_o is the force constant, and $(\Theta_1^{ref}, \Phi_1^{ref}, \Psi_1^{ref})$ are the reference angle values of the bound crystallographic structure. The axial constraint for the ligand is taken as:

$$u_a\left(\theta_1, \phi_1\right) = k_a\left[\left(\theta_1 - \theta_1^{ref}\right)^2 + \left(\phi_1 - \phi_1^{ref}\right)^2\right], \quad (3)$$

with analogously defined constants.

3.3 Umbrella Sampling MD Simulations

Umbrella sampling [14] MD simulations are used to obtain the potential of mean force (PMF) as a function of a reaction coordinate for a number of steps in Fig. 1.

The relevant reaction coordinate space is subdivided into a discrete set of intervals, each of which corresponds to different windows of the umbrella sampling. MD simulations are run for each window in the presence of harmonic constraints with minima located at the given offset value of the reaction coordinate. The dynamics are run starting from the solvated initial structure built as described in Sect. 3.1 in constant pressure (1 atm) and temperature (300 K) in periodic boundary conditions. The electrostatic interactions are treated with the particle mesh Ewald method [15], using the grid size of approximately 1 Å in each dimension. The van der Waals interactions are cut off at a suitable distance (10 Å). The hydrogen–heavy atom bond lengths are fixed using the SHAKE algorithm [16], and the dynamics are run with 2-fs time steps.

3.4 Calculation of Free Energy Terms for a Ligand in the Bulk

3.4.1 Conformational Free Energy $G_c^{(bulk)}$

The term $G_c^{(bulk)}$ corresponds to the free energy change of reversibly turning on the conformational constraint for a flexible ligand in the bulk. The harmonic constraint is taken as:

$$u_c = k_c(\xi - \xi_0)^2, \tag{4}$$

where k_c is the force constant, ξ is the root mean square deviation (RMSD) of the ligand with respect to the reference conformational state (taken as the crystallographic structure of the bound complex) with ξ_0 as the offset value (equal to zero in this case). The RMSD is calculated using a fixed set of atoms, taken here as the heavy (nonhydrogen) atoms with known crystallographic coordinate for the ligand. The PMF as a function of the RMSD reaction coordinate ξ is calculated using umbrella sampling MD. A suitable number of windows (20 windows) are used to divide the range of ξ values ($0 < \xi < 10$ Å). Constrained dynamics simulations in the presence of harmonic constraints are run for each windows (for up to 2 ns), preferably with varying strengths of force constants (1 kcal/mol Å^2 and 10 kcal/mol Å^2). The umbrella sampling data is recombined using the weighted histogram analysis method (WHAM) [17, 18] with different sets of simulation time series unbiased using respective constraints. Figure 4a shows the resulting PMF $w_c^{(bulk)}(\xi)$ for the ligand in the bulk [4]. The free energy $G_c^{(bulk)}$ is calculated from the PMF by the formula:

$$\exp\left(-\beta G_c^{(bulk)}\right) = \frac{\int d\xi \exp\left[-\beta w_c^{(bulk)}(\xi) - \beta u_c(\xi)\right]}{\int d\xi \exp\left[-\beta w_c^{(bulk)}(\xi)\right]}, \tag{5}$$

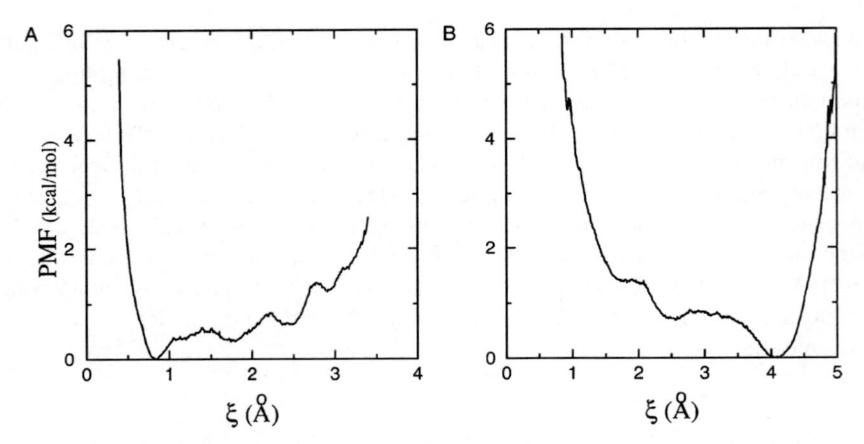

Fig. 4 The PMF $w_c{}^{(bulk)}(\xi)$ (**a**) and $w_c{}^{(site)}(\xi)$ (**b**) of the ligand conformational states in the bulk and the binding site, respectively. Reproduced from Ref. [4]. Copyright © 2005 by the National Academy of Sciences

with each integral in the fraction calculated numerically using the PMF $w_c^{(bulk)}(\xi)$. The conformational constraint u_c in Eq. 5 is taken as Eq. 4 with $\xi_0 = 0$ and the fixed value of $k_c(1.192\,\text{kcal/mol}\,\text{Å}^2)$, which is used in other parts of calculations where the ligand conformation remains constrained.

3.4.2 Orientational Free Energy $G_o^{(bulk)}$

The free energy of imposing the orientational constraint u_o is calculated by the numerical evaluation of the integral:

$$\exp\left(-\beta G_o^{(bulk)}\right) = \frac{1}{8\pi^2}\int_0^{\pi} d\Theta_1 \sin\Theta_1 \int_0^{2\pi} d\Phi_1 \int_0^{2\pi} d\Psi_1 \exp\left(-\beta u_o\right) \qquad (6)$$

with Eq. 2.

3.5 Calculation of Free Energy Terms for the Ligand–Receptor Complex

3.5.1 Conformational Free Energy $G_c^{(site)}$

The free energy change of imposing the conformational constraint to the ligand bound to receptor is calculated analogous to the case of the ligand in the bulk. The solvated protein–ligand complex is simulated by umbrella sampling MD (for up to

1 ns) to yield the PMF $w_c^{(site)}(\xi)$ (Fig. 4B). A numerical integration of the equation analogous to Eq. 5 yields $G_c^{(site)}$.

3.5.2 Orientational/Axial Free Energy Terms $G_o{}^{(site)}$ and $G_a{}^{(site)}$

The free energy perturbation [5, 6] is used to calculate the free energy terms of constraining the direction and orientation of the ligand in the binding site (using the PERT module of CHARMM). The MD simulation is made more efficient by modifying the solvated complex built in Sect. 3.1, making the simulation box geometry cubic (56 Å in each dimension). The two constraints, Eqs. 2 and 3, are turned on with suitable force constant values ($k_o = k_a = 100\,\text{kcal/mol rad}^2$) using a thermodynamic coupling parameter $\lambda(0 \leq \lambda \leq 1)$ divided into 10 intermediate intervals. The set of windows are simulated (for up to 0.5 ns), and the free energy terms are calculated by adding the free energy perturbation contributions from each window (see **Note 5**). All simulations need to be run in the presence of the conformational constraint, Eq. 4, with $\xi_0 = 0$ and the chosen value of force constant ($k_c = 1.192\,\text{kcal/mol Å}^2$).

3.6 Unbinding of the Ligand from the Receptor

3.6.1 PMF as a Function of r_1

The PMF as a function of the receptor–ligand center of mass distance r_1 is calculated by umbrella sampling MD. The radial constraint is taken as:

$$u_r = k_r(r_1 - r_1{}^0)^2, \tag{7}$$

where k_r is the force constant ($1\,\text{kcal/mol Å}^2$) and $r_1{}^0$ is the offset distance for each window. The relevant range of distances ($10\,\text{Å} < r_1 < 40\,\text{Å}$) is subdivided into a number (28 windows) that are simulated with the constraint, Eq. 7, with the offset distance at the center of the window. MD simulations are performed (up to 2 ns), all in the presence of conformational, axial, and orientational constraints, and the resulting time-series are recombined using the WHAM algorithm to construct the PMF $W(r_1)$ (Fig. 5) (see **Note 6**).

3.6.2 Prefactor S^*

The prefactor S^* is calculated by the numerical integration of:

$$S^* = (r_1{}^*)^2 \int_0^{\pi} d\theta_1 \sin\theta_1 \int_0^{2\pi} d\phi_1 \exp(-\beta u_a), \tag{8}$$

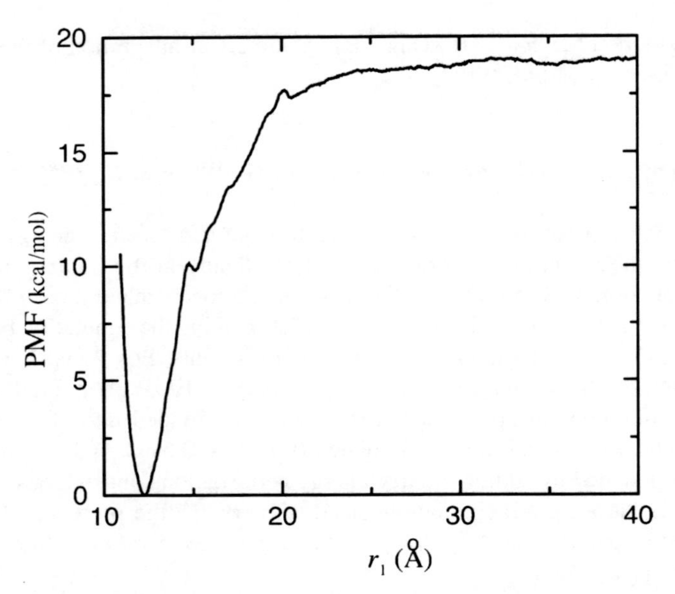

Fig. 5 The radial PMF $W(r_1)$ for the displacement of the ligand from the receptor binding site in the presence of conformational, orientational, and axial constraints. Reproduced from Ref. [4]. Copyright © 2005 by the National Academy of Sciences

where $r_1{}^*$ is chosen as a distance (30 Å) sufficiently large such that the ligand can be regarded as in an isotropic bulk solution. The radial factor I^* is calculated from the PMF by:

$$I^* = \int_0^{r_1{}^*} dr_1 \exp\left[-\beta W\left(r_1\right) + \beta W\left(r_1{}^*\right)\right] \tag{9}$$

(see **Note 7**).

The binding constant is calculated from Eq. 1.

4 Notes

1. Other choices can be made, most notably the use of implicit solvent treatments, such as the Generalized Born or Poisson–Boltzmann methods [19–21], which can reduce the computational cost significantly for larger systems.
2. The box size is to be large enough so that, with periodic boundary conditions, the minimum distances between any protein atoms in the primary cell and their images would not be smaller than the nonbonding interaction cutoff. In addition, the box size along the x-direction has to be large enough so that the center of mass of the ligand can be displaced along the x-axis away from the receptor up to a certain distance (30 Å) without getting too close to the protein image atoms.

3. For the umbrella sampling of radial PMF (Step 3), it is useful to impose an additional constraint that prevents the receptor–ligand complex from making rotational drifts with the protein near the center. The MMFP command of CHARMM, for example, can be used for the center of mass of the ligand with a potential of cylindrical geometry centered on the x-axis of Fig. 2 and a moderate force constant (1 kcal/mol $Å^2$) (see Fig. 2).
4. The choice of groups used for defining the orientational and axial constraints needs to be made such that none of the groups used for dihedral angles are close to collinear. In addition, it is best to choose the group P1 and the direction of pulling (θ_1, ϕ_1) so that any possible steric hindrances the ligand might feel while being "pulled away" from the binding site via the radial PMF calculation would be minimized.
5. The orientational and axial constraints on the ligand in the binding site can be turned on for the free energy perturbation calculations either at the same time or one after the other. The consistency of the resulting numbers can serve as a cross check. Because the orientation of the bound ligand is already stable, imposing the constraints does not lead to significant changes in the free energy, and the magnitudes of the two free energy terms should be relatively small.
6. The radial PMF calculation is the most computationally intensive part of the scheme in Fig. 1. It is helpful to vary the force constant of the radial constraint (1 and 10 kcal/mol $Å^2$) in multiple sets of umbrella sampling data to enhance the quality of statistics. The range of distances near the binding region ($10\,Å < r_1 < 20\,Å$) is more difficult to sample sufficiently. Extra sets of windows can be assigned to this region with larger force constant and/or shorter interwindow spacing. The overall convergence of the PMF calculation can be assessed by performing the WHAM analysis using partial sets of time-series data, and comparing the results [4].
7. Although in Eqs. 8 to 9, the absolute values of S^* and I^* do seem to depend on the particular choice of the "bulk" distance $r_1{}^*$, the binding free energy $G_{bind} = -k_B T \ln\left(K_{eq} C^0\right)$, where $C_0 = 1/1661\,Å^3$, is insensitive to the choice as long as the value chosen is sufficiently larger than the binding site "well" (near 12 Å) in Fig. 5. In calculating I^* by Eq. 9, one notes that $W(r_1{}^*) - W(r_1)$ has a functional form resembling a Gaussian function peaked at the well ($r_1 = 12\,Å$; Fig. 5), and, therefore, its peak height can be factored out of the integral, whereas the remaining (small) correction can be easily calculated either with direct numerical integration or by approximating the function as a Gaussian. The dominant contribution of Step 3 to the binding free energy, therefore, comes from the height of the radial PMF.

Acknowledgements The author thanks Benoît Roux for reading the manuscript and suggestions.

References

1. Gane, P. J. and Dean, P. M. (2000) Recent advances in structure-based rational drug design. *Curr. Opinion Struct. Biol.* **10**, 401–404.
2. Massova, I. and Kollman, P. A. (1999) Computational alanine scanning to probe protein-protein interactions: a novel approach to evaluate binding free energies. *J. Am. Chem. Soc.* **121**, 8133–8143.
3. Wang, J., Morin, P., Wang, W., et al. (2001) Use of MM-PBSA in reproducing the binding free energies to HIV-1 RT of TIBO derivatives and predicting the binding mode to HIV-1 RT of efavirenz by docking and MM-PBSA. *J. Am. Chem. Soc.* **123**, 5221–5230.
4. Woo, H.-J. and Roux, B. (2005) Calculation of absolute protein–ligand binding free energy from computer simulations. *Proc. Natl. Acad. Sci. USA* **102**, 6825–6830.
5. Kollman, P. A. (1993) Free energy calculations: Applications to chemical and biochemical phenomena. *Chem. Rev.* **93**, 2395–2417.
6. Simonson, T., Archontis, G., and Karplus, M. (1997) Continuum treatment of long-range interactions in free energy calculations: Application to protein–ligand binding. *J. Phys. Chem. B* **101**, 8349–8362.
7. Hermans, J. and Wang, L. (1997) Inclusion of loss of translational and rotational freedom in theoretical estimates of free energies of binding: Application to a complex of benzene and mutant T4 lysozyme. *J. Am. Chem. Soc.* **119**, 2707–2714.
8. Boresch, S., Tettinger, F., Leigeb, M., et al. (2003) Absolute binding free energies: A quantitative approach for their calculation. *J. Phys. Chem. B* **107**, 9535–9551.
9. Lee, M. S. and Olson, M. A. (2006) Calculation of absolute protein–ligand binding affinity using path and endpoint approaches. *Biophys. J.* **90**, 864–877.
10. Brooks, B. R., Bruccoleri, R. E., Olafson, B. D., et al. (1983) CHARMM: A program for macromolecular energy, minimization, and dynamics calculations. *J. Comp. Chem.* **4**, 187–217.
11. MacKerell, Jr., A. D. Bashford, D., Bellott, M., et al. (1998) All-atom empirical potential for molecular modeling and dynamics studies of proteins. *J. Phys. Chem. B* **102**, 3586–3616.
12. Waksman, G., Shoelson, S. E., Pant, N., et al. (1993) Binding of a high affinity phosphotyrosyl peptide to the Src SH2 domain: Crystal structures of the complexed and peptide-free forms. *Cell* **72**, 779–790.
13. Jorgensen, W. L., Blake, J. F., Buckner, J. K., et al. (1983) Comparison of simple potential functions for simulating liquid water. *J. Chem. Phys.* **79**, 926–935.
14. Torrie, G. M. and Valleau, J. P. (1977) Non-physical sampling distributions in Monte Carlo free-energy estimation: Umbrella sampling. *J. Comp. Phys.* **23**, 187–199.
15. Essman, U., Perera, L., Berkowitz, M. L., et al. (1995) A smooth particle mesh Ewald method. *J. Chem. Phys.* **103**, 8577–8593.
16. van Gunsteren, W. F. and Berendsen, H. J. C. (1977) Algorithms for macromolecular dynamics and constraint dynamics. *Mol. Phys.* **34**, 1311–1327.
17. Ferrenberg, A. M. and Swendsen, R. H. (1989) Optimized Monte Carlo data analysis. *Phys. Rev. Lett.* **63**, 1195–1198.
18. Roux, B. (1995) The calculation of the potential of mean force using computer simulations. *Comp. Phys. Comm.* **91**, 275–282.
19. Bashford, D. and Case, D. A. (2000) Generalized Born models of macromolecular solvation effects. *Ann. Rev. Phys. Chem.* **51**, 129–152.
20. Simonson, T. (2001) Macromolecular electrostatics: Continuum models and their growing pains. *Curr. Opinion Struct. Biol.* **11**, 243–252.
21. Chen, J. H., Im, W. and Brooks, C. L. (2006) Balancing solvation and intramolecular interactions: Toward a consistent generalized Born force field. *J. Am. Chem. Soc.* **128**, 3728–3736.

Chapter 7
Free Energy Calculations Applied to Membrane Proteins

Christophe Chipot

Summary Selected applications of free energy calculations to the realm of membrane proteins are reviewed. The theoretical underpinnings of these calculations are described, focusing on free energy perturbation and the use of thermodynamic integration to determine free energy changes along well–delineated order parameters. Current strategies for improving the reliability of free energy calculations, while making them somewhat more affordable are outlined. Application of the free energy methodology to understand the structure and function of membrane proteins is illustrated in three concrete examples: The binding of an agonist ligand to a G protein–coupled receptor, the assisted transport of a small permeant through a membrane channel, and the recognition and association of transmembrane α–helical domains.

Keywords: Free energy calculations · molecular dynamics simulations · membrane proteins · transport · recognition and association · signal transduction

1 Introduction

The paucity of structural information available for membrane proteins has imparted a new momentum to the computational investigations of complex biological systems. The grand challenge of molecular modeling is to attain the microscopic detail that is often inaccessible to conventional experimental techniques. Among the 30,000–some protein structures that can be found in the protein data bank [1] (PDB), slightly more than a hundred correspond to unique membrane proteins, and in this subset, a single one belongs to the family of G protein–coupled receptors (GPCRs) — a prominent class of targets for *de novo* drug design. This situation, which may seem paradoxical considering that about 30% of the human genome actually code for membrane proteins, can be easily understood by realizing how difficult the expression and the purification of these proteins are, especially in quantities compatible with x-ray crystallography. Such technical obstacles are further magnified by

From: Methods in Molecular Biology, vol. 443, Molecular Modeling of Proteins
Edited by Andreas Kukol © Humana Press, Totowa, NJ

the imperious necessity to extract the protein from its natural lipid environment and solubilize it gingerly in a proper detergent that will preserve its three–dimensional structure and, hence, its function.

Access to massively parallel computational resources has undeniably pushed back the limits of molecular simulations, allowing larger assemblies of atoms to be investigated over longer times. Theoretical studies of biological systems, in general, and membrane proteins, in particular, have benefited from this increase in both size and time scales. Such advances on the hardware front — and to a lesser extent on the methodological front, have without a doubt helped decipher at the atomic level how membrane proteins operate in lipid bilayers. Yet, to understand in depth the molecular mechanisms responsible for their function, a close examination of the underlying free energy behavior is necessary [2, 3]. For instance, assisted transport phenomena across the membrane, or recognition and association of transmembrane (TM) protein segments cannot be fully appreciated without the knowledge of the constituent free energy changes. Furthermore, free energy represents a tangible link between experimental and computational investigations, and, hence, a quantitative tool for appraising the quality of the designed models. The ability to determine *a priori* and with a reasonable level of accuracy free energy differences through statistical simulations is within reach. Relentless developments over the past twenty years have contributed to bring free energy calculations at the level of similarly robust and well–characterized modeling tools, while widening their field of applications. It is fair to recognize, however, that in spite of bolstering results, the accurate estimation of free energy changes in large, biologically realistic molecular assemblies still constitutes a challenge for the modeler. Taking advantage of the newest, fastest architectures, cost–effective and precise free energy calculations can provide a convincing answer to help rationalize experimental observations. In some instances, they may even play a predictive role — for instance, in the development of new leads for a specific target.

In the first section of this chapter, the theoretical underpinnings of free energy calculations are recapped, focusing on the methods that are currently utilized to determine free energy differences. Next, the reader is invited to delve into three biologically relevant examples that correspond to distinct facets of free energy simulations applied to membrane proteins. In the first application, molecular dynamics (MD) simulations and free energy methodology are utilized to probe the three–dimensional structure of a complex formed by a GPCR and an agonist ligand. Arguably enough, understanding how the agonist ligand interacts with its receptor constitutes the first step in deciphering the complex mechanism whereby the cellular signal is transduced across the cell by means of GPCRs. Next, we examine how Nature has designed specific and extremely efficient carriers for transporting water as well as small, linear polyalcohols like glycerol across the cell membrane. Last, the intricate mechanism responsible for the reversible association of TM α–helices in a membrane–like environment is dissected. Conclusions on the role played by free energy calculations in the molecular modeling of membrane proteins are drawn with a glimpse into their promising future.

2 Methods

Molecular simulations provide the modeler with ensembles of configurations, from which detailed structural information can be gained. Such configurations, or sets of Cartesian coordinates, $\{\mathbf{x}\}$, can also be used to determine thermodynamic averages, which, in turn, can be confronted directly to experimental observables. In this section, we will show that free energy can be expressed as an ensemble average, and, hence, can be inferred from computer simulations. In the following subsections, we shall assume that these simulations are carried out in the canonical, NVT, ensemble.

2.1 Statistical Mechanics Background

In the canonical ensemble, the Helmholtz free energy for an N–particle system writes [4]:

$$A = -\frac{1}{\beta} \ln Q_{\mathrm{NVT}} \tag{1}$$

where $\beta = 1/k_B T$. k_B is the Boltzmann constant and T is the temperature of the system. Q_{NVT} denotes its $6N$–dimensional partition function:

$$Q_{\mathrm{NVT}} = \frac{1}{h^{3N} N!} \int \int \exp\left[-\beta \mathcal{H}(\mathbf{x}, \mathbf{p}_x)\right] \, \mathrm{d}\mathbf{x} \, \mathrm{d}\mathbf{p}_x \tag{2}$$

Here, $\mathcal{H}(\mathbf{x}, \mathbf{p}_x)$ is the classical Hamiltonian describing the system. In equation (2), integration is carried out over all atomic coordinates, $\{\mathbf{x}\}$, and momenta, $\{\mathbf{p}_x\}$. The normalization factor reflects the measure of the volume of the phase space through the Planck constant, h, and the indistinguishable nature of the particles, embodied in the factorial term, $N!$.

The definition of the partition function may be utilized to introduce the concept of probability distribution to find the system in the unique microscopic state characterized by positions $\{\mathbf{x}\}$ and momenta $\{\mathbf{p}_x\}$:

$$\mathcal{P}(\mathbf{x}, \mathbf{p}_x) = \frac{1}{h^{3N} N!} \frac{1}{Q_{\mathrm{NVT}}} \exp\left[-\beta \mathcal{H}(\mathbf{x}, \mathbf{p}_x)\right] \tag{3}$$

A logical consequence of this expression is that low–energy regions of the phase space will be sampled predominantly, according to their respective Boltzmann weight [3, 5].

To a large extent, the canonical partition function constitutes the corner stone of the statistical mechanical description of the assembly of particles. From a phenomenological point of view, it can be seen as a measure of the thermodynamic states accessible to the system in terms of spatial coordinates and momenta.

It seems rather obvious from the above that the estimation of Q_{NVT} and, hence, A, will be an extremely challenging task — virtually impossible from the perspective of finite molecular simulations. In practice, however, the modeler is generally interested in free energy differences, ΔA, between well–delineated thermodynamic states. In this case, the free energy difference can be expressed in terms of a ratio of partition functions. Using a rather straightforward transform, this ratio can be further restated in terms of energies:

$$\Delta A = -\frac{1}{\beta} \ln \int \varrho[\Delta \mathscr{H}(\mathbf{x}, \mathbf{p}_x)] \exp[-\beta \Delta \mathscr{H}(\mathbf{x}, \mathbf{p}_x)] \, \mathrm{d}\Delta \mathscr{H}(\mathbf{x}, \mathbf{p}_x) \tag{4}$$

where $\varrho[\Delta \mathscr{H}(\mathbf{x}, \mathbf{p}_x)]$ is the so–called density of states accessible to the system of interest. As will be seen momentarily, the concept of density of states is particularly useful to understand the convergence properties of free energy calculations.

2.2 Free Energy Perturbation

Let us assume that we are interested in estimating the free energy difference between a reference system, a, described by Hamiltonian $\mathscr{H}_a(\mathbf{x}, \mathbf{p}_x)$, and a target system, b, described by Hamiltonian $\mathscr{H}_b(\mathbf{x}, \mathbf{p}_x)$, such that:

$$\mathscr{H}_b(\mathbf{x}, \mathbf{p}_x) = \mathscr{H}_a(\mathbf{x}, \mathbf{p}_x) + \Delta \mathscr{H}(\mathbf{x}, \mathbf{p}_x) \tag{5}$$

Here, $\Delta \mathscr{H}(\mathbf{x}, \mathbf{p}_x)$ represents a perturbation between the initial and the final states of the transformation. As has been hinted in the previous subsection, the difference in the Helmholtz free energy can be expressed in terms of a ratio of the corresponding partition function (2):

$$\Delta A_{a \to b} = -\frac{1}{\beta} \ln \frac{Q^b_{\mathrm{NVT}}}{Q^a_{\mathrm{NVT}}} \tag{6}$$

Substituting equation (2) to equation (6), it follows that:

$$\Delta A_{a \to b} = -\frac{1}{\beta} \ln \frac{\displaystyle\int\!\!\int \exp[-\beta \mathscr{H}_b(\mathbf{x}, \mathbf{p}_x)] \, \mathrm{d}\mathbf{x} \, \mathrm{d}\mathbf{p}_x}{\displaystyle\int\!\!\int \exp[-\beta \mathscr{H}_b(\mathbf{x}, \mathbf{p}_x)] \, \mathrm{d}\mathbf{x} \, \mathrm{d}\mathbf{p}_x} \tag{7}$$

$$= -\frac{1}{\beta} \ln \frac{\displaystyle\int\!\!\int \exp[-\beta \Delta \mathscr{H}(\mathbf{x}, \mathbf{p}_x)] \exp[-\beta \mathscr{H}_a(\mathbf{x}, \mathbf{p}_x)] \, \mathrm{d}\mathbf{x} \, \mathrm{d}\mathbf{p}_x}{\displaystyle\int\!\!\int \exp[-\beta \mathscr{H}_b(\mathbf{x}, \mathbf{p}_x)] \, \mathrm{d}\mathbf{x} \, \mathrm{d}\mathbf{p}_x}$$

Recalling the definition (3) of the probability to find the system in the unique microscopic state characterized by positions $\{\mathbf{x}\}$ and momenta $\{\mathbf{p}_x\}$, the free energy difference between the reference and the target states becomes:

$$\Delta A_{a \to b} = -\frac{1}{\beta} \ln \int\!\!\int \exp[-\beta \Delta \mathscr{H}(\mathbf{x}, \mathbf{p}_x)] \, \mathscr{P}(\mathbf{x}, \mathbf{p}_x) \, \mathrm{d}\mathbf{x} \, \mathrm{d}\mathbf{p}_x \tag{8}$$

or equivalently:

$$\boxed{\Delta A_{a \to b} = -\frac{1}{\beta} \ln \langle \exp [-\beta \, \Delta \mathscr{H}(\mathbf{x}, \mathbf{p}_x)] \rangle_a} \tag{9}$$

where $\langle \cdots \rangle_a$ stands for an ensemble average over configurations representative of the reference system. Equation (9), the fundamental free energy perturbation (FEP) formula [6], thus, states that $\Delta A_{a \to b}$ can be determined by sampling only equilibrium configurations of the initial state, a. In principle, this equation is "exact" in the sense that it is expected to converge regardless of $\Delta \mathscr{H}(\mathbf{x}, \mathbf{p}_x)$ — which is true in the limit of infinite sampling. In practice, however, validity of the perturbation formula (9) only holds for small changes between a and b, for obvious numerical reasons.

At this stage, the condition of *small changes* ought to be clarified, as it is often misconstrued. It does not imply that the free energies characteristic of the reference and the target systems be sufficiently close, but rather that the corresponding configurational ensembles overlap appropriately to guarantee the desired accuracy [7, 8]. In other words, it is expected that the density of states, $\varrho[\Delta \mathscr{H}(\mathbf{x}, \mathbf{p}_x)]$, describing the transformation between these systems be narrow enough — *viz.* typically on the order of $1/\beta$, to ascertain that, when multiplied by the exponential term of equation (4), the resulting distribution be located in a region where ample statistical data has been accrued.

Under most circumstances, however, single–step transformations between rather orthogonal states only seldom fulfill this requirement. To circumvent this difficulty, the reaction pathway connecting the reference and the target systems is broken down into a number of intermediate, states, so that between any two contiguous states, the condition of overlapping ensembles is satisfied [9]. To achieve this goal, the Hamiltonian, $\mathscr{H}(\mathbf{x}, \mathbf{p}_x)$, is made a function of the order parameter, or "coupling parameter", λ, characterizing the transformation [10]. Conventionally, λ varies between 0 and 1 when the system goes from the initial state, a, to the final state, b. In practice, λ can correspond to a variety of order parameters — possibly a true reaction coordinates, *e.g.* non–bonded parameters in the so–called "alchemical transformations" or *in silico* point mutations [11, 12].

The interval separating the intermediate states of the transformation between the reference and the target systems, which corresponds to selected fixed values of the coupling parameter, λ, is often referred to as "window". It should be reminded that the vocabulary *window* adopted in perturbation theory is distinct from that utilized in "umbrella sampling" (US) simulations [13], where it denotes a range of values taken by the order parameter. For a series of a N intermediate states, the total free energy change for the transformation from a to b is expressed as a sum of $N - 1$ free energy differences [9]:

$$\Delta A_{a \to b} = -\frac{1}{\beta} \sum_{k=1}^{N-1} \ln \Big\langle \exp \big\{ -\beta \left[\mathscr{H}(\mathbf{x}, \mathbf{p}_x; \lambda_{k+1}) - \mathscr{H}(\mathbf{x}, \mathbf{p}_x; \lambda_k) \right] \big\} \Big\rangle_{\lambda_k} \tag{10}$$

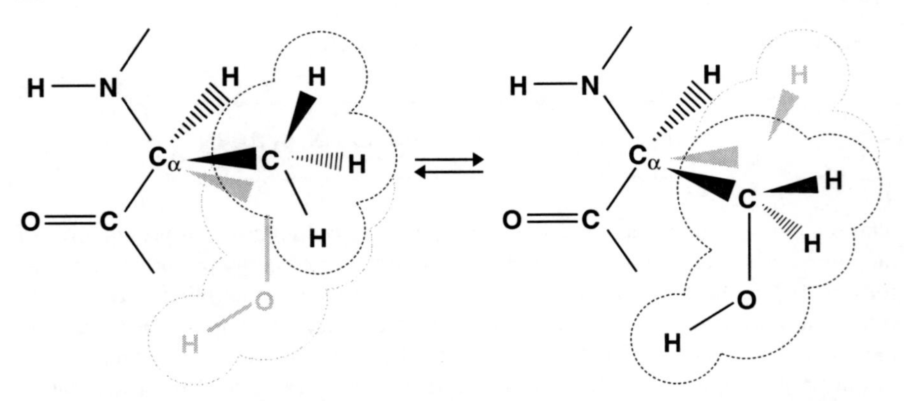

Fig. 1 Dual–topology paradigm illustrated in the case of a serine to alanine "alchemical transformation". The initial and the final states coexist , yet without "seeing" each other. The interaction energy of these topologies with their environment is scaled as λ goes from 0 to 1, so that $\mathscr{H}(\mathbf{x}, \mathbf{p}_x; \lambda) = \lambda \mathscr{H}_b(\mathbf{x}, \mathbf{p}_x) + (1 - \lambda)\mathscr{H}_a(\mathbf{x}, \mathbf{p}_x)$

Assessing an ideal number of intermediate states, N, between the initial and the final states is evidently dependent upon the nature of the system that undergoes the transformation. The condition of overlapping ensembles should be kept in mind when setting N, remembering that the choice of $\Delta\lambda = \lambda_{k+1} - \lambda_k$ ought to correspond to a finite perturbation of the system. A natural choice consists in using a number of windows that guarantees reasonably similar free energy changes between contiguous intermediate states. The consequence of this choice is that the width of the consecutive windows connecting a to b will be different.

Performing "alchemical transformations" calls for the definition of topologies that describe the initial and the final states of the mutation. In the dual–topology approach [14, 15], shown in Figure 1, the topologies representative of the reference and the target states are defined concomitantly. Yet, these topologies do not interact with each other in the course of the simulation. Their interaction energy with the surroundings is scaled as λ goes from 0 to 1. The dual–topology paradigm has been recognized to be sensitive to the so–called "end–point catastrophes", when λ tends towards 0 or 1, because ghost particles can appear where solvent molecules are already present, thereby causing severe van der Waals clashes, and, thus, numerical instabilities. A number of schemes have been devised to circumvent this problem, among which the use of windows of decreasing width as λ tends towards 0 or 1. Introduction of a soft–core potential [16] to eliminate the singularities at 0 or 1 perhaps constitutes the most elegant method proposed hitherto.

2.3 Thermodynamic Integration

Closely related to the FEP formalism (9), thermodynamic integration (TI) restates the free energy difference between the reference and the target systems as a finite difference [10, 17]:

$$\Delta A_{a \to b} = A(\lambda_b) - A(\lambda_a) \tag{11}$$

$$= \int_{\lambda_a}^{\lambda_b} \frac{\mathrm{d}A(\lambda)}{\mathrm{d}\lambda} \, \mathrm{d}\lambda$$

Substituting the definition of the canonical partition function (2) to the above equation, it follows that:

$$\frac{\mathrm{d}A(\lambda)}{\mathrm{d}\lambda} = \frac{\displaystyle\int \frac{\partial \mathscr{H}(\mathbf{x}, \mathbf{p}_x; \lambda)}{\partial \lambda} \exp -\beta \mathscr{H}(\mathbf{x}, \mathbf{p}_x; \lambda) \, \mathrm{d}\mathbf{x} \, \mathrm{d}\mathbf{p}_x}{\displaystyle\int \exp -\beta \mathscr{H}(\mathbf{x}, \mathbf{p}_x; \lambda) \, \mathrm{d}\mathbf{x} \, \mathrm{d}\mathbf{p}_x} \tag{12}$$

We recognize here again the expression of the probability (3) of finding the system in the unique microscopic state characterized by positions $\{\mathbf{x}\}$ and momenta $\{\mathbf{p}_x\}$. The free energy difference between the reference and the target states then becomes:

$$\Delta A_{a \to b} = \int_{\lambda_a}^{\lambda_b} \frac{\partial \mathscr{H}(\mathbf{x}, \mathbf{p}_x; \lambda)}{\partial \lambda} \mathscr{P}(\mathbf{x}, \mathbf{p}_x; \lambda) \, \mathrm{d}\lambda \tag{13}$$

and the integrand can be written as an ensemble average:

$$\boxed{\Delta A_{a \to b} = \int_{\lambda_a}^{\lambda_b} \left\langle \frac{\partial \mathscr{H}(\mathbf{x}, \mathbf{p}_x; \lambda)}{\partial \lambda} \right\rangle_{\lambda} \mathrm{d}\lambda} \tag{14}$$

In sharp contrast with the FEP method, the criterion of convergence here is the appropriate smoothness of the free energy as a function of λ.

2.4 Unconstrained Molecular Dynamics and Average Forces

Assuming that the variation of the kinetic energy between the reference and the target systems can be neglected, it is apparent from equation (14) that the derivative of $\Delta A_{a \to b}$ with respect to some order parameter, ξ, is equal to $-\langle F_\xi \rangle_\xi$, the average of the force exerted along ξ, hence, the concept of potential of mean force (PMF) [18].

Traditionally, the PMF, which can be used to quantify the reversible work required to bring two particles in a solvent bath from infinity to a contact distance, is expressed from the corresponding pair correlation function, $g(r)$. Generalization of this classical definition [18] is, however, far from straightforward. For this reason, the free energy as a function of order parameter ξ will be restated as:

$$A(\xi) = -\frac{1}{\beta} \ln \mathscr{P}(\xi) + A_0 \tag{15}$$

where $\mathscr{P}(\xi)$ is the probability distribution to find the system at a given value, ξ, along that order parameter:

$$\mathscr{P}(\xi) = \int \delta[\xi - \xi(\mathbf{x})] \; \exp[-\beta \mathscr{H}(\mathbf{x}, \mathbf{p}_x)] \; \mathbf{dx} \; \mathbf{dp}_x \tag{16}$$

Equation (15) corresponds to the classical definition of the free energy in methods like US, in which external biasing potentials are included to ensure a uniform distribution $\mathscr{P}(\xi)$. To improve sampling efficiency, the complete reaction pathway is broken down into "windows", or ranges of ξ, wherein individual free energy profiles are determined. The latter are subsequently pasted together using, for instance, the self–consistent weighted histogram analysis method (WHAM) [19].

For a number of years, the first derivative of the free energy with respect to the order parameter has been written as [20]:

$$\frac{\mathrm{d}A(\xi)}{\mathrm{d}\xi} = \left\langle \frac{\partial \mathscr{V}(\mathbf{x})}{\partial \xi} \right\rangle_{\xi} \tag{17}$$

This description is, however, erroneous because ξ and $\{\mathbf{x}\}$ do evidently not constitute independent variables [21, 22]. Furthermore, it assumes that kinetic contributions can be omitted, which may not always be necessarily the case. Rigorous separation of the variables imposes a transformation of the metric, so that:

$$\mathscr{P}(\xi) = \int |J| \; \exp[-\beta \mathscr{V}(\mathbf{q}; \xi)] \; \mathbf{dq} \int \exp[-\beta \mathscr{T}(\mathbf{p}_x)] \mathbf{dp}_x \tag{18}$$

Introducing probability $\mathscr{P}(\xi)$ in the first derivative (17), it follows that the kinetic contribution vanishes in $\mathrm{d}A(\xi)/\mathrm{d}\xi$:

$$\frac{\mathrm{d}A(\xi)}{\mathrm{d}\xi} = -\frac{1}{\beta} \frac{1}{\mathscr{P}(\xi)} \int \exp[-\beta \mathscr{V}(\mathbf{q}; \xi^*)] \; \delta(\xi^* - \xi) \times \left\{ -\beta |J| \frac{\partial \mathscr{V}(\mathbf{q}; \xi^*)}{\partial \xi} + \frac{\partial |J|}{\partial \xi} \right\} \mathbf{dq} \; \mathrm{d}\xi^* \tag{19}$$

After back transformation into Cartesian coordinates, the derivative of the free energy with respect to ξ can be expressed as a sum of configurational averages at constant ξ [23]:

$$\boxed{\frac{\mathrm{d}A(\xi)}{\mathrm{d}\xi} = \left\langle \frac{\partial \mathscr{V}(\mathbf{x})}{\partial \xi} \right\rangle_{\xi} - \frac{1}{\beta} \left\langle \frac{\partial \ln |J|}{\partial \xi} \right\rangle_{\xi} = -\langle F_{\xi} \rangle_{\xi}} \tag{20}$$

In this approach, only the average $\langle F_{\xi} \rangle_{\xi}$ is the physically meaningful quantity, unlike the instantaneous components, F_{ξ}, from which it is evaluated. From a computational perspective, F_{ξ} is accrued in bins of finite size, $\delta\xi$. After a predefined number of observables are accumulated in each bin, the adaptive biasing force

(ABF) [24, 25] is applied along the order parameter:

$$\mathbf{F}^{\mathrm{ABF}} = \nabla \widetilde{A} = -\langle F_\xi \rangle_\xi \, \nabla \xi \tag{21}$$

which, in turn, yields a Hamiltonian in which no average force is exerted along ξ. It follows that the evolution of the system in that direction is governed mainly by its self–diffusion properties. It is apparent from the present description that the ABF method is significantly more effective than US or its variants, because no knowledge of the free energy hypersurface is required beforehand to define the necessary biasing potentials that guarantee uniform sampling along ξ. Determining the form of such external biases may easily become intricate in the case of qualitatively new problems, in which variation of the free energy behavior cannot be guessed with the appropriate accuracy. Yet, it should be clearly understood that, whereas ABF undoubtedly improves sampling dramatically along the order parameter, efficiency still suffers, like in any other free energy method, from slowly relaxing, orthogonal degrees of freedom.

2.5 Convergence Properties of Free Energy Calculations and Error Analysis

When is enough sampling really enough to assume safely that the free energy calculation has converged? constitutes a classical conundrum that often leaves the modelers performing free energy calculations discomfited. Assessing the convergence properties and the error associated to a free energy calculation often turns out to be a challenging task. Sources of errors likely to be at play are diverse, and, hence, can modulate the results differently.

One usually distinguishes between systematic and statistical errors. In the first category, the choice of the force field parameters undoubtedly affects the results of the simulation, albeit this contribution can be largely concealed by the statistical error arising from insufficient sampling. Paradoxically, exceedingly short free energy calculations employing inadequate non–bonded parameters may, nonetheless, yield the correct answer [26]. Under the hypothetical assumption of an optimally designed potential energy function, quasi non–ergodicity scenarios constitute a common pitfall towards fully converged simulations [3].

Quasi non–ergodicity originates from different sources. In a vast number of cases, however, it is a manifestation of slow degrees of freedom relaxing over time scales that cannot be easily embraced by MD simulations. As a consequence, sampling is impeded and the system is trapped in regions of configurational space of lesser relevance for the determination of the targeted free energy change. This unfortunate situation may be the result of a poorly chosen order parameter for characterizing a process of interest. Under most circumstances, finding an adequate order parameter, let alone a true reaction coordinate, constitutes a daunting task, because the degrees of freedom at play as the system progresses are not known *a priori*. Furthermore, obvious order parameters along which the system is envisioned to glide in an unhampered fashion can be strongly coupled to degrees of freedom from the slow manifolds. As illustrated in Figure 2, these shortcomings may be revealed in multistage simulations, wherein the reaction pathway is split into ranges of the order parameter, or windows. Non–Boltzmann sampling methods, like the US and the ABF schemes, are designed to flatten the free energy landscape in the direction of the selected order parameter. If the latter is coupled to orthogonal, slow degrees of freedom, exploration of the reaction pathway over the range of interest can be severely jeopardized, because sampling of these degrees of freedom is disrupted by large free energy barriers.

Appreciation of the statistical error has been devised following different schemes. Historically, the free energy changes for the $\lambda \rightarrow \lambda + \delta\lambda$ and the $\lambda \rightarrow \lambda - \delta\lambda$ perturbations were computed simultaneously to provide the hysteresis between the forward and the reverse transformations. In practice, it can be shown that when $\Delta\lambda$ is sufficiently small, the hysteresis of such "double–wide sampling" simulation [27] becomes negligible, irrespective of the amount of sampling generated in each window — as would be the case in a "slow–growth" calculation [28].

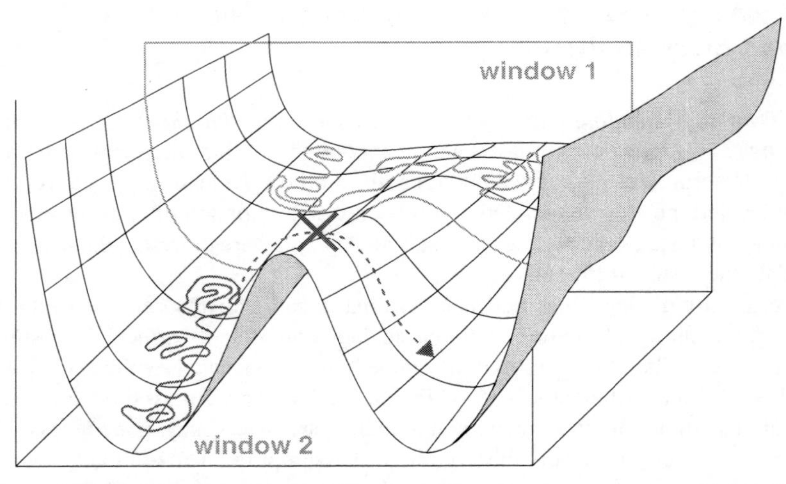

Fig. 2 Illustration of a quasi non–ergodicity scenario: Progress along the order parameter, ξ, is enhanced by means of properly designed external biases that ensure sampling uniformity. To improve the efficiency of the calculation further, the reaction pathway can be broken down into windows, in which sampling is confined. Whereas the external biases help overcome the free energy barriers along ξ, sampling may be hampered by coupled, orthogonal degrees of freedom, $\zeta \neq \xi$. In window 1, the smooth free energy landscape in both the ξ– and the ζ–directions suggests that averaging over ζ as the system progresses along ξ is likely to be very effective. In sharp contrast, sampling along ξ in window 2 is expected to be incomplete because the system is trapped along ζ, which precludes proper averaging in that direction. It is anticipated that a free energy calculation performed using a single, broad window will yield a free energy landscape distinct from that obtained by pasting the profiles of window 1 and window 2 [47]

A somewhat less arguable point of view consists in performing the transformation in the forward, $a \rightarrow b$, and in the reverse, $b \rightarrow a$, directions. Micro–reversibility imposes that, in principle, $\Delta A_{b \rightarrow a} = -\Delta A_{a \rightarrow b}$. Unfortunately, forward and reverse transformations do not necessarily share the same convergence properties. Case in point, the insertion and deletion of a particle in a liquid [29]: Whereas the former simulation converges rapidly towards the expected excess chemical potential, the latter never does. This shortcoming can be ascribed to the fact that configurations in which a cavity does not exist where a real atom is present are never sampled. In terms of density of states, this scenario would translate into $\varrho_a[\Delta \mathscr{H}(\mathbf{x}, \mathbf{p}_x)]$ embracing $\varrho_b[\Delta \mathscr{H}(\mathbf{x}, \mathbf{p}_x)]$ entirely, thereby ensuring a proper convergence of the forward simulation, whereas the same cannot be said for the reciprocal, reverse transformation. Estimation of errors based on forward and reverse simulations should, therefore, be considered with great care. Yet, appropriate combination of the two can be used profitably to improve the accuracy of free energy calculations [7].

In FEP calculations, convergence may be probed by monitoring the evolution of the ensemble average (9) as a function of time. This rather coarse test constitutes, however, a necessary, albeit not sufficient condition for convergence, because apparent plateaus of the ensemble average often conceal anomalous overlap of the density of states characterizing the reference and the target systems [7, 8]. The latter should be the key–criterion to ascertain the local convergence of the simulation for those degrees of freedom that are effectively sampled.

In addition, statistical errors in FEP calculations can be estimated by means of a first–order expansion of the free energy:

$$\Delta A = -\frac{1}{\beta}\left\{\ln \langle \exp[-\beta\, \Delta \mathscr{H}(\mathbf{x}, \mathbf{p}_x; \lambda)] \rangle_\lambda \pm \frac{\delta\varepsilon}{\langle \exp[-\beta\, \Delta \mathscr{H}(\mathbf{x}, \mathbf{p}_x; \lambda)] \rangle_\lambda}\right\} \quad (22)$$

where $\delta\varepsilon$ is the statistical error on the ensemble average, $\langle \exp[-\beta \Delta \mathscr{V}(\mathbf{x}; \lambda)] \rangle_\lambda$, defined as:

$$\begin{aligned} \Delta\varepsilon^2 = {} & \frac{1+2\tau}{N} \\ & \times \left\{ \langle \exp[-2\beta \Delta \mathscr{H}(\mathbf{x}, \mathbf{p}_x; \lambda)] \rangle_\lambda - \langle \exp[-\beta\, \Delta \mathscr{H}(\mathbf{x}, \mathbf{p}_x; \lambda)] \rangle_\lambda^2 \right\} \end{aligned} \quad (23)$$

Here, N is the number of samples accrued in the FEP calculation and $(1 + 2\tau)$ is the sampling ratio of the latter [30].

In the idealistic cases where a thermodynamic cycle can be defined, closure of the latter imposes that individual free energy contributions add up to zero [31]. In principle, any deviation from this target should provide a valuable guidance to improve sampling efficiency. In practice, however, discrimination of the faulty transformation, or transformations, becomes rapidly intricate on account of possible mutual compensation or cancelation of errors.

As has been commented on previously, visual inspection of $\varrho_a[\Delta \mathscr{H}(\mathbf{x}, \mathbf{p}_x)]$ and $\varrho_b[\Delta \mathscr{H}(\mathbf{x}, \mathbf{p}_x)]$ indicates whether the free energy calculation has converged [7, 8]. Deficiencies in the overlap of the two distributions is also suggestive of possible errors, but it should be kept in mind that approximations like (22) only reflect the *statistical precision* of the computation, and evidently do not account for fluctuations in the system occurring over long time scales. In sharp contrast, the *statistical accuracy* is expected to yield a more faithful picture of the degrees of freedom that have been actually sampled. The safest route to estimate this quantity consists in performing the same free energy calculation, starting from different regions of the phase space — *viz.* the error is then defined as the root mean square deviation over the different simulations [32]. Semantically speaking, the error measured from one individual run yields the statistical precision of the free energy calculation, whereas that derived from the ensemble of simulations provides its statistical accuracy.

3 A First Step Towards Understanding Cellular Signal Transduction

As has been mentioned earlier in this chapter, the lack of structural information available for membrane proteins has encouraged the theoretical and computational biophysics community to investigate these systems by means of large–scale statistical simulations. Of topical interest are seven TM domain GPCRs [33], which correspond to the third largest family of genes in the human genome, and, therefore, represent privileged targets for the pharmaceutical industry. Full resolution by x-ray crystallography of the three–dimensional structure of bovine rhodopsin [34], the only GPCR structure known to this date, has paved the way for the modeling of other, related membrane proteins. Unfortunately, crystallization of this receptor in its dark, inactive state does not constitute the best possible basis for homology modeling of GPCR–ligand activated complexes [35].

With over ten years of hindsight, it has become apparent that neither theory nor experiment *alone* can provide atomic–level, three–dimensional structures of activated GPCRs. It would seem, however, that their synergistic combination offers an interesting perspective to reach this goal. Such a self–consistent strategy between experimentalists and modelers has been applied rather successfully to elucidate the structure of the human receptor of cholecystokinin (CCK1R) in the presence of an agonist ligand [36] — *viz.* a nonapeptide (CCK9) [37] of sequence Arg–Asp–S-Tyr–Thr–Gly–Trp–Met–Asp–Phe–NH_2, where S-Tyr stands for a sulfated tyrosyl amino acid. Design of a consistent *in vacuo* construct of the complex involved site–directed mutagenesis experiments targeted at highlighting key receptor–ligand interactions, which helped position the constituent TM α–helices and dock CCK9 in its designated binding pocket.

In vacuo models reflect the geometrical constraints enforced in the course of their construction — *e.g.* TM segments are necessarily coerced in their putative orientation by means of appropriately chosen restraints. It is far from clear, however, whether these constructs will behave as anticipated when immersed in a realistic membrane environment. Accordingly, the model formed by CCK1R and CCK9 was inserted in a fully hydrated palmitoyloleylphosphatidylcholine (POPC) bilayer, and the complete assembly was scrutinized over a period of 30 ns, using MD simulation [38]. Thorough analysis of the trajectory reveals no apparent loss of secondary structure in the TM domain, and the distance root mean square deviation (RMSD) for the backbone atoms never exceeded 2 Å. More importantly, all crucial receptor–ligand interactions brought to light by site–directed mutagenesis experiments are preserved throughout the simulation — *e.g.* Arg^{336} with Asp^{8} [39], and Met^{195} and Arg^{197} with S-Tyr^{3} [40,41].

Arguably enough, such MD simulations only supply a qualitative picture of the molecular assembly. Integrity of the complex is probed over the time scale amenable to MD, which evidently cannot capture large spatial rearrangements. Beyond the qualitative view, free energy calculations quantify intermolecular interactions according to their importance, and, as such, form a bridge with experiment to assess the accuracy of the proposed model. Moreover, free energies are directly comparable to site–directed mutagenesis experiments utilized to build the receptor, thereby closing the loop of the modeling process.

Performing free energy calculations in such large molecular assemblies may be viewed as a bold and perhaps foolish leap of faith, considering the variety of sources of errors likely to affect the final result. Among the latter, attempting to reproduce free energy differences using a three–dimensional model in lieu of a well–resolved, experimentally determined structure casts tremendous doubts on the chances of success of this venture. Of equal concern, “alchemical transformations” involving charged amino acids are driven primarily by the hydration of the appearing, or the vanishing ionic moieties, which usually yields large free energies, the difference of which, between the free and the bound states, is expected to be small.

Assuming a valid, consistent model, which appears to be confirmed by the preliminary MD simulation, the key–question, already mentioned in this chapter, remains: *When is “enough sampling” really enough?* This question should be, in fact, rephrased here as: *Are the time scales characteristic of the slowest degrees of freedom in the system crucial for the free energy changes that are being estimated?* For instance, is the mutation of the penultimate amino acid of CCK9 — *viz.* Asp^{8} into alanine (see Figure 3), likely to be affected by the slow collective motions of lipid molecules, or possible translational motions of TM α–helices? Nanosecond MD simulations obviously cannot capture these events, which occur over significantly longer times. Yet, under the assumption that the replacement of an agonist ligand by an alternate one does not entail any noticeable rearrangement of the TM domain, the present free energy calculations are likely to be appropriate for ranking ligands according to their affinity towards a given receptor.

The theoretical free energy estimates were obtained from two runs of 3.4 ns each, in bulk water and in CCK1R, respectively, breaking the reaction path into 114 consecutive windows of uneven width, and using the dual–topology paradigm. The error was estimated from two distinct runs performed at 5.0 and 10.5 ns of the MD simulation. In contrast with an error derived from a first–order expansion of the free energy, which only reflects the statistical *precision* of the calculation — here, ± 0.3 kcal/mol, repeating the simulation from distinct initial conditions accounts for

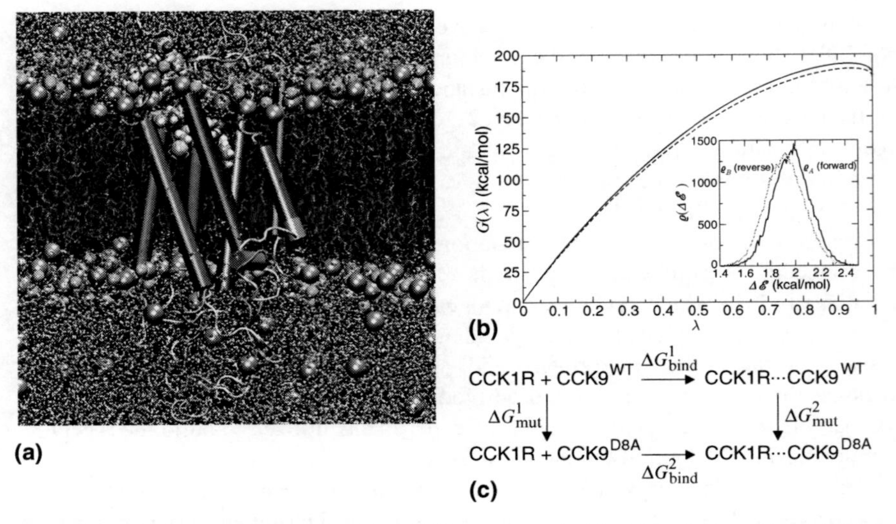

Fig. 3 Human receptor of cholecystokinin, CCK1R, embedded in a fully hydrated POPC bilayer. The agonist ligand, the CCK9 nonapeptide, is shown in a space–filling representation (**a**). Free energy change for the mutation in CCK9 of Asp^8 into alanine: Transformation in the receptor (solid line) and in water (dashed line). Inset: Overlapping density of states characterizing adjacent states, at $\lambda = 0.5$ (**b**). Thermodynamic cycle utilized to estimate the relative receptor–ligand binding free energy for the D8A point mutation in CCK9 (**c**)

fluctuations of the structure over longer time scales, thereby providing a measure of its *accuracy*.

The FEP estimate of $+3.0 \pm 0.7$ kcal/mol for the Asp^8 to alanine transformation agrees very well with the site–directed mutagenesis experiments that yielded a free energy change equal to $+.2 \pm 0.3$ kcal/mol. Replacement of the eight, sulfated–tyrosyl residue of CCK9 by tyrosine yielded a free energy change of $+1.9 \pm 0.4$ kcal/ mol, which does not compare as nicely with the experimental estimate of $+2.7 \pm 0.1$ kcal/mol. Disagreement between theory and experiment may be ascribed to, at least, two distinct sources — First, the flexibility of the extracellular loop with which S-Tyr^8 interacts, which evidently cannot be modeled fully over finite MD simulations. Second, possible imperfections in the parametrization of the non–standard sulfated tyrosyl residue.

To conclude, while it is difficult to ascertain without ambiguity the correctness of the three–dimensional structure of CCK1R:CCK9 in the sole light of a limited number of numerical experiments, it still remains that the variety of observations accrued in these simulations coincide nicely with the host of available experimental data. *De novo* development of new drug candidates for targets of unknown structure constitutes one of the greatest challenges faced today by the pharmaceutical industry. It is envisioned that the very encouraging results presented herein for CCK1R will pave the way towards a more self–contained approach to drug design, virtually emancipated from the requirement of well–resolved structures.

4 Deciphering Transport Phenomena Using Free Energy Methods

Considerable effort has been invested in recent years to understand the assisted transport of small molecules across the cell membrane. Of particular interest, the conduction events of water and small, linear polyalcohols through aquaporins like the *Escherichia coli* glycerol transport facilitator [42–44] (GlpF) have been explored by means of classical atomistic MD simulations [45, 46]. Owing to the significant time scales covered by the very slow permeation of the four channels forming the homotetrameric membrane protein [44], evidently not amenable to all–atom MD, investigation of such rare events constitutes a paradigmatic application for free energy calculations along an appropriately chosen order parameter. These calculations are expected to provide a realistic picture of the conformational and orientational relaxation phenomena in the TM channels.

Conduction of glycerol in GlpF was investigated at thermodynamic equilibrium, using the ABF scheme outlined above. Solvation of the GlpF homotetramer in a fully hydrated palmitoyloleylphosphatidylethanolamine (POPE) bilayer is described in reference [46]. To enhance the statistical information supplied by the simulations, the assisted transport of glycerol was investigated along the z–direction of Cartesian space, normal to the water–membrane interface, in the four channels of GlpF, through the concomitant definition of four independent order parameters. The order parameter was chosen as the distance separating the center of mass of glycerol from the centroid of the channel in which it was confined, projected onto z. The efficiency of the calculation was further increased by dividing the pathway connecting the cytoplasm and the periplasm sides of the membrane into seven non–overlapping windows, in which up to 20 ns of MD trajectory was generated, amounting to a total simulation time of 70 ns. Quasi non–ergodicity scenarios prone to occur in multi–stage approaches [47] were circumvented by means of an additional, 20–ns simulation performed in a single, large window spanning 25 Å. The set of initial biases for this simulation were inferred from the windowed free energy calculations.

In the hypothetical limit of infinite sampling, the free energy profiles characterizing the permeation of GlpF by glycerol in its four constituent channels should superimpose perfectly. 90 ns of sampling proved, however, that such may not necessarily be true. The similarity of the different curves suggests, nonetheless, that convergence is within reach. This assertion is reflected in the moderate RMSD of the average force, $\langle F_z \rangle_z$, which peaks at *ca.* 2.5 kcal/mol/Å in the constriction section of the conduction pathway, as shown in Figure 4.

The average free energy profile determined from the different channels is remarkably simple. The constriction region is preceded by a shallow vestibular minimum. The selectivity filter (SF) is embodied in a single free energy barrier, the average height of which is *ca.* 9.0 kcal/mol, matching closely the experimental activation energy of Borgnia and Agre, of 9.6 kcal/mol [44]. The remainder of the free energy landscape along z, in particular near the sequence formed by the three consecutive asparagine, proline and alanine amino acids (NPA motif), is essentially flat.

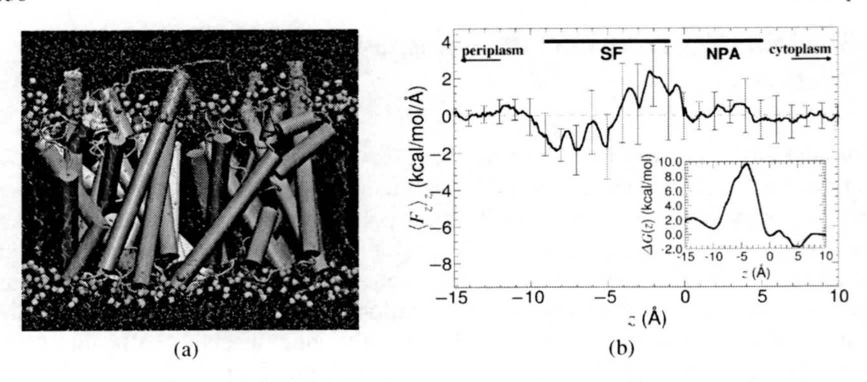

Fig. 4 Molecular assembly formed by the homotetrameric facilitator of glycerol transport, GlpF, embedded in a fully hydrated POPE lipid bilayer (**a**). Average force acting along the order parameter, the z–direction of Cartesian space, which coincides with the normal to the water–membrane interface (**b**). The error bars represent the RMSD computed over the four channels Inset: Free energy profile delineating the permeation of GlpF by glycerol

At the experimental level, the activation energy measured for glycerol conduction through GlpF [44] consists of an average over an ensemble of permeant molecules entering the constriction region of the aquaglyceroporin with distinct conformations and orientations. A closer look at the glycerol molecules as they enter the SF sheds new light on the intimate relationship between orientation, isomerization, and free energy.

Orientation of glycerol in the midst of the SF follows a two–state regime, whereby the vector joining the first and the last carbon of the molecule is either parallel or antiparallel to the normal to the water–membrane interface. It is worth noting that the preferred parallel orientation is conducive to the emergence of the *gauche–gauche* conformer, hence, suggesting that orientation and conformation are closely coupled. The marked propensity towards *gauche–anti* conformers for antiparallel orientations further illustrates the stereoselectivity of the channel, in which conformation is dictated by prochirality. Furthermore, the average orientation of the dipole moment follows the expected mechanism for the permeation process: The dipole moment is roughly antiparallel to the normal of the aqueous interface, before it tilts to a parallel orientation near the NPA motif [48].

On the biological time scale, transport of one glycerol molecule in GlpF occurs within *ca.* 55 ms [44], which evidently cannot be described by current atomistic simulations. This "blue moon" event from the perspective of theoretical and computational biophysics can, nonetheless, be modeled by accelerating the natural process to a time scale compatible with the contingencies of MD–related approaches. Capturing the full permeation event still remains limited by isomerization of the permeant in the channel. The latter occurs on the multi–nanosecond time scale, flipping of the two torsional angles, φ_1 and φ_2, of glycerol being concerted. This phenomenon spans a somewhat shorter, yet appreciable time scale when glycerol is solvated in bulk water. Not too unexpectedly, reorientation of the permeant is reasonably fast near the cytoplasm, but is hampered dramatically in the SF region, where it can be congealed either parallel or antiparallel to the normal to the water–membrane interface for as long as *ca.* 10 ns.

Glycerol conduction in GlpF has been investigated from the perspective of equilibrium, sub–hundred–nanosecond free energy calculations. Compared to shorter, irreversible pulling experiments, unable to capture relaxation phenomena embracing significant time scales, the length of the present simulations and their reversible character allow the permeant to reorient and isomerize freely as it diffuses slowly through the conduction pathway. They illuminate that orientational and conformational relaxation of glycerol and its ABF–assisted transport in the channels span comparable time scales. The heights of the free energy barrier separating the periplasmic vestibule from the NPA motif, initially modulated by the original orientation of glycerol in the channel, appears to converge after appropriate sampling towards the experimentally determined activation energy [44]. The reported free energy calculations, therefore, constitute an important, albeit still incomplete step towards the full understanding of glycerol diffusion in GlpF. To reconcile fully theoretical and experimental results, a better characterization of the slow degrees of freedom that thwart diffusion in the channels is highly desirable.

5 Recognition and Association in Membrane Proteins

It is fair to recognize that our current knowledge of how membrane protein domains recognize and associate into functional, three–dimensional entities is still fragmentary. Whereas the structure of membrane proteins can be particularly complex, their TM region is often simple, consisting in general of a bundle of α–helices, or barrels of β–strands. A fundamental result brought to light by deletion experiments reveals that some membrane proteins can retain their biological function even when large fractions of the protein are removed [49–52]. This suggests that rudimentary models, like simple α–helices, can be utilized profitably to understand how TM segments recognize and associate into complex membrane proteins.

The "two–stage" model of Popot and Engelman [53], a pioneering attempt to reach this goal, represents an interesting view for rationalizing the folding of membrane proteins. According to this model, elements of the secondary structure — *viz.* under most circumstances, α–helices — are first formed and inserted into the lipid bilayer, prior to specific inter–helical interactions that drive the TM segments towards well–ordered, native structures.

Capturing the atomic detail of the underlying mechanisms of α–helix recognition and association requires model systems supported by robust experimental data to appraise the accuracy of the computations endeavored. Glycophorin A (GpA), a glycoprotein ubiquitous to the human erythrocyte membrane, represents one such system. It forms non–covalent dimers through the reversible association of its membrane–spanning domain — *i.e.* residues 62 to 101, albeit only residues 73 to 96 actually adopt an α–helical conformation [54–56]. Inter–helical association

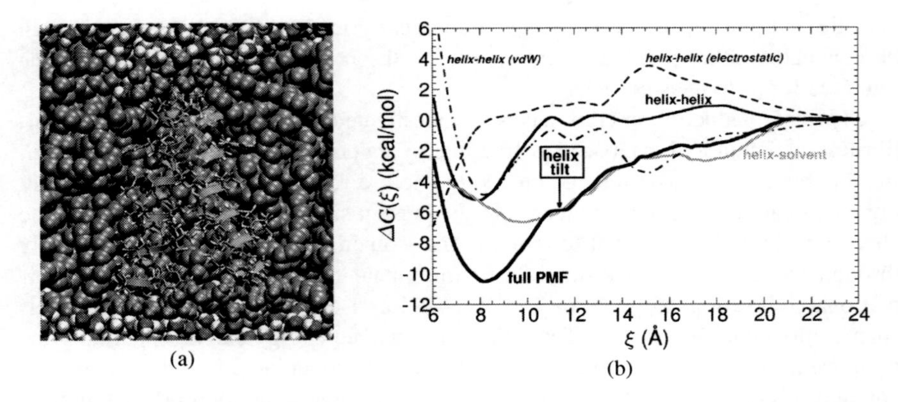

Fig. 5 TM domain of GpA formed by a homodimer of α–helices, embedded in a lipid membrane mimetic. The heptad of residues involved in the association of the TM segments are shown as transparent van der Waals spheres. Note the crossing angle between the two α–helices, equal to *ca.* 40° (**a**). Free energy profile delineating the reversible association of the TM segments (**b**). From the generated MD trajectory, it is possible to determine *a posteriori* the different contributions to the total free energy change by evaluating the associated force and projecting the latter onto the order parameter, ξ

has been shown to result from specific interactions involving a heptad of residues, essentially located on one face of each TM segment, as may be seen in Figure 5.

The reversible association of GpA in a lipid bilayer was examined using its dimeric, α–helical TM segments immersed in a membrane mimetic resulting from the assembly of a dodecane lamella placed between two lamellae of water. The ABF method was employed to allow the TM segments to diffuse freely along an order parameter, ξ, chosen to be the distance separating the centers of mass of the two TM α–helices. Such a free energy calculation is not only challenging methodologically, but it is also of paramount importance from a biophysical standpoint, because it provides a tangible link between the structural data obtained from nuclear magnetic resonance (NMR) [54–56] and the thermodynamic data derived from analytical ultracentrifugation [57, 58] and fluorescence resonance energy transfer (FRET) [59, 60] measurements, while providing a dynamic view of the recognition and association stages.

Reversible association of the α–helices was examined over a period of 125 ns [61]. The PMF derived from this simulation is shown to be qualitatively simple, featuring a single minimum characteristic of the native dimer — see Figure 5. As ξ increases, so does the free energy, progressing by steps that correspond to the successive breaking of all inter–helical contacts. Beyond 21 Å, the TM segments are sufficiently separated to assume that they no longer interact. It is noteworthy that as the interaction of the two α–helices vanish, the latter progressively tilt towards an upright orientation, which is suggestive that the formation of the signature right–handed 40° crossing angle is concomitant with the creation of short–range inter–helical contacts.

Table 1 Distinction between the short–range and long–range regimes that govern α–helix recognition and association in the TM region of GpA

	Short–range interactions	Long–range interactions
Contacts	native	non–native
Crossing angle	native (40°)	upright (<15°)
Driving force	helix–helix	helix–solvent

The association constant of the TM segments may be recovered by integrating the PMF in the limit of α–helix association. The corresponding standardized free energy of dimerization is equal to $+11.5 \pm 0.4$ kcal/mol. Yet, direct and precise comparison of this value with experiment is not possible, because measurements were carried in different environments, namely hydrocarbon *vs.* detergent micelles. It can, nonetheless, be inferred that the value in dodecane probably constitutes an upper bound to the experimental estimates determined in micelles, for two reasons, namely (i) the greater order imposed by the detergent chains, and, (ii) the hydrophobic fraction of the system increasing with the length of the chain [60].

Valuable information may be derived from the PMF Deconvolution of the PMF into free energy components illuminates two distinct regimes controlling recognition and association, which are summarized in Table 1. At large separations, as inter–helical contacts vanish, the helix–helix term becomes progressively negligible, resulting essentially from the $1/\xi^3$ interaction of two macro–dipoles — see Figure 5. The TM segments are stabilized by favorable helix–solvent contributions. In contrast, at short separations, helix–helix interactions are prominent and govern the change in the free energy near the global minimum. Association proceeds through the transient formation of early, non–native contacts involving residues that act as recognition sites. These contacts are subsequently replaced by contacts in the heptad of residues responsible for association, concomitantly with the tilt of the two α–helices from an upright position to that characteristic of the native dimer.

6 Conclusion

Beyond the qualitative structural information inferred from molecular simulations, free energy calculations offer a robust link between theory and experiment by quantifying the modeled processes at the thermodynamic level. In addition, these calculations provide a rigorous and discriminating assessment of the quality of the model, while suggesting how the latter might be improved.

The simulations reported herein demonstrate that with twenty years of hindsight gained from methodological development and characterization, a variety of biologically relevant problems can now be tackled with confidence. Whereas the foundation of the FEP machinery have been established many years ago, significant progress has been achieved only recently in the calculation of free energies along a chosen order parameter. Employing the concept of an average force acting along this parameter [24, 25], free energy barriers may be overcome without any *a priori* knowledge of the actual free energy landscape, from which the necessary biases could be determined to ensure a uniform sampling. This methodology is now available in the popular program NAMD [62], well suited for the simulation of complex biological system on large arrays of processors.

By and large, free energy calculations have greatly benefited from the increased access to massively parallel architectures as the price/performance ratio of computer processors continues to fall inexorably. They have also benefited from advances in the understanding of the methods and how the latter should be applied [63], as well as in the characterization of the error affecting the simulations [7,8,64]. Put together, they have come of age to emancipate from their traditional role of a mere proof of concept [65], and progressively become a predictive tool. As has been illustrated in this chapter, they can be applied profitably to a host of problems in theoretical and computational biophysics, involving large assemblies of atoms. In the field of membrane proteins, whenever a structural information is available, free energy calculations are expected to help decipher the mechanisms whereby these systems operate in the cell machinery, and, thus, help relate the concepts of *structure* and *function*.

Free energy calculations, however, cannot yet be considered as "black box" routine jobs. A robust and reliable methodology does not necessarily imply that it can be used blindly, as the nature of the problem essentially dictates the choice of the method and the associated sampling strategy. In the investigation of assisted transport phenomena, for instance, the definition of a suitable order parameter has proven to be particularly problematic on account of its often equivocal nature. If the order parameter is too strongly coupled to degrees of freedom in the slow manifolds, convergence of the simulation will be unavoidably plagued by these slowly relaxing degrees of freedom. Circumventing such quasi non–ergodicity scenarios is generally difficult, because we do not know beforehand what are the fast and the slow degrees of freedom in the system.

It still remains that the results reported here are not only very encouraging, but they also illustrate how much progress has been accomplished since the pioneering calculations of Tembe and McCammon, who demonstrated, using a very rudimentary model, that the FEP machinery could be applied successfully to model ligand–receptor assemblies [66]. Reproduction of relative binding free energies within chemical accuracy for protein–ligand complexes formed by several thousands of atoms evidently opens new vistas for the rational design of novel drug candidates aimed at membrane proteins. Moreover, comparing the free energy profiles characterizing glycerol diffusion in GlpF and α–helix dimerization in GpA with that derived by Berne and coworkers back in 1979 to investigate the hydrophobic effect [67], using a multistage strategy and a model system formed by two Lennard–Jones spheres in a water bath, one can appreciate the latest advances in the computation of free energy changes along an order parameter. Although these recent results hold great promise for the future of free energy calculations, exploring reaction pathways in complex biological systems remains thwarted by the recurrent difficulty to select an adequate order parameter capable of describing fully a process of interest. Bridging theory and experiment inevitably implies that such order parameters can be determined, but it also suggests that we can discriminate between the slow and the fast degrees of freedom coupled to the latter. Advancing with this objective in mind constitutes a priority for the years to come [3].

Acknowledgements The author warmly thank Dr. Jérôme Hénin, Profs. Andrew Pohorille, Emad Tajkhorshid and Klaus Schulten, who were deeply involved in the computational investigations compiled in this chapter. The Centre Informatique National de l'Enseignement Supérieur (CINES) and the centre de Calcul Réseaux et Visualisation Haute Performance (CRVHP) are gratefully acknowledged for generous provision of CPU time on their SGI Origin 3000 architectures.

References

1. Sussman, J. L., Lin, D., Jiang, J., Manning, N. O., Prilusky, J., Ritter, O., and Abola E. E., (1998), Protein data bank (PDB): Database of three–dimensional structural information of biological macromolecules, *Acta Cryst.*, **D54**, 1078–1084.
2. Kollman, P. A., (1993), Free energy calculations: Applications to chemical and biochemical phenomena, *Chem. Rev.*, **93**, 2395–2417.
3. Chipot, C. and Pohorille, A., Eds., *Free energy calculations. Theory and applications in chemistry and biology*, Springer Verlag, 2007. (in press).
4. McQuarrie, D. A., *Statistical mechanics*, Harper and Row: New York, 1976.
5. Allen, M. P. and Tildesley, D. J., *Computer Simulation of Liquids*, Clarendon Press: Oxford, 1987.
6. Zwanzig, R. W., (1954), High–temperature equation of state by a perturbation method. I. Nonpolar gases, *J. Chem. Phys.*, **22**, 1420–1426.
7. Lu, N., Adhikari, J. and Kofke, D. A., (2003), Variational formula for the free energy based on incomplete sampling in a molecular simulation, *Phys. Rev. E*, **68**, 026122–1–026122–7.
8. Lu, N., Kofke, D. A. and Woolf, T. B., (2004), Improving the efficiency and reliability of free energy perturbation calculations using overlap sampling methods, *J. Comput. Chem.*, **25**, 28–39.
9. Mark, A. E. Free Energy Perturbation Calculations. In *Encyclopedia of computational chemistry*, Schleyer, P. v. R.; Allinger, N. L.; Clark, T.; Gasteiger, J.; Kollman, P. A.; Schaefer III, H. F.; Schreiner, P. R., Eds., vol. 2. Wiley and Sons, Chichester, (1998), pp. 1070–1083.
10. Kirkwood, J. G., (1935), Statistical mechanics of fluid mixtures, *J. Chem. Phys.*, **3**, 300–313.
11. Bash, P. A., Singh, U. C., Brown, F. K., Langridge, R. and Kollman, P. A., (1987), Calculation of the relative change in binding free energy of a protein–inhibitor complex, *Science*, **235**, 574–576.
12. Bash, P. A., Singh, U. C., Langridge, R. and Kollman, P. A., (1987), Free energy calculations by computer simulation, *Science*, **236**, 564–568.
13. Torrie, G. M. and Valleau, J. P., (1977), Monte Carlo study of phase separating liquid mixture by umbrella sampling, *J. Chem. Phys.*, **66**, 1402–1408.
14. Gao, J., Kuczera, K., Tidor, B. and Karplus, M., (1989), Hidden thermodynamics of mutant proteins: A molecular dynamics analysis, *Science*, **244**, 1069–1072.
15. Pearlman, D. A., (1994), Free energy derivatives: A new method for probing the convergence problem in free energy calculations, *J. Comput. Chem.*, **15**, 105–124.
16. Beutler, T. C., Mark, A. E., van Schaik, R. C., Gerber, P. R. and van Gunsteren, W. F., (1994), Avoiding singularities and neumerical instabilities in free energy calculations based on molecular simulations, *Chem. Phys. Lett.*, **222**, 529–539.

17. Straatsma, T. P. and Berendsen, H. J. C., (1988), Free energy of ionic hydration: Analysis of a thermodynamic integration technique to evaluate free energy differences by molecular dynamics simulations, *J. Chem. Phys.*, **89**, 5876–5886.
18. Chandler, D., *Introduction to modern statistical mechanics*, Oxford University Press, 1987.
19. Kumar, S., Bouzida, D., Swendsen, R. H., Kollman, P. A. and Rosenberg, J. M., (1992), The weighted histogram analysis method for free energy calculations on biomolecules. I. The method, *J. Comput. Chem.*, **13**, 1011–1021.
20. Pearlman, D. A., (1993), Determining the contributions of constraints in free energy calculations: Development, characterization, and recommendations, *J. Chem. Phys.*, **98**, 8946–8957.
21. Carter, E, A., Cicotti, G., Hynes, J. T. and Kapral, R., (1989), Constrained reaction coordinate dynamics for the simulation of rare events, *Chem. Phys. Lett.*, **156**, 472–477.
22. den Otter, W. K. and Briels, W. J., (1998), The calculation of free–energy differences by constrained molecular dynamics simulations, *J. Chem. Phys.*, **109**, 4139–4146.
23. den Otter, W. K., (2000), Thermodynamic integration of the free energy along a reaction coordinate in Cartesian coordinates, *J. Chem. Phys.*, **112**, 7283–7292.
24. Darve, E. and Pohorille, A., (2001), Calculating free energies using average force, *J. Chem. Phys.*, **115**, 9169–9183.
25. Hénin, J. and Chipot, C., (2004), Overcoming free energy barriers using unconstrained molecular dynamics simulations, *J. Chem. Phys.*, **121**, 2904–2914.
26. Pearlman, D. A. and Kollman, P. A., (1991), The overlooked bond–stretching contribution in free energy perturbation calculations, *J. Chem. Phys.*, **94**, 4532–4545.
27. Jorgensen, W. L. and Ravimohan, C., (1985), Monte Carlo simulation of differences in free energies of hydration, *J. Chem. Phys.*, **83**, 3050–3054.
28. Chipot, C., Kollman, P. A. and Pearlman, D. A., (1996), Alternative approaches to potential of mean force calculations: Free energy perturbation versus thermodynamic integration. Case study of some representative nonpolar interactions, *J. Comput. Chem.*, **17**, 1112–1131.
29. Widom, B., (1963), Some topics in the theory of fluids, *J. Chem. Phys.*, **39**, 2808–2812.
30. Straatsma, T. P., Berendsen, H. J. C. and Stam, A. J., (1986), Estimation of statistical errors in molecular simulation calculations, *Mol. Phys.*, **57**, 89–95.
31. Chipot, C. and Pohorille, A., (1998), Conformational equilibria of terminally blocked single amino acids at the water–hexane interface. A molecular dynamics study, *J. Phys. Chem. B*, **102**, 281–290.
32. Chipot, C., Millot, C., Maigret, B. and Kollman, P. A., (1994), Molecular dynamics free energy perturbation calculations. Influence of nonbonded parameters on the free energy of hydration of charged and neutral species, *J. Phys. Chem.*, **98**, 11362–11372.
33. Takeda, S. Kadowaki, S., Haga, T., Takaesu, H. and Mitaku, S., (2002), Identification of G protein–coupled receptor genes from the human genome sequence, *FEBS Lett.*, **520**, 97–101.
34. Palczewski, K., Kumasaka, T., Hori, T., Behnke, C. A., Motoshima, H., Fox, B. A., Le Trong, I., Teller, D. C., Okada, T., Stenkamp, R. E., Yamamoto, M. and Miyano, M., (2000), Crystal structure of rhodopsin: A G protein–coupled receptor, *Science*, **289**, 739–745.
35. Archer, E., Maigret, B., Escrieut, C., Pradayrol, L. and Fourmy, D., (2003), Rhodopsin crystal: New template yielding realistic models of G–protein–coupled receptors?, *Trends Pharmacol. Sci.*, **24**, 36–40.
36. Talkad, V. D., Fortune, K. P., Pollo, D. A., Shah, G. N., Wank, S. A. and Gardner, J. D., (1994), Direct demonstration of three different states of the pancreatic cholecystokinin receptor, *Proc. Natl. Acad. Sci. U. S. A.*, **91**, 1868–1872.
37. Moroder, L., Wilschowitz, L., Gemeiner, M., Göhring, W., Knof, S., Scharf, R., Thamm, P., Gardner, J. D., Solomon, T. E. and Wünsch, E., (1981), Zur Synthese von Cholecystokinin–Pankreozymin. Darstellung von [28–Threonin, 31–Norleucin]– und [28–Threonin, 31–Leucin]– Cholecystokinin–Pankreozymin–(25–33)–Nonapeptid, *Z. Physiol. Chem.*, **362**, 929–942.
38. Hénin, J., Maigret, B., Tarek, M., Escrieut, C., Fourmy, D. and Chipot, C., (2006), Probing a model of a GPCR/ligand complex in an explicit membrane environment. The human cholecystokinin–1 receptor, *Biophys. J.*, **90**, 1232–1240.

39. Gigoux, V., Escrieut, C., Fehrentz, J. A., Poirot, S., Maigret, B., Moroder, L., Gully, D., Martinez, J., Vaysse, N. and Fourmy, D., (1999), Arginine 336 and Asparagine 333 of the human cholecystokinin–A receptor binding site interact with the penultimate aspartic acid and the C–terminal amide of cholecystokinin, *J. Biol. Chem.*, **274**, 20457–20464.
40. Gigoux, V., Escrieut, C., Silvente-Poirot, S., Maigret, B., Gouilleux, L., Fehrentz, J. A., Gully, D., Moroder, L., Vaysse, N. and Fourmy, D., (1998), Met–195 of the cholecystokinin–A interacts with the sulfated tyrosine of cholecystokinin and is crucial for receptor transition to high affinity state, *J. Biol. Chem.*, **273**, 14380–14386.
41. Gigoux, V., Maigret, B., Escrieut, C., Silvente-Poirot, S., Bouisson, M., Fehrentz, J. A., Moroder, L., Gully, D., Martinez, J., Vaysse, N. and Fourmy, D., (1999), Arginine 197 of the cholecystokinin–A receptor binding site interacts with the sulfate of the peptide agonist cholecystokinin, *Protein Sci.*, **8**, 2347–2354.
42. Stahlberg, H., Braun, T., de Groot, B., Philippsen, A., Borgnia, M. J., Agre, P., Kühlbrandt, W. and Engel, A., Nov 2000, The 6.9-Å structure of GlpF: A basis for homology modeling of the glycerol channel from *Escherichia coli*, *J. Struct. Biol.*, **132**, 133–141.
43. Fu, D., Libson, A., Miercke, L. J., Weitzman, C., Nollert, P., Krucinski, J. and Stroud, R. M., (2000), Structure of a glycerol-conducting channel and the basis for its selectivity., *Science*, **290**, 481–6.
44. Borgnia, M. J. and Agre, P., (2001), Reconstitution and functional comparison of purified GlpF and AqpZ, the glycerol and water channels from *Escherichia coli*, *Proc. Natl. Acad. Sci. USA*, **98**, 2888–2893.
45. Jensen, M. Ø., Park, S., Tajkhorshid, E. and Schulten, K., (2002), Energetics of glycerol conduction through aquaglyceroporin GlpF, *Proc. Natl. Acad. Sci. USA*, **99**, 6731–6736.
46. Jensen, M. O., Tajkhorshid, E. and Schulten, K., (2001), The mechanism of glycerol conduction in aquaglyceroporins, *Structure*, **9**, 1083–1093.
47. Chipot, C. and Hénin, J., (2005), Exploring the free energy landscape of a short peptide using an average force, *J. Chem. Phys.*, **123**, 244906.
48. Wang, Y., Schulten, K. and Tajkhorshid, E., (2005), What makes an aquaporin a glycerol channel? A comparative study of AqpZ and GlpF, *Structure*, **13**, 1107–1118.
49. Duff, K. C. and Ashley, R. H., (1992), The transmembrane domain of influenza A M2 protein forms amantidine sensitive proton channels in planar lipid bilayers, *Virology*, **190**, 485–489.
50. Oblatt-Montal, M., Buhler, L., Iwamoto, T., Tomich, J. and Montal, M., (1993), Synthetic peptides and four-helix bundle proteins as model systems for the pore–forming structure of channel proteins. I. Transmembrane segment M2 of the nicotinic cholinergic receptor channel is a key pore–lining structure, *J. Biol. Chem.*, **268**, 14601–14607.
51. Montal, M., (1995), Molecular mimicry in channel–protein structure, *Curr. Opin. Struct. Biol.*, **5**, 501–506.
52. Montal, M., (1995), Design of molecular function: Channels of communication, *Annu. Rev. Biophys. Biomol. Struct.*, **24**, 31–57.
53. Popot, J. L. and Engelman, D. M., (1990), Membrane protein folding and oligomerization: The two–stage model, *Biochemistry*, **29**, 4031–4037.
54. MacKenzie, K. R., Prestegard, J. H. and Engelman, D. M., (1997), A transmembrane helix dimer: Structure and implications, *Science*, **276**, 131–133.
55. MacKenzie, K. R. and Engelman, D. M., (1998), Structure–based prediction of the stability of transmembrane helix–helix interactions: The sequence dependence of glycophorin A dimerization, *Proc. Natl. Acad. Sci. USA*, **95**, 3583–3590.
56. Smith, S. O., Song, D., Shekar, S., Groesbeek, M., Ziliox, M. and Aimoto, S., (2001), Structure of the transmembrane dimer interface of glycophorin A in membrane bilayers, *Biochemistry*, **40**, 6553–6558.
57. Fleming, K. G., Ackerman, A. L. and Engelman, D. M., (1997), The effect of point mutations on the free energy of transmembrane α–helix dimerization, *J. Mol. Biol.*, **272**, 266–275.
58. Fleming, K. G., (2002), Standardizing the free energy change of transmembrane helix–helix interactions, *J. Mol. Biol.*, **323**, 563–571.
59. Fisher, L. E., Engelman, D. M. and Sturgis, J. N., (1999), Detergents modulate dimerization, but not helicity, of the glycophorin A transmembrane domain, *J. Mol. Biol.*, **293**, 639–651.

60. Fisher, L. E., Engelman, D. M. and Sturgis, J. N., (2003), Effects of detergents on the association of the glycophorin A transmembrane helix, *Biophys. J.*, **85**, 3097–3105.
61. Hénin, J., Pohorille, A. and Chipot, C., (2005), Insights into the recognition and association of transmembrane α–helices. The free energy of α–helix dimerization in glycophorin A, *J. Am. Chem. Soc.*, **127**, 8478–8484.
62. Phillips, J. C., Braun, R., Wang, W., Gumbart, J., Tajkhorshid, E., Villa, E., Chipot, C., Skeel, R. D. Kalé, L. and Schulten, K., (2005), Scalable molecular dynamics with NAMD, *J. Comput. Chem.*, **26**, 1781–1802.
63. Dixit, S. B. and Chipot, C., (2001), Can absolute free energies of association be estimated from molecular mechanical simulations? The biotin–streptavidin system revisited, *J. Phys. Chem. A*, **105**, 9795–9799.
64. Rodriguez-Gomez, D., Darve, E. and Pohorille, A., (2004), Assessing the efficiency of free energy calculation methods, *J. Chem. Phys.*, **120**, 3563–3578.
65. Simonson, T., Archontis, G. and Karplus, M., (2002), Free energy simulations come of age: Protein–ligand recognition, *Acc. Chem. Res.*, **35**, 430–437.
66. Tembe, B. L. and McCammon, J. A., (1984), Ligand–receptor interactions, *Comp. Chem.*, **8**, 281–283.
67. Pangali, C., Rao, M. and Berne, B. J., (1979), A Monte Carlo simulation of the hydrophobic interaction, *J. Chem. Phys.*, **71**, 2975–2981.

Part III
Molecular Modeling of Membrane Proteins

Chapter 8
Molecular Dynamics Simulations of Membrane Proteins

Philip C. Biggin and Peter J. Bond

Summary Membrane protein structures are underrepresented in the Protein Data Bank (PDB) because of difficulties associated with expression and crystallization. As such, it is one area in which computational studies, particularly molecular dynamics (MD), can provide useful additional information. Recently, there has been substantial progress in the simulation of lipid bilayers and membrane proteins embedded within them. Initial efforts at simulating membrane proteins embedded within a lipid bilayer were relatively slow and interactive processes, but recent advances now mean that the setup and running of membrane protein simulations is somewhat more straightforward, although not without its problems. In this chapter, we outline practical methods for setting up and running MD simulations of a membrane protein embedded within a lipid bilayer and discuss methodologies that are likely to contribute future improvements.

Keywords: Computational · Ion channels · Membrane proteins · Molecular Dynamics · Simulation

1 Introduction

Membrane proteins are thought to constitute approximately 30% of genomes [1]. Furthermore, it has been estimated that more than half of all drug targets are membrane proteins [2]. However, because of problems associated with expression and crystallization, the number of high resolution crystal structures is less than 1% of the total number of structures (see http://blanco.biomol.uci.edu/Membrane_Proteins_ xtal.html for a maintained list of membrane proteins). The situation is further complicated by the fact that many membrane proteins undergo very large conformational changes to complete their function (for example, transporter proteins [3–5], which cycle between at least two distinct states). Crystallography will, at best, only be able to capture a time- and space-averaged snapshot

From: Methods in Molecular Biology, vol. 443, Molecular Modeling of Proteins
Edited by Andreas Kukol © Humana Press, Totowa, NJ

of these states. Computer simulations, on the other hand, and, in particular, molecular dynamics (MD), are useful tools that, in addition to providing information regarding the stability of a membrane protein, can also provide insight into the manner in which these conformational changes can proceed. Thus, there has been a large increase in the application of computer simulation methods to membrane proteins [6, 7], ranging from ion channels [8–10] to outer membrane proteins [11]. MD can also be used to test hypotheses in idealized systems in which one can explore underlying biophysical principals governing a process [12], and through to systems that represent in vivo systems as closely as possible [13].

The field of membrane protein simulation has matured during recent years, and worldwide there are now many groups performing simulations. One reason for the recent increase in the number of research groups is that the computational facilities required to perform membrane protein simulations are now accessible to more people. Because the most common computational method used is MD, in this chapter, we discuss how to set up and run a membrane protein simulation, focusing on the more practical aspects. Until recently, the setup required a large amount of interactive input from the researcher. Now, principally because of increases in computational power, the setup and running of such simulations is much simpler. We divide the process into four distinct steps: the preparation of the protein itself, the preparation of the lipid (although this is discussed only briefly because the main thrust of this chapter is the setup and running of a membrane protein simulation), the actual insertion and establishing of a stable system, and, finally, running the simulation. Of these steps, it is perhaps the preparation of the protein itself that requires the most care and interactive input from the researcher.

2 Theory

The underlying theory for MD simulations is discussed in detail in Chap. 1, and, therefore, in this section, we will briefly discuss some specific considerations that researchers should bear in mind when performing simulations of membrane proteins. Perhaps the most important of these considerations are the timescale of the problem that is under consideration and the resources that are available. The many different aspects of membrane dynamics span a large timescale, ranging from a few picoseconds (for a protein side chain to rotate) through minutes and longer (for flip-flop motion of lipids). Indeed, where resources are minimal and only an approximate representation of the bilayer is required, one may be content with using a slab of octane to represent the hydrophobic core of the bilayer [14]. More recently, efforts have been made to approximate the lipid molecules in a different way, by using a coarse-grained (CG) approach, where, typically, four atoms are represented by one particle [15]. These methods have become popular because they allow much longer timescale events to be explored, but, of course, they are less detailed than a fully atomistic simulation.

Typically, for simulations of membrane proteins, a stable simulation is required, usually reflecting some sort of equilibrium of the system. In these simulations, we have two components to worry about, the protein and the lipid bilayer. Thus, some metrics of stability are required. For the protein, the most common of these metrics is the root mean square deviation (RMSD) of Cα atoms from the initial (usually x-ray) starting structure. In the case of the lipid, a good indicator is the mean surface area per lipid [16].

The basic theory underlying the insertion process below is very simple. We place the protein in the bilayer and remove overlapping atoms. We then allow the whole system to relax and equilibrate as the lipids adjust conformation around the protein. The positioning of the protein is still somewhat subjective, especially with respect to its displacement along the membrane normal axis. However, structural bioinformatics analysis [17] has demonstrated that nearly all membrane proteins have two "aromatic girdles" (although not phenylalanine), separated by approximately 30 Å. These girdles are thought to interact with the interface region of the lipid bilayer and, thus, provide an approximate indication of how to position the protein. Another approach is to treat the protein as a rigid body and optimize the transfer free energy between water and a hydrophobic slab that represents the location of the lipid bilayer [18–20]. Some of these methods have been developed into web-servers where one can obtain a pre-oriented inserted protein. Toward the end of the chapter, we discuss how this issue may also be addressed via a CG simulation approach. If one is not interested in the interaction of the protein with the lipid at the particle level at all, then a more suitable approach may be to use an implicit membrane, which is discussed in detail in Chap. 10.

3 Methods

There are obviously two main components to a membrane protein simulation: the actual protein and the lipid bilayer in which it is to be embedded. Although we focus on the issues concerning the whole system, it is worth briefly reviewing practical considerations for these individual components.

3.1 Preparation of the Protein

Typically, the starting point for the protein will be a structure deposited in the protein data bank (PDB; www.rcsb.org). Often, however, these structures will need a certain amount of preparation before production level MD can be run. The most severe of these considerations might be missing atoms, which can range from entire loops to a couple of side chain atoms. How one deals with this problem depends on the question one is trying to address with the simulation. For the case in which only a few atoms are missing from a small number of side chains, one can manually build in the missing atoms using an interactive modeling program such as PyMOL [21] or What-If [22] (see **Note 1**). For the more complicated case in which whole loops are missing, typically, one has to resort to programs that can build random structures that are geometrically correct, such as Modeller [23]. Indeed, in some cases, it may be that construction of an entire homology model is required (Chap. 11). Another related consideration is how to deal with the termini in the structure. Frequently, the structure is not the whole sequence of the protein, and, therefore, charged termini may not be appropriate. One common procedure has been to build on capping groups that help to best mimic the continuing protein chain (see **Note 2**). A simpler approach involves simply protonating the C terminus and deprotonating the N terminus.

In all but the very high-resolution structures, one will still have to add hydrogen atoms because these will not be present in the PDB file. Although this is a very simple process, there are decisions to be made even for this process: 1) the choice of force field—all atom versus a united-atom model in which only polar hydrogens are explicit, and 2) the protonation states of ionizable side chains. United-atom force fields will give the benefit of reduced computational effort because of reduction in the number of particles, but all-atom models might be preferred in some cases in which greater accuracy is required. Various programs exist to calculate the pK_a of ionizable side chains (PROPKA [24], H++ [25], and WHAT IF [22]). Most rely on calculating an estimate of the free energy (via the thermodynamic cycle) of protonating the residue within its proteinaceous environment. It may be the case that the protonation state is not important, in which case, default ionization states at pH 7.0 are assumed. However, there are examples in which the protonation state may be critical, as exemplified by the protonation state of Glu71 in KcsA [26–29]. The position of the hydrogens on histidine residues should also be considered carefully, usually by simple visual inspection to optimize local hydrogen bonding.

Finally, although solvation of the system is generally automated, oxygen atoms from water molecules are often included in protein crystal structures. These reflect low-energy minima for a water molecule and, thus, it is usual to include these before "bulk solvation." Deciding whether a water molecule belongs to the subunit of interest from the PDB file is a problem and usually one simply chooses an arbitrary cut-off within which to include these crystallographic waters in the simulation. A cut-off that we have used in the past has been 4 Å [30–32]. The remainder of the simulation box is solvated with pre-equilibrated water boxes. Although we are typically interested in the protein–lipid interactions, it is important to have adequate solvation of the entire protein (see **Note 3**).

3.2 Preparation of the Lipid

Not all of the methods below rely on a preformed lipid bilayer. However, one invariably will need a lipid-only system for control purposes, therefore, simulation of the pure system should be done at some point. The simulation of lipid bilayers has matured during the past 15 to 20 years and it is beyond the scope of this chapter to cover this in depth. The interested reader is referred to several excellent reviews [33–36]. Some groups have generously made equilibrated conformations of some lipid bilayer systems freely available (see **Note 4**), which provide a good starting point for the main procedure we outline below in Sect. 3.3. Sometimes it will be necessary to generate a new lipid bilayer system from scratch, and one then needs a measure of how stable or good that pure system is before proceeding to insert a protein into it. The most commonly used measure of equilibration or stability is to analyze the mean area per lipid; a quantity for which there frequently exists experimental data with which to make a direct comparison. Furthermore, if this is incorrect, then it is likely that most other properties will also be inaccurate [16].

3.3 Setup of the Protein in the Membrane

We will focus here on methods currently used in our laboratory, but it is worthwhile to briefly mention alternative methods. Earlier setup methods were developed with the limitations imposed by available computer power at the time. The approach adopted by Woolf and Roux was to build up lipids around the protein by placing isolated lipid molecules randomly selected from a library of 2,000 conformations. The system was adjusted to remove as many overlaps as possible, followed by a constrained minimization procedure [37, 38]. Although one does not start from a preformed lipid bilayer in this case, the conformations in the library will be derived from a simulation of pure lipids.

A different approach was proposed by Faraldo-Gómez and colleagues [39], who used preformed lipid bilayers as the starting point. Their method relies on creating a cavity in a pre-equilibrated lipid bilayer using the solvent-accessible surface area of the protein as a template. Lipid molecules whose head groups fall within the volume are removed while remaining lipids are subjected to an ever-increasing force acting perpendicular to the surface of the cavity template until the cavity is empty. The protein itself can then simply be inserted into the cavity. This method has the advantage that a pre-existing lipid bilayer can be used and in such a way that non-interfacial lipid molecules are not significantly perturbed, which results in a faster equilibration time.

With the advent of more powerful computers, a more direct approach to the setup has become possible, and we will focus on this approach. We have implemented this procedure using GROMACS [40] in combination with VMD [41], but there is also a plug-in available at http://www.ks.uiuc.edu/Research/vmd/plugins/membrane/ to do the entire process in VMD. This plug-in, however, is more useful if you intend to use NAMD [42] as your MD program. Both methodologies exploit the fact that enough simulation time is available to adequately equilibrate the system. The procedure is very interactive and can be summarized by the following steps:

1. Obtain a pre-equilibrated lipid bilayer (see **Note 4**)
2. Align protein in the lipid bilayer

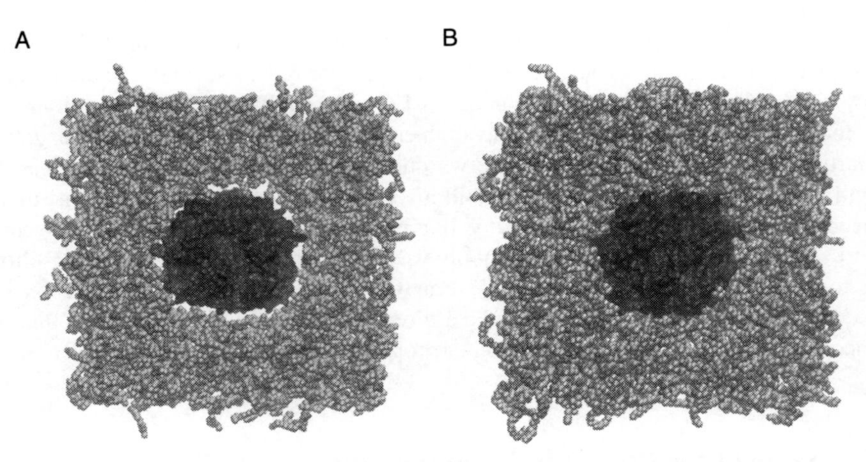

Fig. 1 **(A)** The protein BtuB (*dark molecular surface*), embedded in the bilayer after the removal of overlapping lipids (only protein and lipid are shown in this figure for clarity). Lipid atoms are shown as van der Waals spheres. During the course of the equilibration phase, lipid molecules will move in around the protein as shown in **(B)**, which is an equilibrated system

3. Remove overlapping lipid molecules
4. Equilibrate new system

The alignment of the protein with the pre-equilibrated lipid bilayer in this process is essentially something that is performed by eye. As mentioned in the introduction, some guidance is afforded by the presence of the tryptophan/tyrosine girdles that are associated with membrane proteins. Removal of whole overlapping lipid molecules means that the resulting system will have a vacuum in between the protein and the lipid molecule (see Fig. 1a). During the first stage of the equilibration, this will be removed as the lipid molecules relax around the protein. Typically, for this stage, it is important to keep the protein conformation as close to the starting coordinates as possible. Thus, it is common for positional restraints to be imposed on the protein atoms (or a subset thereof) during this stage. An NPT ensemble (see **Note 5**) MD simulation is then performed to allow the lipid molecules to equilibrate around the newly inserted membrane protein (Fig. 1b). During this stage, water may penetrate slightly into the vacuum between lipids and the protein. These will be expelled during the course of the simulation as the lipids move toward the protein and the system equilibrates. The length of this equilibration phase is usually determined by monitoring the area per lipid as a function of time. After a period of time (typically between 1 and 3 ns for systems with 512 lipids), one should see this plateau. This value can be checked against experimental data. Before unconstrained production or further dynamics can be performed, it is best to allow the protein to relax in stages. There are many different approaches reported in the literature, which can seem to be rather subjective, but the underling philosophy is to work back from the backbone of the protein (see **Note 6**).

3.4 An Alternative CG Method

A more recent approach has been to simulate the whole system *de novo* from a random arrangement of molecules in the system. Such an approach is made possible via the use of the CG methods [15] in which small groups of atoms (typically four) are treated as single particles. Because of the associated reduction in the number of particles, much longer time and length scales can be addressed. This presents the opportunity to simulate large-scale changes in protein–lipid interactions, such as membrane protein bilayer insertion. Thus, at that point, decide whether the problem requires a switch to a fully atomistic description or whether the CG description is adequate. There are currently ongoing efforts to integrate multiscale methods into a self-consistent representation [43–45]. A full discussion of these methods is not possible here, but it seems likely that these methods offer the greatest potential in terms of flexibility across time and length scales for membrane systems.

CG simulations have previously proved useful in modeling the dynamics of lipids and detergents [46–51], DNA [52], proteins [53], and "toy" peptides [54–56]. A semiquantitative model for lipid systems was devised based on thermodynamic data as well as structural and dynamic properties of atomistic simulations [57]. We recently adapted this model for application to membrane proteins [58, 59], and a similar model has been developed for related systems [60]. In our CG model, instead of representing every atom in a protein or lipid molecule, approximately four atoms are grouped together into one particle, and are parameterized to capture the hydrophobicity/hydrophilicity, charge, and hydrogen-bonding properties of their constituent atoms. Bonds, angles, and the overall backbone secondary/tertiary structure are preserved through soft harmonic potentials [58].

The initial CG model for a protein is derived by extracting the coordinates for all Cα atoms and selected side chain atoms from the corresponding all-atom protein file. The overall shape and surface area of a lipid or protein molecule is preserved in the model (Fig. 2). Subsequently, lipid molecules taken from a library (derived from, e.g., a pure lipid simulation) are randomly placed in a box containing the protein, before solvation with a pre-equilibrated box of water particles, and neutralizing ions (Fig. 3a). A number of factors must be considered when solvating the protein. First, as with atomistic models, the number of lipid molecules should be such that the bilayer formed is sufficiently large enough to allow plenty of space between the embedded protein and its periodic image within the membrane plane (see **Note 3**). This number may be estimated by considering the equilibrium area per lipid of interest. Second, a ratio of CG water particles to lipid molecules of approximately 10 to 25 should be used to favor the formation of a bilayer, rather than, e.g., hexagonal, micellar, or other nonlamellar phases [57].

Once the system has been prepared, a short energy minimization procedure is carried out, before one or more production runs are performed. A typical such CG simulation will normally be approximately two orders of magnitude faster than the corresponding atomistic simulation, as a consequence of the reduced number of particles, softer potentials (and, thus, longer MD integration timestep), and consideration of only short-range nonbonded interactions. Our experience with approximately

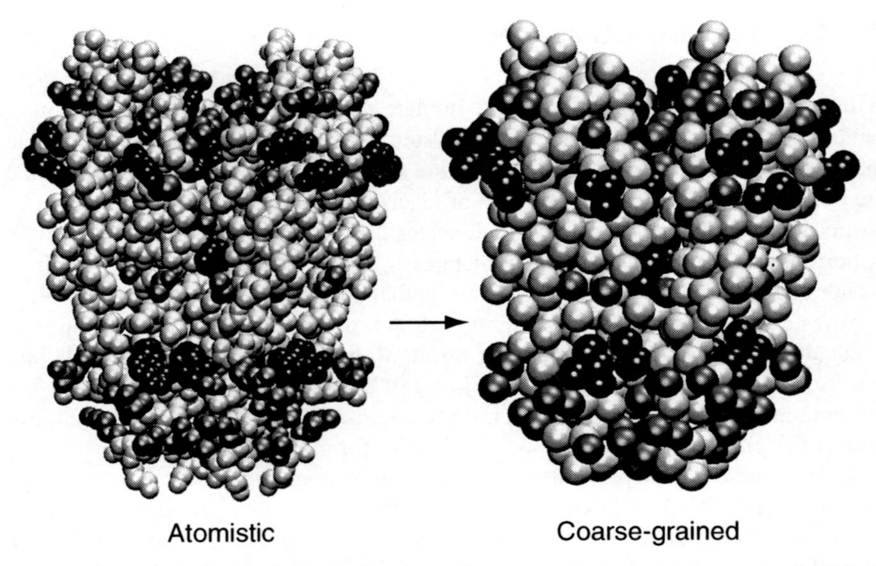

Fig. 2 Illustration of the how the atomistic model translates into the CG model for the KcsA potassium channel. Aromatic particles are shown as *black* van der Waals spheres, hydrophobic or backbone particles are shown in *light grey*, and polar/charged particles are shown in *dark grey*

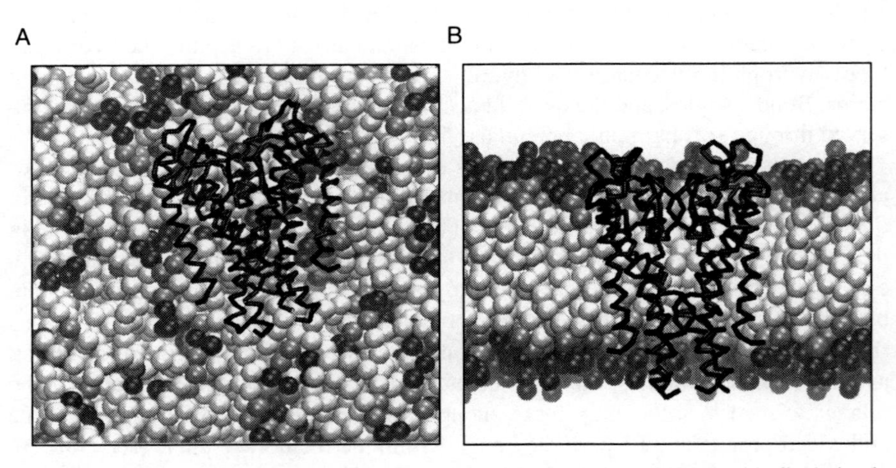

Fig. 3 **(A)** The random starting configuration of the CG simulation of KcsA with dipalmitoyl-phosphatidylcholine (DPPC). KcsA is drawn as a *black backbone trace*. Lipid acyl chain particles are drawn as *light grey* van der Waals spheres, glycerol backbone particles are show in *dark grey*, and lipid head groups (including the phosphates) are drawn as *black spheres*. Water molecules are not shown for clarity. **(B)** The configuration after 200 ns, which clearly shows that the system has evolved into a bilayer arrangement with KcsA embedded within it

40 different membrane proteins suggests that a period of approximately 0.1 to 0.2 μs is normally sufficient to allow the self-assembly of a phospholipid bilayer around the protein of interest. For a typical system of approximately 6,000 CG particles, this translates to a CPU time of approximately 1 to 2 days on a typical workstation computer [58]. The simulation proceeds via an initial assembly of lipids into a continuous lamellar phase, with a lipid "stalk" that bridges between the bilayer and its periodic image. Eventually this "stalk" is broken to form a defect-free bilayer with the membrane protein correctly inserted (Fig. 3b).

This CG method for bilayer insertion has been successfully applied to a number of membrane proteins, and shown to agree in terms of lipid–protein interactions with extended atomistic simulations of an eight-stranded β-barrel, OmpA; a transmembrane α-helical dimer, Glycophorin A [58]; and a 12-transmembrane α-helix bundle, LacY [59]. It has also been extensively tested against a number of α-helical membrane peptides and proteins and shown to be in good agreement with experimental data in terms of orientation within the bilayer [59]. The primary advantage of the CG approach to membrane insertion is the elimination of the need for user input, e.g., using the aromatic girdles to guide placement. This is particularly useful for proteins that might be tilted with respect to the bilayer normal, such as the Vpu α-helical fragment from HIV-1 [59]. This is also true of proteins that are nonuniform in their transmembrane distribution, e.g., the coat protein from fd phage, which contains an amphipathic in-plane helix thought to reside at the membrane/water interface [59]; monotopic proteins that sit on the surface of the bilayer; or proteins with large extracellular regions, such as the multidomain ABC transporter family. Finally, self-assembly simulations of complex, atypical membrane proteins, such as the highly charged voltage sensor domain from a potassium channel, reveal that considerable local bilayer deformation may be necessary for insertion, rather than a bilayer of fixed and uniform thickness surrounding the protein [61].

Once the protein has stably inserted within the membrane, it may be desirable to convert from the CG representation back to an atomistic level of detail. The most obvious method for achieving this is to use the CG results as a "rough guide" for positioning the atomistic protein into a bilayer via the method detailed above (Sect. 3.3) involving placement into a pre-equilibrated bilayer before removal of overlapping lipid molecules. For example, the peaks in densities of the head groups along the membrane normal in the atomistic preformed bilayer may be matched up with the CG system, before least-squares fitting the atomistic protein Cα atoms onto the backbone of the CG model, or, alternatively, using homology modeling techniques. For proteins that are more complex, such as those that are significantly tilted or that induce local bilayer deformation, a more direct matching of atomistic and CG coordinates may be necessary. Therefore, a method is currently being developed whereby atomistic lipid structures from a pure lipid simulation library are iteratively least-squares fitted onto their CG lipid counterparts, with each of the best matching molecules being retained. Thus, an atomistic bilayer resembling the CG system is gradually built up around the atomistic protein model, which is again obtained by fitting the Cα atoms onto the backbone of the CG protein.

3.5 Running the Simulation

The last step is to actually run your atomistic simulation. The primary emphasis has been on using parameters and ensembles that best reproduce the properties of lipid bilayers in the absence of proteins. A full review of these considerations is beyond the scope of this chapter, but the interested reader is referred to several recent articles that discuss sources of error and the best choice of parameters in membrane simulations [16, 62–64].

There are many properties that one could check in the simulation, but probably the most useful is the area per lipid, which gives an indication of molecular packing and the membrane fluidity. It is also a property that is sensitive to simulation setup while also being a reasonably reliable indicator that other properties will also be correct. It is important to remember here what your question is—large undulations across large membrane patches will require much longer simulation time than a study of water–head group interactions, for example.

Finally, there are practical considerations, such as disk space and storage of very large trajectories (see **Note 7**), a problem that is presumably going to parallel the increase in computer power.

4 Conclusions

We have discussed two approaches that can be used to set up and perform MD simulations of membrane proteins. The advantage of the first atomistic approach is that it is easy to use and generally applicable. A disadvantage of this approach is that, to some extent, it depends on a subjective positioning of the protein within the bilayer in terms of its overall tilt and its disposition along the bilayer normal. The second approach, via the use of CG methodologies, allows one to circumvent these problems. The combination of both of these methodologies allows one to explore a wide range of time and length scales with respect to membrane proteins, and should provide valuable information regarding their structure and function.

5 Notes

1. There is also an online server version of the What-If program (http://www.cmbi.kun.nl:1100/WIWWWI/) that provides useful tools features to rebuild missing atoms in side chains. Stereochemical checking tools are also available at this site (useful if you are starting from a model).
2. Typical capping groups are an acetyl on the N terminus (to give CH_3-CO-NH_2–protein) or amidation at the C terminus (to give protein–CO-NH_2). These can be added with a molecule-building program such as Pymol [21]. These additional

Table 1 Lipid configurations available for download

Principal investigator	URL	Lipids
Scott Feller	http://persweb.wabash.edu/facstaff/fellers/	POPC, DOPC, DPPC, SDPC
Helmut Heller	http://www.lrz-muenchen.de/~heller/membrane/membrane.html	POPC
Mikko Karttunen	http://www.lce.hut.fi/research/ polymer/downloads.shtml	DMTAP, DMP, DPPC
Peter Tieleman	http://moose.bio.ucalgary.ca/index.php?page=Structures_and_Topologies	DPC micelles, POPC, DMPC, DPPC, PLPC

groups are either treated as separated residues (as is case for GROMACS [40]) or as patches (as is the case for CHARMM [65]).

3. In periodic systems (nearly all lipid bilayer simulations will be periodic), it is important to make sure that the parts of the protein that are not in the bilayer are adequately solvated to ensure that the protein near one edge of the box does not "see" itself in the nearest periodic image. To avoid such problems, we have typically set up the system such that there is 10 Å of water between the protein and its nearest box edge.
4. Some groups have made freely available their coordinates of pre-equilibrated lipid bilayers, and these provide a useful start point. Some that are available at the time of writing are summarized in Table 1.
5. NPT refers to the thermodynamic ensemble used. In this case, a constant number of particles (N), constant pressure (P), and constant temperature (T). This allows the volume of the system to change and, hence, the surface area of the lipid, which can then be compared back with experiment as a measure of simulation quality.
6. Typically, the protein is relaxed in steps. For example, there may be a period during which backbone atoms only are constrained, followed by just Cα atoms, followed by no constraints during the production phase of the simulation.
7. An issue that requires constant revisiting is how often one writes simulation frames to the trajectory file. The problem is compounded by two factors: the ever-increasing size of the system that can be reasonably addressed (currently routinely up to 200,000 atoms), and the length of simulation time (of the order of tens of nanoseconds). A reasonable value for atomistic simulations is to write to disk every 5 ps, but, again, this will depend on what question you are trying to address.

Acknowledgements We thank the Wellcome Trust for support and Dr. Jorge Pikunic for the BtuB coordinates and useful discussions.

References

1. Wallin, E., and von Heijne, G. (1998) Genome-wide analysis of integral membrane proteins from eubacterial, archean, and eukaryotic organisms. *Prot. Sci.* **7,** 1029–1038.
2. Terstappen, G. C., and Reggiani, A. (2001) In silico research in drug discovery. *Trends Pharmacol. Sci.* **22,** 23–26.
3. Lemieux, M. J., Huang, Y., and Wang, D. N. (2004) The structural basis of substrate translocation by the Escherichia coli glycerol-3-phosphate transporter: a member of the major facilitator superfamily. *Curr. Opin. Struct. Biol.* **14,** 405–412.
4. Guan, L., and Kaback, H. R. (2006) Lessons from lactose permease. *Annu. Rev. Biophys. Biomol. Struct.* **35,** 67–91.
5. Gether, U., Andersen, P. H., Larsson, O. M., and Schousboe, A. (2006) Neurotransmitter transporters: molecular function of important drug targets. *Trends Pharmacol. Sci.* **27,** 375–383.
6. Ash, W. L., Zlomislic, M. R., Oloo, E. O., and Tieleman, D. P. (2004) Computer simulations of membrane proteins. *Biochem. Biophys. Acta* **1666,** 158–189.
7. Sperotto, M. M., May, S., and Baumgaertner, A. (2006) Modelling of proteins in membranes. *Chem. Phys. Lipids.* **141,** 2–29.
8. Beckstein, O., Biggin, P. C., Bond, P., Bright, J. N., Domene, C., Grottesi, A., Holyoake, J., and Sansom, M. S. P. (2003) Ion channel gating: insights via molecular simulations. *FEBS Lett.* **555,** 85–90.
9. Gumbart, J., Wang, Y., Aksimentiev, A., Tajkhorshid, E., and Schulten, K. (2005) Molecular dynamics simulations of proteins in lipid bilayers. *Curr. Opin. Struct. Biol.* **15,** 423–431.
10. Roux, B. (2005) Ion conduction and selectivity in K(+) channels. *Annu. Rev. Biophys. Biomol. Struct.* **34,** 153–171.
11. Bond, P. J., and Sansom, M. S. P. (2004) The simulation approach to bacterial outer membrane proteins. *Mol. Memb. Biol.* **21,** 151–161.
12. Beckstein, O., Biggin, P. C., and Sansom, M. S. P. (2001) A hydrophobic gating mechanism for nanopores. *J. Phys. Chem. B.* **105,** 12902–12905.
13. Beckstein, O., and Sansom, M. S. P. (2006) A hydrophobic gate in an ion channel: the closed state of the nicotinic acetylcholine receptor. *Phys. Biol.* **3,** 147–159.
14. Arinaminpathy, Y., Biggin, P. C., Shrivastava, I. H., and Sansom, M. S. P. (2003) A prokaryotic glutamate receptor: homology modelling and molecular dynamics simulations of GluR0. *FEBS Lett.* **553,** 321–327.
15. Nielsen, S. O., Lopez, C. F., Srinivas, G., and Klein, M. L. (2004) Coarse grain models and the computer simulation of soft materials. *J. Phys. Cond. Matt.* **16,** R481–R512.
16. Anézo, C., de Vries, A. H., Hoeltje, H.-D., Tieleman, D. P., and Marrink, S. J. (2003) Methodological issues in lipid bilayer simulations. *J. Phys. Chem. B.* **107,** 9424–9433.
17. Ulmschneider, M. B., Sansom, M. S. P., and Di Nola, A. (2005) Properties of integral membrane protein structures: derivation of an implicit membrane potential. *Proteins* **59,** 252–265.
18. Basyn, F., Charloteaux, B., Thomas, A., and Brasseur, R. (2001) Prediction of membrane protein orientation in lipid bilayers: a theoretical approach. *J. Mol. Graph. Model.* **20,** 235–244.
19. Tusnady, G. E., Dosztanyi, Z., and Simon, I. (2005) PDB_TM: selection and membrane localization of transmembrane proteins in the protein data bank. *Nucleic Acids Res.* **33,** D275–D78.
20. Lomize, M. A., Lomize, A. L., Pogozheva, I. D., and Mosberg, H. I. (2006) OPM: Orientations of proteins in membranes database. *Bioinformatics* **22,** 623–625.
21. DeLano, W. L. (2004) The PyMOL molecular graphics system. *DeLano Scientific LLC, San Carlos, CA.*
22. Vriend, G. (1990) A molecular modelling and drug design program. *J. Mol. Graph.* **8,** 52–56.
23. Fiser, A., and Sali, A. (2003) Modeller: generation and refinement of homology-based protein structure models. *Meths. Enzym.* **374,** 461–491.
24. Li, H., Robertson, A. D., and Jensen, J. H. (2005) Very fast empirical prediction and interpretation of protein pKa values. *Proteins* **61,** 704–721.

25. Gordon, J. C., Myers, J. B., Folta, T., Shoja, V., Heath, L. S., and Onufriev, A. (2005) H++: a server for estimating pKas and adding missing hydrogens to macromolecules. *Nucleic Acids Res.* **33,** W368–W71.
26. Luzhkov, V. B., and Åqvist, J. (2000) A computational study of ion binding and protonation states in the KcsA potassium channel. *Biochim. Biophys. Acta* **1481,** 360–370.
27. Ranatunga, K. M., Shrivastava, I. H., Smith, G. R., and Sansom, M. S. P. (2001) Side-chain ionization states in a potassium channel. *Biophys. J.* **80,** 1210–1219.
28. Bernèche, S., and Roux, B. (2002) The ionization state and the conformation of Glu-71 in the KcsA K(+) channel. *Biophys. J.* **82,** 772–780.
29. Cordero-Morales, J. F., Cuello, L. G., Zhao, Y., Jogini, V., Cortes, D. M., Roux, B., and Perozo, E. (2006) Molecular determinants of gating at the potassium channel selectivity filter. *Nat. Struct. Biol.* **13,** 319–322.
30. Arinaminpathy, Y., Sansom, M. S. P., and Biggin, P. C. (2002) Molecular dynamics simulations of the ligand-binding domain of the ionotropic glutamate receptor GluR2. *Biophys. J.* **82,** 676–683.
31. Arinaminpathy, Y., Sansom, M. S. P., and Biggin, P. C. (2006) Binding site flexibility: Molecular simulation of partial and full agonists within a glutamate receptor. *Mol. Pharm.* **69,** 11–18.
32. Kaye, L. S., Sansom, M. S. P., and Biggin, P. C. (2006) Molecular dynamics simulations of an NMDA Receptor. *J. Biol. Chem* **281,** 12736–12742.
33. Mouritsen, O. G., and Jorgensen, K. (1997) Small-scale lipid-membrane structure: simulation versus experiment. *Curr. Opin. Struct. Biol.* **7,** 518–527.
34. Feller, S. E. (2000) Molecular dynamics simulations of lipid bilayers. *Curr. Opin. Coll. Interface Sci.* **5,** 217–223.
35. Scott, H. L. (2002) Modeling the lipid component of membranes. *Curr. Opin. Struct. Biol.* **12,** 495–502.
36. Tieleman, D. P., Marrink, S. J., and Berendsen, H. J. C. (1997) A computer perspective of membranes: Molecular dynamics studies of lipid bilayer systems. *Biochim. Biophys. Acta* **1331,** 235–270.
37. Belohorcova, K., Davis, J. H., Woolf, T. B., and Roux, B. (1997) Structure and dynamics of an amphiphilic peptide in a lipid bilayer: a molecular dynamics study. *Biophys. J.* **73,** 3039–3055.
38. Woolf, T. B., and Roux, B. (1996) Structure, energetics, and dynamics of lipid-protein interactions - a molecular-dynamics study of the gramicidin-A channel in a DMPC bilayer. *Proteins: Struc. Func. Genet.* **24,** 92–114.
39. Faraldo-Gómez, J. D., Smith, G. R., and Sansom, M. S. P. (2002) Setting up and optimization of membrane protein simulations. *Eur. Biophys. J.* **31,** 217–227.
40. Lindahl, E., Hess, B., and van der Spoel, D. (2001) GROMACS 3.0: A package for molecular simulation and trajectory analysis. *J. Mol. Model* **7,** 306–317.
41. Humphrey, W., Dalke, A., and Schulten, K. (1996) VMD—Visual molecular dynamics. *J. Molec. Graph.* **14,** 33–38.
42. Phillips, J. C., Braun, R., Wang, W., Gumbart, J., Tajkhorshid, E., Villa, E., Chipot, C., Skeel, R. D., Kale, L., and Schulten, K. (2005) Scalable molecular dynamics with NAMD. *J. Comp. Chem.* **26,** 1781–17802.
43. Christen, M., and Van Gunsteren, W. F. (2006) Multigraining: An algorithm for simultaneous fine-grained and coarse-grained simulation of molecular systems. *J. Chem. Phys.* **124,** 154106.1-06.7.
44. Chang, R., Ayton, G. S., and Voth, G. A. (2005) Multiscale coupling of mesoscopic and atomistic-level lipid bilayer simulations. *J. Chem. Phys.* **122,** 244716.
45. Shi, Q., Izvekov, S., and Voth, G. A. (2006) Mixed atomistic and coarse-grained molecular dynamics: Simulation of a membrane-bound ion channel. *J. Phys. Chem. B.* **110,** 15045–15048.
46. Smit, B., Hilbers, A. J., Esselink, K., Rupert, L. A. M., Van Os, N. M., and Schlijper, G. (1990) Computer simulations of a water/oil interface in the presence of micelles. *Nature* **348,** 624–625.
47. Murtola, T., Falck, E., Patra, M., Karttunen, M., and Vattulainen, I. (2004) Coarse-grained model for phospholipid/cholesterol bilayer. *J. Chem. Phys.* **121,** 9156–9165.

48. Stevens, M. J., Hoh, J. H., and Woolf, T. B. (2003) Insights into the molecular mechanism of membrane fusion from simulation: Evidence for the association of splayed tails. *Phys. Rev. Lett.* **91,** 188102.1-02.4.
49. Shelley, J. C., Shelley, M. Y., Reeder, R. C., Bandyopadhyay, S., and Klein, M. L. (2001) A coarse grain model for phospholipid simulations. *J. Phys. Chem. B* **105,** 4464–4470.
50. Goetz, R., and Lipowsky, R. (1998) Computer simulations of bilayer membranes: Self-assembly and interfacial tension. *J. Chem. Phys.* **108,** 7397–73409.
51. Whitehead, L., Edge, C. M., and Essex, J. W. (2001) Molecular dynamics simulation of the hydrocarbon region of a biomembrane using a reduced representation model. *J. Comput. Chem.* **22,** 1622–1633.
52. Tepper, H. L., and Voth, G. A. (2005) A coarse-grained model for double-helix molecules in solution: Spontaneous helix formation and equilibrium properties. *J. Chem. Phys.* **122,** 124906.1-06.11.
53. Tozzini, V. (2005) Coarse-grained models for proteins. *Curr. Opin. Struct. Biol.* **15,** 144–150.
54. Nielsen, S. O., Lopez, C. F., Ivanov, I., Moore, P. B., Shelley, J. C., and Klein, M. L. (2004) Transmembrane peptide-induced lipid sorting and mechanism of Lalpha-to-inverted phase transition using coarse-grain molecular dynamics. *Biophys. J.* **87,** 2107–2115.
55. Venturoli, M., Smit, B., and Sperotto, M. M. (2005) Simulation studies of protein-induced bilayer deformations, and lipid-induced protein tilting, on a mesoscopic model for lipid bilayers with embedded proteins. *Biophys. J.* **88,** 1778–1798.
56. Lopez, C. F., Nielsen, S. O., Moore, P. B., and Klein, M. L. (2004) Understand nature's design for a nanosyringe. *Proc. Nat. Acad. Sci. USA* **101,** 4431–4434.
57. Marrink, S. J., de Vries, A. H., and Mark, A. E. (2004) Coarse grained model for semiquantitative lipid simulations. *J. Phys. Chem. B* **108,** 750–760.
58. Bond, P. J., and Sansom, M. S. P. (2006) Insertion and assembly of membrane proteins via simulation. *J. Am. Chem. Soc.* **128,** 2697–26704.
59. Bond, P. J., Holyoake, J., Ivetac, A., Khalid, S., and Sansom, M. S. P. (2006) Coarse-grained molecular dynamics simulations of membrane proteins and peptides. *J. Struct. Biol.* **157,** 593–605.
60. Shih, A. Y., Arkhipov, A., Freddolino, P. L., and Schulten, K. (2006) Coarse grained protein-lipid model with application to lipoprotein particles. *J. Phys. Chem. B* **110,** 3674–3684.
61. Bond, P. J., and Sansom, M. S. P. (2006) Bilayer deformation by the Kv channel voltage sensor domain revealed by self-assembly simulations. *Proc. Nat. Acad. Sci. USA* **104,** 2631–2636.
62. Patra, M., Karttunen, M., Hyvönen, M. T., Falck, E., Lindqvist, P., and Vattulainen, I. (2003) Molecular dynamics simulations of lipid bilayers: major artifacts due to truncating electrostatic interactions. *Biophys. J.* **84,** 3636–3645.
63. Patra, M., Karttunen, M., Hyvönen, M. T., Falck, E., and Vattulainen, I. (2004) Lipid bilayers driven to a wrong lane in molecular dynamics simulations by subtle changes in long-range electrostatic interactions. *J. Phys. Chem. B.* **108,** 4485–4494.
64. de Vries, A. H., Chandraskhar, I., van Gunsteren, W. F., and Hunenberger, P. H. (2005) Molecular dynamics simulations of phospholipid bilayers: Influence of artificial periodicity, system size, and simulation time. *J. Phys. Chem. B.* **109,** 11643–11652.
65. MacKerrell, A. D., Bashford, D., Bellott, M., Dunbrack, R. L., Evanseck, J. D., Field, M. J., Fischer, S., Gao, J., Guo, H., Ha, S., Joseph-McCartney, D., Kuchnir, L., Kuczera, K., Lau, F. T. K., Mattos, C., Michnick, S., Ngo, T., Nguyen, D. T., Prodhom, B., Reiher, W. E., Roux, B., Schlenkrich, M., Smith, J. C., Stote, R., Straub, J., Watanabe, M., Wiorkiewicz-Kuczera, J., Yin, D., and Karplus, M. (1998) All-atom empirical potential for molecular modeling and dynamics studies of proteins. *J. Phys. Chem. B* **102,** 3586–3616.

Chapter 9
Membrane-Associated Proteins and Peptides

Marc F. Lensink

Summary This chapter discusses the practical aspects of setting up molecular dynamics simulations for membrane-associated proteins and peptides. Special emphasis lies on the analysis of such systems. The main focus is the association between a cationic peptide and an anionic lipid bilayer—a peptide/lipid–bilayer system—but the extension onto more complicated systems is discussed. Topology files for selected lipids and several new analysis tools relevant for protein–membrane simulations are presented, the most important ones of which are: g_helixaxis, to calculate the axis of a helix and its angle with the bilayer; g_arom, to calculate aromatic order parameters; and g_under, to calculate which lipids interact with the protein. A procedure is explained to calculate properties involving peptide-interacting lipids only, as opposed to all lipids.

Keywords: Cell-penetrating peptide · GROMACS · Helix axis · Molecular dynamics · Order parameter · Penetratin · Peptide–lipid interaction · Peptide-membrane association · POPA · POPC · POPG · POPS

1 Introduction

Membrane-associated proteins are relatively little studied by modeling or simulation techniques because of the traditional difficulty in getting high-quality starting structures and the related underrepresentation in the Protein Data Bank [1] (PDB), despite their critical contribution to cell functioning. The simulation of such systems is not necessarily more involved than any other simulation, but the lipid bilayer adds a complexity that requires an elevated level of bookkeeping to keep the analysis of the simulations tangible. On the other hand, the normal to the bilayer plane—especially when it aligns with one of the primary axes—offers an easy frame of reference. Aspects of the simulation setup and analyses will be demonstrated using a peptide–bilayer (as opposed to protein–bilayer) system, more specifically, the association of

From: Methods in Molecular Biology, vol. 443, Molecular Modeling of Proteins
Edited by Andreas Kukol © Humana Press, Totowa, NJ

the cationic helix penetratin to a partially negatively charged bilayer. Please refer to other chapters in this volume for simulation setup and analysis details that are not specific for the simulation of membrane-associated peptides.

The main sections of this chapter are system setup and analysis. In the system setup section, I show how to create a new lipid topology from an existing one by varying only the head group function, how to incorporate these new lipids into an existing (and equilibrated) bilayer, how to add one (or more) peptide(s) in the solvent phase of this bilayer system, and how to mutate residues (to alanine) without solvating the hydrophobic core of the bilayer. In the analysis section, the analysis of the simulation of these systems is discussed and several new tools are presented. Both sections end with a downloadable files subsection. All files relevant for this chapter are freely available to the scientific community.

1.1 Penetratin

The 16-residue helix named penetratin originates from the Antennapedia homeodomain of Drosophila melanogaster. Homeodomain proteins are DNA-binding proteins in which the third helix binds in the major groove (Fig. 1).

This helix, with sequence RQIKIWFQNRRMKWKK (residues 43–58), contains a relatively high number of positively charged residues, and interaction with—specifically the phosphate groups of—a cellular membrane is, therefore,

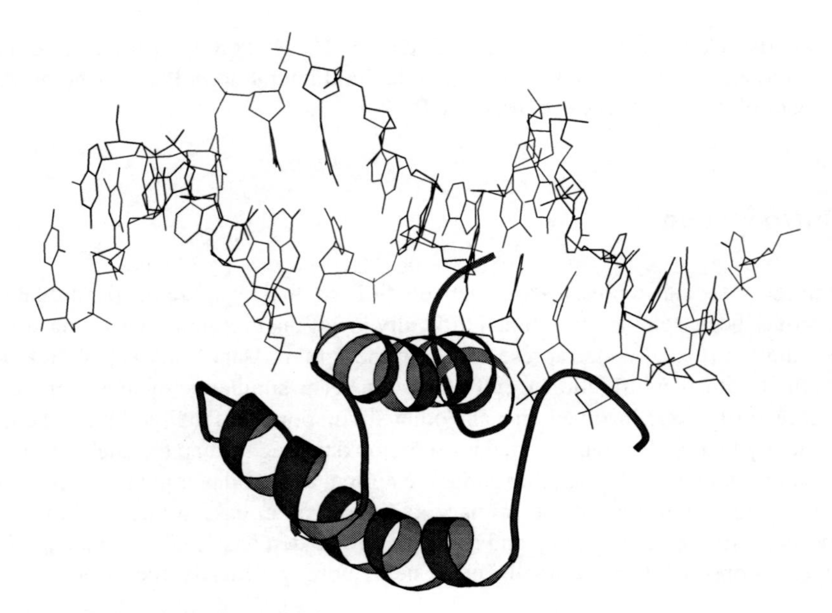

Fig. 1 Homeodomain binding to DNA

not surprising. However, the Antennapedia homeodomain was found to not only bind to, but also be able to translocate across a cellular membrane [2], with its discriminating factor the third helix [3]. This so-called pAntp peptide, or penetratin, was the first member of an increasingly large family of cell-penetrating peptides [4] that have been proposed and reported as carriers for cellular uptake of proteins, oligonucleotides, and drugs [5,6]. Penetratin differentiates itself from most membrane-associating peptides through its non-amphipathic character [7]; the positively charged residues are distributed evenly over the helix, leaving no single side of the helix hydrophobic enough to allow the formation of pores.

The translocation mechanism of penetratin is, to date, not known [8]. Penetratin translocation is thought to be a receptor-independent process [9], but different pathways seem to exist [10] that may be direct [11] or energy dependent [12], involve endocytosis [13], or require a transmembrane potential [14, 15]. Its potential ability to permeate pure lipid bilayers is a matter of controversy and more and more evidence to the contrary is presented [16–18]. Additional simulations, using a transmembrane potential and using the replica exchange molecular dynamics technique, seem to confirm this observation (results not published).

Although the penetratin system is indeed an interesting system to study on its own, the principles that govern the peptide–membrane interaction are easily applicable to protein–membrane systems as well. Structure, orientation, and bilayer–peptide and lipid–residue interaction are all determined by their three-dimensional coordinates and (a combination of) atom–atom contacts. Once the correct correspondence definitions have been set up, traversing the trajectories and processing each frame with the desired analysis function is straightforward (see **Note 1**).

2 Methods

2.1 System Setup

2.1.1 Material

The simulations and analyses presented in this chapter were performed with the GROMACS [19] suite of programs (see **Note 2**), version 3.1.4. Additional UNIX tools such as awk, sed, grep, and cat, as well as programs written by the author of this chapter, using the GROMACS C programming libraries, are used. The latter programs are made available to the scientific community (see **Note 3**). The principles of the simulation setup and analysis explained in this chapter apply to any simulation package, and the presented analysis programs can also be used to analyze simulations done by other simulations packages.

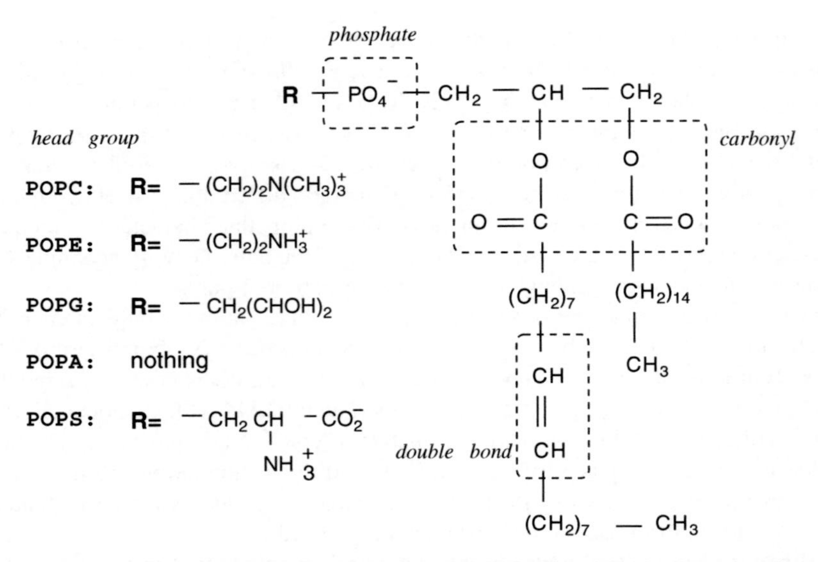

Fig. 2 Structure of several of the members of the POPx family of lipids. Structure of several of the members of the 1-palmitoyl-2-oleoyl-*sn*-glycero-3-phospho-family (POPx) of lipids. The different head groups determine the overall charge of the molecule: POPC and POPE are effectively neutral, whereas POPG, POPA, and POPS carry a negative charge. The dashed boxes indicate commonly defined groups of atoms. These definitions are also used throughout the text. A typical calculation involving these groups would be the calculation of their density throughout the bilayer, thus, investigating the bilayer structure

2.1.2 Lipid Bilayer

As a starting point, a 128-lipid bilayer of POPC molecules was taken, and equilibrated in the GROMOS96 43a2 force field [20] with modified lipid parameters [21]. The bilayer structure and the topology files needed for simulating this system are freely available (see **Note 4**). Figure 2 shows the structure of a selected set of lipids belonging to the glycerol family that have both a 15-carbon 1-palmitoyl chain and a 17-carbon 2-oleoyl chain containing a single double bond in the middle. The head group function determines the specifics of the lipid as well as its overall charge.

To obtain a physiologically relevant bilayer, a portion of the neutral lipids can be replaced by negatively charged ones. This requires modification of the bilayer structure and the lipid topology.

2.1.3 Modification of Lipid Topology

The creation of force field parameters is a field in its own. Here, I will use the principle of parameter transfer and copy parameters from well-calibrated residue parts onto the newly created lipid head group function to make a POPG lipid topology from the existing POPC topology files. Topology files for POPA and POPS are also available (see **Note 3**). The only varying entity is the head group, both in the topology file and the structure. This means that all coordinates, bond definitions, and interaction terms of the glycerol group can simply be copied to the new lipid definition.

1. Remove all atoms, bonds, pairs, angles, etc. involving the head group from the POPC lipid topology file.
2. Increase or decrease all references to atom numbers in what remains of this topology file by the difference in number of atoms of the head group function (see **Note 5**).
3. Add the new head group function to the section listing the atoms, renaming lipid atoms to avoid overlap (see **Note 6**).
4. Add the parameters for the new head group function, copying from chemically equivalent groups in residues (see **Note 7**).
5. Add bonded terms (bond, angle, dihedral, and 1-4 pair interaction definitions) for the atoms involving both the phosphate and head group (see **Note 8**).
6. Perform a simulation in a vacuum to check the correctness of the newly created topology (see **Note 9**).

2.1.4 Modification of Bilayer Structure

Any number of lipids can be replaced from the base POPC bilayer structure, as long as the varying entity—in this case, the head group—occupies more or less the same space. The following procedure will replace eight POPC molecules at either side of the bilayer by POPG (see **Note 3**).

1. Choose, at random, eight lipids on either side of the bilayer.
2. Replace these lipids in reverse order:
   (a) Fit (least-squares root mean square deviation [RMSD]) the phosphate groups of POPG and POPC, correctly aligning the oxygens that point toward the head group and the acyl chains.
   (b) Extract the head group coordinates from the fit and the phosphate and tail coordinates from the reference.
   (c) Rename the tail atoms to correspond to the previously created POPG topology.
   (d) Copy the coordinates of remaining lipids, and add new head, phosphate, and tail coordinates, with POPG as the residue name.
3. Restore the original box, and reorder the file: POPC first, then POPG, then SOL (see **Note 10**).

2.1.5 Incorporation of Peptide Structure

Although the structure of the third helix in the homeodomain is α-helical, this need not be the case in solution, or when interacting with a lipid bilayer. Previous investigations have shown a random structure in solution [22]. In contrast, binding to lipid bilayers promotes the formation of secondary structure [23], both α- or β-like, through backbone hydrogen bond shielding [24], and a decreased dielectric constant in the membrane core [25]. Although an increase in both peptide concentration or negatively charged lipid content was shown to induce an $\alpha \rightarrow \beta$ conversion [26], at high lipid-to-peptide ratio and low negatively charged lipid concentration, penetratin was shown to be α-helical [27].

The following procedure will place the penetratin helix at a distance of approximately 2 nm above the bilayer, horizontally aligned with the bilayer surface (see **Note 11**). The solvation step will add solvent molecules both around the helix and in any vacuum that was formed by the replacement of lipids (previous section) without placing water in the hydrophobic core of the bilayer.

1. Extract the coordinates of the third helix from PDB entry 1ahd, residues 43 to 58.
2. Create a topology for the helix, capping the C terminus with an amine group (see **Note 12**).
3. Rotate the helix to align itself perpendicular to the z-axis, i.e., the bilayer surface, and increase its z coordinates by approximately 10 nm (see **Note 13**).
4. Add the helix coordinates to the solvated bilayer box.
5. Calculate the minimum box size needed, combine this z-axis of this box with the x- and y-axes of the original bilayer (see **Note 14**).
6. Remove counterions and solvate the box, i.e., the resulting vacuum around the peptide and wherever modified lipids have a smaller head group volume.
7. Remove added waters that have their z coordinate in between the membrane layer (see **Notes 15** and **3**).
8. Add counterions to make the system electrostatically neutral.
9. Combine the topology files for the peptide, POPC, POPG, solvent, and counterions (see **Note 16**).

Figure 3 shows a representation of the resulting peptide–bilayer system. The same procedure can be repeated to incorporate additional peptide molecules (see **Note 17**). Overlap of these molecules with solvent is not a problem because the solvation step in the above procedure will eliminate these water molecules.

2.1.6 Mutation of Selected Residues to Alanine

Single-residue mutations are often used in biochemical experiment because a change in reactivity can be directly associated with a single residue. Likewise, mutation in simulation can provide an important insight into the physical interactions that lie at the basis of this reactivity.

Mutating selected residues to alanine is a straightforward step that only involves deletion of the side chain. This results in a local vacuum that can be filled by a few water molecules, depending on the size of the initial side chain. By allowing only water molecules to be inserted that show overlap with the original side chain, one avoids the step of having to remove membrane-inserted water molecules again. The

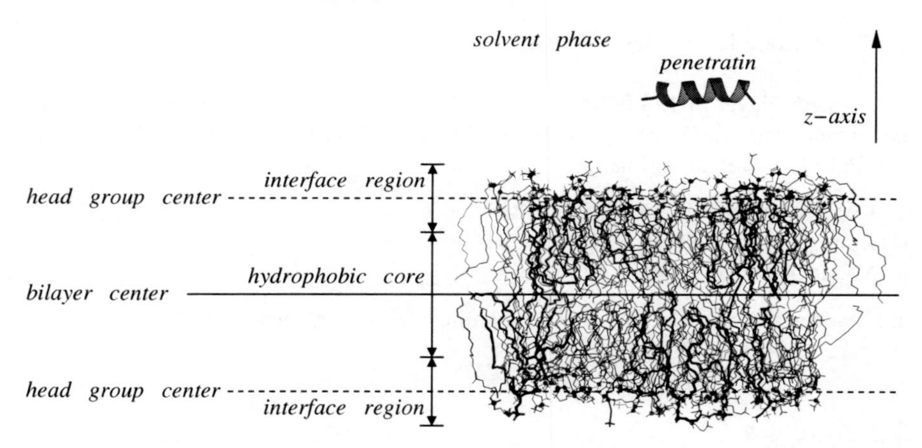

Fig. 3 Initial penetratin–bilayer structure. Neutral POPC and negatively charged POPG lipids are displayed in *thin* and *thick* wireframe, respectively. The helix is displayed as a ribbon structure. Solvent and counterions are not shown. The head group centers, defined as the average coordinate in the z direction (perpendicular to the bilayer surface) of all lipid phosphorus atoms of each bilayer half, represent the center of the interface regions. The center of the hydrophobic core of the membrane is the average of these

easiest approach to this is a two-step procedure, extracting in a first step the added solvent molecules and adding them manually in a second step. This procedure is especially desirable when mutating residues that are located in the bilayer interface region.

1. Pass one:

   (a) Convert the system box to force all atoms to be in the box. This will break molecules but it is easier to visualize the overlap with added water molecules (see **Note 18**).
   (b) Mutate residues by removing their side chain and renaming the residue name of the remaining atoms (see **Note 3**).
   (c) Solvate the box, copy the added solvent molecules to the unmutated system box.
   (d) Extract the solvent molecules that overlap with the original unmutated side chain.

2. Pass two:

   (a) Mutate the residues in the original box, but keep the molecules whole (see **Note 19**).
   (b) Add the selected solvent molecules from the last step of pass one.
   (c) Add counterions if charged residues were mutated.

3. Create a new topology (see **Note 20**).

2.1.7 Downloadable Files

A solvated bilayer of 128 POPC lipids, and topology files for POPC and POPE can be downloaded from http://moose.bio.ucalgary.ca/. Topology files for POPG, POPA, and POPS, as well as scripts to:

- Replace selected POPC lipids
- Remove unwanted water molecules
- Mutate selected residues into alanine

are made available at http://www.scmbb.ulb.ac.be/Users/lensink/lipid/.

2.2 Analysis

The dynamics of a protein in equilibrium may be severely modified by the presence of an external influence. Such external forces may occur on membrane binding and/or ligand release and can trigger a cascade of events that are characterized by slow-motion displacements of secondary structure elements. The dynamics of these elements are readily identified from RMSD plots after fitting to a common reference frame such as a central β-sheet [28]. Whenever a membrane is present, an additional frame of reference exists in the surface of the bilayer (tacitly ignoring curvature effects). The orientation of a protein with respect to the membrane to which it is binding is especially relevant in the case of peptide–membrane association.

2.2.1 Calculation of Helical Axis, g_helixaxis

The calculation of the axis of one or more helices, together with the normal to the bilayer plane, offers a view onto (inter)-helical displacements in combination with their orientation with respect to the bilayer.

The axis of a helix is best determined using a rotational least-squares fitting procedure, mapping the Cα's of the helix onto itself but one residue out of phase, i.e., residue i is mapped onto residue $i + 1$. A quaternion-based method is used to identify the screw transform (translation along and rotation about the helix axis) that will superimpose the two helices. The rotational least-squares method is fast, accurate, and insensitive to noise and, thus, able to deal well with imperfect helices [29].

The method has been implemented in a program using the GROMACS development and analysis libraries: g_helixaxis (see **Note 3**). It can read trajectory and single structure PDB files and will output for each helix its angle with the z-axis as well as their interhelical angles. Optionally, the initial point and vector components and length of each helix are written, in a format following PDB standards and, thus, easily visualized using standard molecule viewers.

2.2.2 Orientation of Aromatic Residues, g_arom

Whereas the calculation of helix axes helps in determining the dynamics of such secondary structure elements, at a more local level, aromatic order parameters are used (Fig. 4) [30]. The aromatic order parameters S(N) and S(L) are calculated relative to the normal to the bilayer plane, through the formula $1/2\,(3\cos^2\theta - 1)$. S(N) relates to the normal to the aromatic ring, whereas S(L) describes the vector from $C\gamma$ through the ring. θ is the angle between the respective vector and the bilayer normal. For $S = 1$, these vectors are aligned, whereas $S = -1/2$ means orthogonality. Because these vectors cannot both simultaneously be aligned with the bilayer normal, the combinations $S(N) = 1$ and $S(L) = 1$ are mutually exclusive, and an increase in the one induces a decrease of the other. They can, however, both equal $-1/2$, meaning that both vector are orthogonal to the bilayer normal. The mutually exclusive behavior is illustrated in Fig. 5.

Calculation of the aromatic order parameters has been implemented in an analysis tool that can read both trajectories and PDB files: g_arom (see **Note 3**). Detection of aromatic residues is automatic, but it is also possible to select the residues of interest. The vectors $C\gamma \rightarrow \zeta$ for PHE and TYR, and $C\gamma \rightarrow C\zeta_2$ for TRP, are used for the calculation of S(L), whereas the aromatic plane, defined by the atoms $C\gamma$, $C\varepsilon_1$,

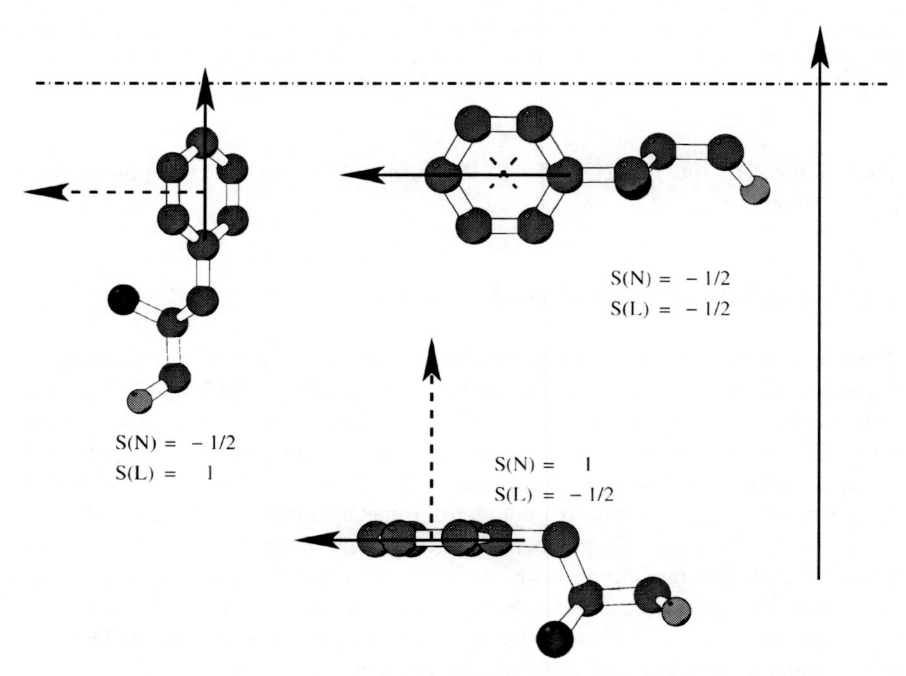

Fig. 4 Aromatic order parameters. Visualization of aromatic order parameters. *Solid* and *dashed arrows* represent S(L) and S(N), respectively. When either arrow is aligned with the normal to the bilayer plane (*long arrow*), the respective order parameter equals 1

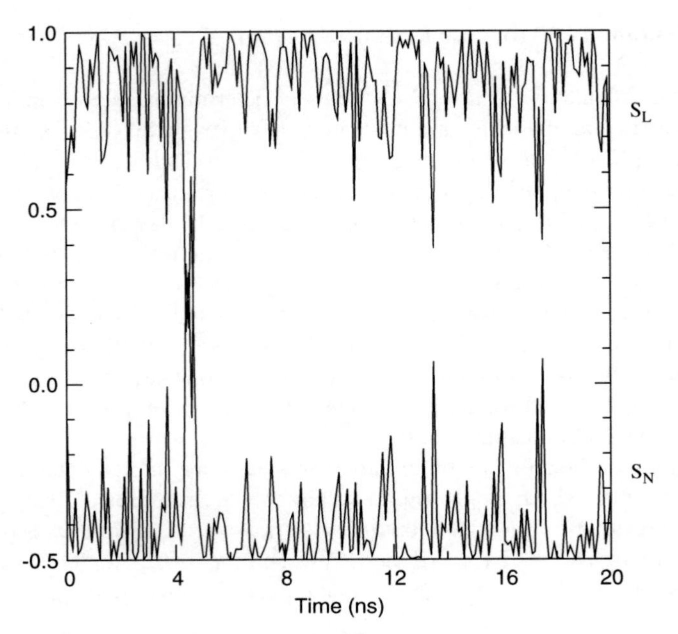

Fig. 5 Semiconcerted behavior of aromatic order parameters. Aromatic order parameters for TRP 48 during a 20-ns simulation of the R53A,K57A double mutant, in which mutation occurred after a 100-ns simulation of the association of penetratin with a bilayer consisting of POPC/POPG lipids in a 7:1 ratio. Notice the concomitant *decrease* in S(L) after an *increase* of S(N)

and Cε_2 for PHE and TYR, Cζ_3 for TRP, determines the normal to this plane, used in the calculation of S(N).

2.2.3 Lipids Interacting with Peptide, g_under

The interaction between a peptide or protein and a lipid bilayer is not a static quantity that can be defined as the interaction between residues X and Y and 1, 2, 3, or N specific lipids. Although membrane-interacting residues may be likely to sustain this interaction once established, lateral lipid diffusion will take place, replacing individual lipids, much like a water molecule interacting through hydrogen bonding with a protein residue may be replaced a number of times by another water molecule.

In Fig. 3, a positively charged peptide hovers above a negatively charged bilayer. During the peptide-membrane association, the attractive force on the peptide will effectively change from an almost uniformly distributed electric field at large distance into more specific, close distance, atom–atom interactions that include, besides Coulomb, also Lennard-Jones interaction terms.

Here, I present a pragmatic way of defining lipids that interact with an approaching peptide or protein that is purely distance based. This definition [17] is implemented in the program g_under (see **Note 3**). Essentially, lipids that come within a certain cut-off distance of the peptide are defined as interacting with this peptide. This distance can be calculated between any two peptide and lipid atoms, but also be restricted to use only backbone atoms. The distance criteria need not be the same in the *x*, *y*, or *z* direction. Picturing a cylinder with a radius of 0.1 Å and height of 2 nm under every atom of the helix in Fig. 3 will define a lipid as interacting with the peptide when any of its atoms enters this cylinder. This procedure includes lipids that have their acyl chains under, but the head group beside, the peptide (see **Note 21**).

The resulting time-dependent evolution of peptide-interacting lipids can subsequently be used to calculate properties involving these lipids only, as opposed to the entire bilayer or bilayer half.

Calculating Properties of Peptide-Interacting Lipids

Most analysis programs calculate a quantity from the interaction between the atomic coordinates of one set of atoms versus another. If this happens for every frame in the simulation, one obtains the evolution of this quantity in time. It is usually (also in the case of GROMACS analysis tools) not possible to vary one of these two sets of atoms, as one would need, e.g., in the case of peptide-interacting lipids or when studying a shell of water molecules around an active site.

However, when the quantity to be calculated is cumulative, i.e., the quantity can be calculated a posteriori from each individual lipid molecule at the instantaneous time t, e.g., the average *z* coordinate of the phosphorus atoms of the peptide-interacting lipids, or the order parameters, one simply needs to traverse the trajectory and extract—for the first example—the *z* coordinate for every single lipid. Then, in a second step, these can be combined with the list of peptide-interacting lipids to get the evolution of average *z* coordinate of all peptide-interacting lipids during the course of the simulation (see **Note 22**).

Alternatively, one could cut the trajectory in n pieces of each P ps and scan these individually. This has the advantage that, for a lipid that becomes interesting during only a fraction of the simulation, not the entire trajectory needs to be scanned, but only a (very small) part of it (see **Note 23**).

2.2.4 Bilayer Structure

Lipid deuterium order parameters describe the ordering of the lipid acyl chains with respect to the bilayer normal. They can be measured by nuclear magnetic resonance (NMR) experiments, but can also calculated from the lipid tail C-C dihedral angles [31] and are expressed as a scalar value per lipid carbon atom that typically ranges between 0 for disordered and 0.5 for ordered lipid structure. The following example shows the calculation of lipid order parameters for lipids that are interacting with the peptide:

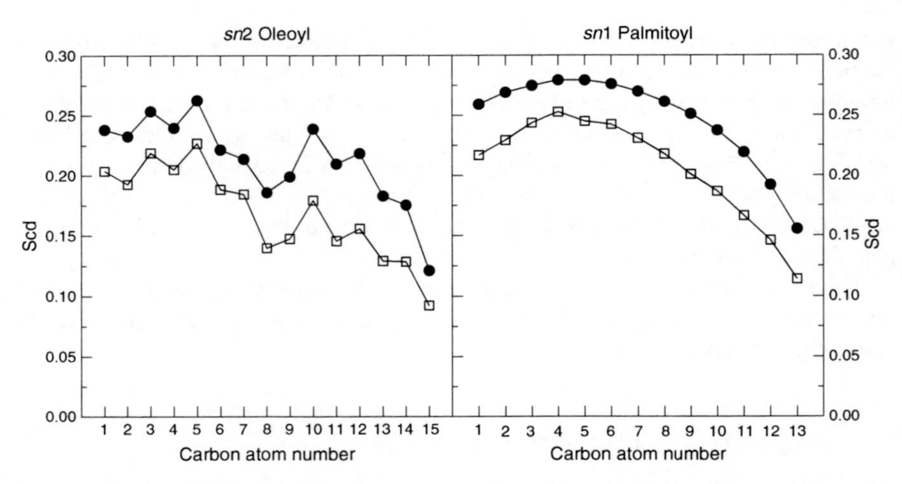

Fig. 6 Calculated lipid deuterium order parameters for a 50-ns simulation of a penetratin molecule bound to a bilayer of 128 lipid molecules in a 7:1 POPC/POPS ratio. *Solid circles* denote order parameters calculated over all lipids, *open squares* are for peptide-interacting lipids only. No difference is made between POPC and POPS lipid molecules

1. Calculate which lipids interact with the peptide for every frame in the trajectory.
2. For each lipid tail:
   (a) Calculate the lipid order parameters for each individual lipid and lipid tail for every frame in the trajectory. These time frames should match the calculation of peptide-interacting lipids.
   (b) For each time frame:
      i. Extract the residue numbers of the peptide-interacting lipids.
      ii. Average the calculated order parameters for the given lipid tail for these lipids at the given time frame.
   (c) Average these averaged order parameters over all time frames.

   The resulting graph is depicted in Fig. 6.

2.2.5 Coordinate Frame in Bilayer Simulations, g_zcoor and g_xycoor

Choosing the appropriate reference axes for the simulation significantly facilitates subsequent analyses because no coordinate transformation is necessary. Taking the example of Fig. 3, the z-axis coincides with the normal to the bilayer surface. Any property involving "distance to bilayer" is calculated in the z-axis only, possibly in relation to the bilayer or head group center. The program g_zcoor (see **Note 3**) does exactly this: extract the (center-of-mass averaged) z coordinate for a combination of molecules and/or atoms.

Combining the extraction of z coordinates relative to the bilayer center with the definition of peptide-interacting lipids produces Fig. 7. It can be seen that the

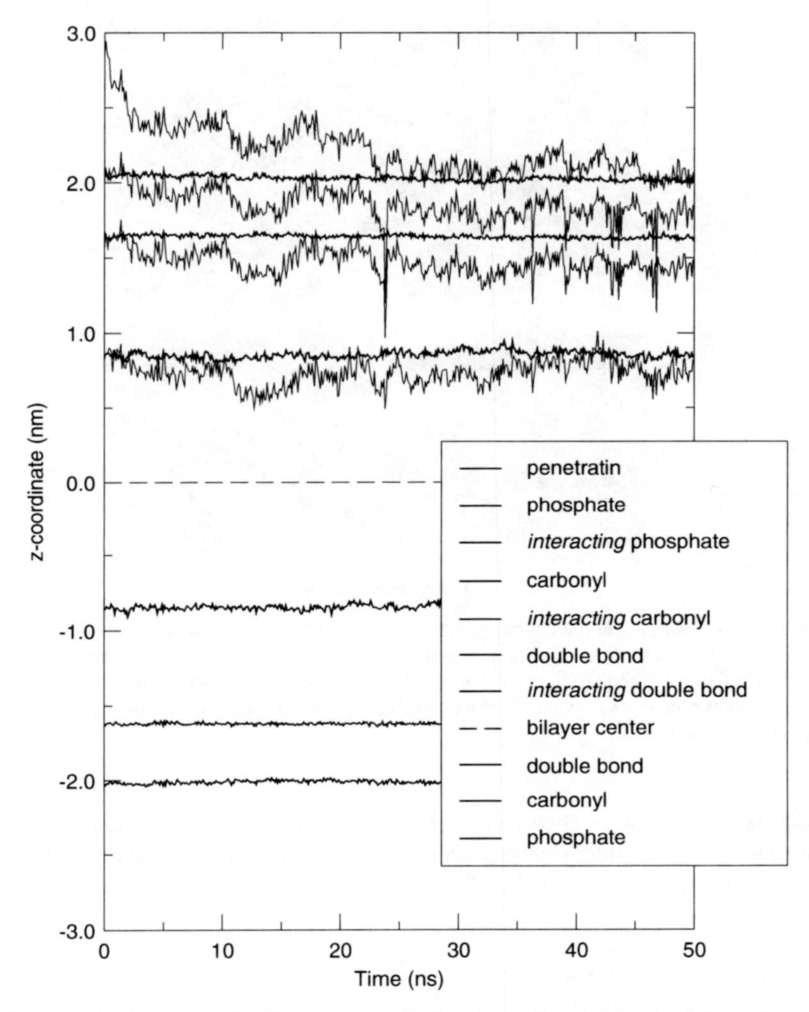

Fig. 7 Average distance to the bilayer center (in the *z* coordinate) for penetratin and selected lipid groups (see also Fig. 3) for a 50-ns simulation of penetratin binding to a bilayer of 128 lipid molecules in a 7:1 POPC/POPA ratio. The legend follows the lines from top to bottom, with *thick lines* drawn for peptide-interacting lipids and a *dashed line* for the bilayer center

association between peptide and bilayer has an immediate effect on the lipids that are defined as interacting with the peptide as they are pushed down toward the bilayer center, whereas other lipids do not seem to be affected.

Orthogonally to this, the *x*- and *y*-axes describe diffusion within the bilayer surface. The program g_xycoor extracts the *x* and *y* coordinates for every single atom in any number of combinations of molecules and/or atoms (see **Note 3**).

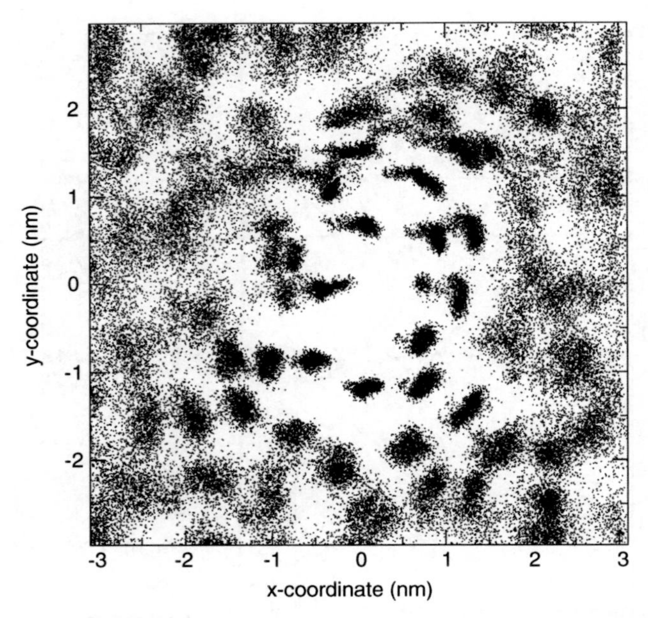

Fig. 8 Diffusion of lipids in the peptide-binding bilayer half, for a 130-ns time frame with a 100-ps increment of a 230-ns simulation of penetratin bound to bilayer of 128 lipid molecules in a 7:1 POPC/POPG ratio. Plotted are the x and y coordinates of the lipid phosphorus atoms, relative to the peptide center-of-mass. Notice the fixating effect that the peptide has on the lipid phosphate groups in its proximity

Modification of these absolute coordinates into relative ones, e.g., relative to a membrane bound peptide, shows the restricted diffusional motion of peptide-bound lipid molecules, as depicted in Fig. 8.

2.2.6 Downloadable Files

The analysis programs:

- g_helixaxis, to calculate the axis of a helix
- g_arom, to calculate aromatic order parameters
- g_under, to calculate which lipids interact with a protein
- g_zcoor, to plot average z coordinates
- g_xycoor, to plot x and y coordinates

are available to the scientific community (see **Note 3**). GROMACS needs to be installed, because these programs dynamically link to the GROMACS libraries, but to be able to use these programs the simulations need not be performed by GROMACS.

3 Notes

1. In the analysis of many different simulations that all consist of almost the same setup but with minor differences, such as a single mutated residue or a certain number of POPC lipids replaced by POPS (or any other), a lot of time can be saved by creating correspondence files that describe which group of atoms of simulation A corresponds to which group of atoms in simulation B. These files can then be parsed in the scripts that call the analysis programs to supply these programs with the correct input.
2. http://www.gromacs.org/.
3. File(s) can be downloaded at http://www.scmbb.ulb.ac.be/Users/lensink/lipid.
4. http://moose.bio.ucalgary.ca/.
5. Renumbering is necessary because the varying entity, the head group, is at the beginning of the topology.
6. Renaming is not necessary as long as newly added atoms have unique names.
7. If chemically equivalent groups are not available for the force field you are using, you will have to go through the whole process of deriving parameters—especially partial charges—from quantum mechanical calculations, following the procedure as described in the literature for that force field. This lies outside of the scope of this chapter.
8. Also here **Note 7** applies, but at this point, the charges are already known. Other parameters are less critical, e.g., angles and dihedrals can be made to follow sp_2 or sp_3 hybridization and bond lengths taken from experimentally determined values (NMR or x-ray). Moreover, in most present-day simulations, bond lengths are constrained.
9. Incorrect topologies will quickly explode or collapse. Check the final structure: if it looks okay, it probably is okay. Remove rotational center-of-mass motion to avoid accelerated spinning. Vacuum simulations should be sufficiently long (on the order of several ns) to allow the dissipation of energy in the limited number of degrees of freedom.
10. The order of the molecules listed in the structure file must correspond to the order of the molecules in the system topology file.
11. Penetratin positioning in lipid membranes or membrane-mimicking systems was found to be with the axis lying in the head group region [32–34]. Moreover, the uniform distribution of positively charged residues around the helix is likely to result in a horizontal approach of the peptide toward the bilayer surface [35, 36].
12. Capping is generally necessary to avoid artifacts from a terminal charge caused by the artificial chain breaking. Here the capping was performed to be in accordance with previous experiments [35]. Capping is easiest performed using the residue topology database by adding a "residue" with the correct name at the terminus, hydrogens are then added automatically. Some RMS fitting may be necessary, but the exact position of the cap atoms is not very important because the energy minimization is likely to correct them.

13. This distance will place the helix approximately 2 nm above the bilayer surface, far enough to allow free rotation of the peptide in any direction (the helix itself is ~2.5 nm long) without the helix binding to both bilayer halves simultaneously—which would be an unwanted artifact from the use of periodic boundary conditions—but also close enough to avoid unfolding of the helical structure before membrane association.
14. If you want the helix to be exactly in between the bilayer halfs (taking periodic boundary conditions into account), you have to calculate (for a bilayer that aligns its normal with the z-axis) the average z coordinates of the helix and use this value for the calculation of the box z length.
15. Because the start is an already-equilibrated bilayer on top of which the peptide is placed, only water molecules that have been added in the last solvation step need to be removed. The limiting z coordinates are arbitrarily chosen to represent the hydrophobic core of the bilayer where we do not want water molecules. They should be sufficiently close to the hydrophobic region that any existing vacuum created by a changed lipid head group (POPA has no head group) is filled with water, without adding water in the hydrophobic core of the bilayer. Because the solvation step in this procedure uses Van der Waals radii to eliminate waters overlapping with the template structure, the same effect can probably also be achieved by (temporarily) increasing the Van der Waals radius of the lipid acyl chain atoms.
16. The molecule force field definitions can be included in any order, but, for the section where the molecules' presence is defined, **Note 10** applies.
17. For each additional peptide, the number of counterions has to be corrected to keep the system electrostatically neutral. With two penetratin peptides, that—because of the periodic boundary conditions—each will bind to a different bilayer half, there are no counterions left (−16e for the bilayer and +8e for each peptide). One should then have in mind that a third peptide will effectively see a neutral bilayer and not a negatively charged one. Association between a cationic peptide and an anionic bilayer is relatively fast and, moreover, subject to lateral diffusion: in the case of penetratin, additional molecules could be placed at exactly the same position without introducing a bias from the peptides that were already present.
18. Keeping the molecules whole—as is usual practice—may result in the helix being partially located out of the box, whereas added water molecules would be placed in the box. Because one is only interested in knowing which water molecules occupy the vacuum after the side chain deletion, this would be an unnecessary complication for visual inspection. Visual inspection is to be preferred over distance calculation because the number of water molecules to add would probably not exceed two.
19. In this step, the molecules must be kept whole. Otherwise, the bond connectivity information is lost in the coordinate file, and automatic topology generation (for the helix) is no longer possible.
20. The mutation into alanine requires the generation of a new (partial) topology. It is not possible to simply delete all interactions involving deleted side chain

atoms. Only a new topology for the helix is required, all other parts of the topology can be kept.

21. Specifically, first, every atom that enters the cylinder is calculated and subsequently this group is expanded into full residues. The average distance of the resulting group of atoms to any other group of atoms can be calculated, with the possibility of excluding itself. More concretely, one could calculate the evolution of the distance between the average position of the phosphorus atoms of interacting and non-interacting lipids of one bilayer half during the course of the molecular dynamics simulation.
22. For a 128-lipid bilayer, this still means that the trajectory has to be traversed 128 times. When only the peptide-interacting lipids are required, a first step would be the identification of these lipids to avoid unnecessary processing of the trajectory.
23. Scanning of a trajectory file containing all coordinates in the system, including water, may become inhibitively slow when the simulation length exceeds 100 ns. When not all coordinates in the system are required, a first step would be the extraction of only the coordinates involved into a new trajectory and then only that trajectory is processed. This step usually already results in a trajectory that is small enough to avoid the necessity of cutting it in pieces.

Acknowledgements I thank Prof. D.P. Tieleman (University of Calgary) and his coworkers for making lipid topologies and bilayer structures available, and the GROMACS development team for sharing their simulation program and analysis routines with the open-source community.

References

1. H. M. Berman, J. Westbrook, Z. Feng, G. Gilliland, T. N. Bhat, H. Weissig, I. N. Shindyalov, and P. E. Bourne. The protein data bank. Nucleic Acids Res., 28:235–242, 2000.
2. A. H. Joliot, A. Triller, M. Volovitch, C. Pernelle, and A. Prochiantz. Alpha-2,8-polysialic acid is the neuronal surface receptor of antennapedia homeobox peptide. New Biol., 3:1121–1134, 1991.
3. D. Derossi, A. H. Joliot, G. Chassaing, and A. Prochiantz. The third helix of the antennapedia homeodomain translocates through biological membranes. J. Biol. Chem., 269:10444–10450, 1994.
4. U. Langel, editor. Cell Penetrating Peptides. CRC Press LLC, Boca Raton, FL, 2002.
5. A. Prochiantz. Peptide nucleic acid smugglers. Nat. Biotechnol., 16:819–820, 1998.
6. M. Lindgren, M. Hallbrink, A. Prochiantz, and U. Langel. Cell-penetrating peptides. Trends Pharmacol., 21:99–103, 2000.
7. G. Drin, H. Demene, J. Temsamani, and R. Brasseur. Translocation of the pAntp peptide and its amphipathic analogue AP-2AL. Biochemistry, 40:1824–1834, 2001.
8. R. Fischer, M. Fotin-Mleczek, H. Hufnagel, and R. Brock. Break on through to the other side—biophysics and cell biology shed light on cell-penetrating peptides. Chem. Bio. Chem., 6:2126–2142, 2005.
9. D. Derossi, S. Calvet, A. Trembleau, A. Brunissen, G. Chassaing, and A. Prochiantz. Cell internalization of the third helix of the antennapedia homeodomain is receptor independent. J. Biol. Chem., 271:18188–18193, 1996.

10. M. Hällbrink, A. Florén, A. Elmquist, M. Pooga, T. Bartfai, and Ü. Langel. Cargo delivery kinetics of cell-penetrating peptides. Biochim. Biophys. Acta, 1515:101–109, 2001.
11. T. Lethoa, S. Gaal, C. Somlai, A. Czajlik, A. Perczel, and B. Penke. Membrane translocation of penetratin and its derivatives in different cell lines. J. Mol. Recog., 16:272—279, 2003.
12. T. Letoha, S. Gaal, C. Somlai, Z. Venkei, H. Glavinas, E. Kusz, E. Duda, A. Czajlik, F. Petak, and B. Penke. Investigation of penetratin peptides. Part 2. In vitro uptake of penetratin and two of its derivatives. J. Pept. Sci., 11:805–811, 2005.
13. J. P. Richard, K. Melikov, E. Vivès, C. Ramos, B. Verbeure, M. J. Gait, L. V. Chernomordik, and B. Lebleu. Cell-penetrating peptides: A reevaluation of the mechanism of cellular uptake. J. Biol. Chem., 278:585–590, 2003.
14. D. Terrone, S. L. Sang, L. Roudaia, and J. R. Silvius. Penetratin and related cell-penetrating cationic peptides can translocate across lipid bilayers in the presence of a transbilayer potential. Biochemistry, 42:13787–13799, 2003.
15. M. Magzoub, A. Pramanik, and A. Gräslund. Modeling the endosomal escape of cellpenetrating peptides: Transmembrane pH gradient driven translocation across phospholipid bilayers. Biochemistry, 44:14890–14897, 2005.
16. M. Magzoub, L. E. Eriksson, and A. Gräslund. Comparison of the interaction, positioning, structure induction and membrane perturbation of cell-penetrating peptides and non-translocating variants with phospholipid vesicles. Biophys. Chem., 103:271–288, 2003.
17. M. F. Lensink, B. Christiaens, J. Vandekerckhove, A. Prochiantz, and M. Rosseneu. Penetratin-membrane association: W48/R52/W56 shield the peptide from the aqueous phase. Biophys. J., 88:939–952, 2005.
18. E. Bárány-Wallje, S. Keller, S. Serowy, S. Geibel, P. Pohl, M. Bienert, and M. Dathe. A critical reassessment of penetratin translocation across lipid membranes. Biophys. J., 89:2513–2521, 2005.
19. E. Lindahl, B. Hess, and D. Van der Spoel. GROMACS 3.0: A package for molecular simulation and trajectory analysis. J. Mol. Mod., 7:306–317, 2001.
20. W. F. Van Gunsteren, S. R. Billeter, A. A. Eising, P. H. Hünenberger, P. Krüger, A. E. Mark, W. R. P. Scott, and I. G. Tironi. Biomolecular Simulation: The GROMOS96 Manual and User Guide. Hochschulverlag AG an der ETH Zürich, Zürich, Switzerland, 1996.
21. O. Berger, O. Edholm, and F. Jähnig. Molecular dynamics simulations of a fluid bilayer of dipalmitoylphosphatidylcholine at full hydration, constant pressure, and constant temperature. Biophys. J., 72:2002–2013, 1997.
22. B. Christiaens, J. Grooten, M. Reusens, A. Joliot, M. Goethals, J. Vandekerckhove, A. Prochiantz, and M. Rosseneu. Membrane interaction and cellular internalization of penetratin peptides. Eur. J. Biochem., 271:1187–1197, 2004.
23. S. White and W. Wimley. Membrane protein folding and stability: Physical principles. Ann. Rev. Biophys. Biomol. Struct., 28:319–365, 1999.
24. A. E. García and K. Y. Sanbonmatsu. α-Helical stabilization by side chain shielding of backbone hydrogen bonds. Proc. Nat. Ac. Sci. USA, 99:2782–2787, 2002.
25. F. Avbelj, P. Luo, and R. L. Baldwin. Energetics of the interaction between water and the helical peptide group and its role in determining helix propensities. Proc. Nat. Ac. Sci. USA, 97:10786–10791, 2000.
26. M. Magzoub, L. E. Eriksson, and A. Gräslund. Conformational states of the cell penetrating peptide penetratin when interacting with phospholipid vesicles: Effects of surface charge and peptide concentration. Biochim. Biophys. Acta, 1563:53–63, 2002.
27. S. Balayssac, F. Burlina, O. Convert, G. Bolbach, G. Chassaing, and O. Lequin. Comparison of penetratin and other homeodomain-derived cell-penetrating peptides: Interaction in a membrane-mimicking environment and cellular uptake efficiency. Biochemistry, 45:1408–1420, 2006.
28. M. F. Lensink, A. M. Haapalainen, J. K. Hiltunen, T. Glumoff, and A. H. Juffer. Response of SCP-2L domain of human MFE-2 to ligand removal: Binding site closure and burial of peroxisomal targeting signal. J. Mol. Biol., 323:99–113, 2002.
29. J. A. Christopher, R. Swanson, and T. O. Baldwin. Algorithms for finding the axis of a helix: Fast rotational and parametric least-squares methods. Comput. Chem., 20:339–345, 1996.

30. D. P. Tieleman, L. R. Forrest, M. S. P. Sansom, and H. J. C. Berendsen. Lipid properties and the orientation of aromatic residues in OmpF, influenza M2, and alamethicin systems: Molecular dynamics simulations. Biochemistry, 37:17554–17561, 1998.
31. K. M. Merz, Jr. and B. Roux, editors. Biological Membranes: A Molecular Perspective from Computation and Experiment. Birkhäuser, Boston, MA, 1996.
32. G. Fragneto, F. Graner, T. Charitat, P. Dubos, and E. Bellet-Amalric. Interaction of the third helix of antennapedia homeodomain with a deposited phospholipid bilayer: A neutron reflectivity structural study. Langmuir, 16:4581–4588, 2000.
33. C. E. Brattwall, P. Lincoln, and B. Nord'en. Orientation and conformation of cell-penetrating peptide penetratin in phospholipid vesicle membranes determined by polarized-light spectroscopy. J. Am. Chem. Soc., 125:14214–14215, 2003.
34. M. Lindberg, H. Biverstahl, A. Gräslund, and L. Maler. Structure and positioning comparison of two variants of penetratin in two different membrane mimicking systems by NMR. Eur. J. Biochem., 270:2055–2063, 2003.
35. B. Christiaens, S. Symoens, S. Vanderheyden, Y. Engelborghs, A. Joliot, A. Prochiantz, J. Vandekerckhove, M. Rosseneu, and B. Vanloo. Tryptophan fluorescence study of the interaction of penetratin peptides with model membranes. Eur. J. Biochem., 269:2918–2926, 2002.
36. L. Zhang, A. Rozek, and R. E. W. Hancock. Interaction of cationic antimicrobial peptides with model membranes. J. Biol. Chem., 276:35714–35722, 2001.

Chapter 10
Implicit Membrane Models for Membrane Protein Simulation

Michael Feig

Summary Implicit models of membrane environments offer computational advantages in simulations of membrane-interacting proteins and peptides. Such methods are especially useful for studies of long time scale processes, such as folding and aggregation, or very large complexes that are otherwise intractable with explicit lipid environments. Implicit models replace explicit solute–solvent interactions with a mean-field approach. In the most physical models, continuum dielectric electrostatics is combined with empirical formulations for the nonpolar components of the free energy of solvation. The practical use of a number of implicit membrane models ranging from the empirical IMM1 method to generalized Born-based methods with two-dielectric and multidielectric representations of biological membrane characteristics is presented.

Keywords: Continuum electrostatics · Dielectric constant · Generalized Born · Gouy-Chapman · Langevin dynamics · Poisson-Boltzmann · Solvent-accessible surface area

1 Introduction

Simulations of proteins and peptides interacting with biological membranes often involve either very large system sizes, e.g., when transport through channels is studied, or very long time scales, e.g., in studies of peptide folding and aggregation. Extensive conformational sampling is also necessary to obtain sufficiently accurate estimates of association and insertion free energies that are discussed in previous chapters. Simulations of membrane-bound proteins and peptides with explicit lipids, water molecules, ions, and other co-solvents are relatively straightforward, but the associated computational expense imposes serious limitations on how much sampling can actually be obtained in practice. To save computer time and to simplify the setup of complex membrane systems, the environment may be

From: Methods in Molecular Biology, vol. 443, Molecular Modeling of Proteins
Edited by Andreas Kukol © Humana Press, Totowa, NJ

represented implicitly while maintaining atomic-level detail only for the biomolecule of interest. Implicit membrane models reduce computational costs, not just as a result of much smaller overall system sizes, but also through instantaneous conformational averaging over solvent and lipid molecules. Such models are particularly attractive in cases in which specific interactions with the phospholipid bilayer play a minor role and general characteristics of a layered hydrophobic/hydrophilic system are sufficient in describing the environment for membrane-interacting proteins or peptides. It is also possible to use implicit membrane models in combination with explicitly modeled water, ions, or lipids [1], for example, by explicitly including water and ions inside an ion channel. However, because technical challenges in implementing such hybrid models have, to date, limited their practical application, only pure implicit models of membrane environments are discussed in this chapter.

Implicit models of membrane environments can be classified either as *knowledge based* or *physics based*, although empirical assumptions are often also made in the latter case. Knowledge-based implicit membrane models typically use experimentally determined residue-specific transfer free energies between water and nonpolar organic solvent to capture the energetics of inserting polypeptides into the hydrophobic membrane interior [2–9]. Implicit membrane models based on physical principles commonly rely on a decomposition of membrane–protein interactions into electrostatic contributions based on continuum dielectric theory and nonpolar contributions based on empirical formalisms [10–15]. Related approaches include dipole lattice membrane models [16]. Implicit membrane models based on continuum electrostatics are at the center of this chapter because, to date, they have had the widest application.

2 Theory

2.1 Electrostatic Interactions

The main physical characteristic of biological membranes is a hydrophobic core layer formed by the fatty acid tails of the phospholipid molecules that constitute the membrane. From an electrostatic perspective, such a membrane can be approximated as a layered dielectric system with a low dielectric constant in the nonpolar interior that gradually rises to the high dielectric aqueous solvent environment on either side of the membrane [14]. Such a model is described in general by Poisson-Boltzmann (PB) theory [17]:

$$\nabla \cdot [\varepsilon(\mathbf{r})\nabla\phi(\mathbf{r})] - \kappa^2(\mathbf{r})\phi(\mathbf{r}) = -4\pi\rho(\mathbf{r}) \tag{1}$$

where $\rho(r)$ is the explicit charge distribution of the biomolecular solute, whereas the dielectric constant $\varepsilon(r)$ and the modified Debye-Hückel screening factor $\kappa(r)$ describe the continuum environment. The $\kappa(r)$ term captures the interaction with free ions in the environment and may be neglected as a first approximation because ions do not commonly penetrate membrane interiors. How to implicitly include the effect of ions interacting with phospholipid head groups is discussed below. Given $\varepsilon(r)$ and $\rho(r)$, the PB equation can be solved for the electrostatic potential $\phi(r)$ from which the electrostatic solvation energy is obtained as:

$$\Delta G_{solvation,elec} = \int_V \rho(\mathbf{r}) \left(\phi_{dielectric}(\mathbf{r}) - \phi_{vacuum}(\mathbf{r})\right) d\mathbf{r}, \tag{2}$$

where $\phi_{dielectric}(r)$ and $\phi_{vacuum}(r)$ are the potentials in the presence of the dielectric environment and vacuum, respectively. If $\Delta G_{solvation}$ is added to a given molecular force field, an implicit representation of the environment is introduced. Molecular dynamics simulations with such an implicit model are straightforward but require an estimate of the gradient $\nabla \Delta G_{solvation}$.

The PB equation can be solved directly with a number of different methods [18–22]. However, high-accuracy PB solvers are generally not fast enough to be used effectively in molecular dynamics simulations in which the PB equation would need to be solved repeatedly at every time step [23–26]. Furthermore, the commonly used finite difference and finite element methods introduce large fluctuations in the calculation of derivatives unless the dielectric interface is smoothed substantially [26–28].

As an alternative, the generalized Born (GB) approximation [29, 30] expresses the electrostatic solvation energy in pairwise form as:

$$\Delta G_{solvation,elec} = -\frac{1}{2}\left(1 - \frac{1}{\varepsilon}\right) \sum_{i,j} \frac{q_i q_j}{\sqrt{r_{ij}^2 + \alpha_i \alpha_j e^{-r_{ij}^2/F\alpha_i\alpha_j}}}, \tag{3}$$

where the q_i are partial atomic charges of the biomolecule from a given force field, r_{ij} are pairwise atomic distances, and F is an adjustable parameter. The key to the GB formalism is the calculation of the GB radii α_i which are related to the electrostatic solvation energy of a single charge at the given atomic site in the presence of the otherwise uncharged biomolecular cavity:

$$\Delta G_{i,elec.solvation} = -\left(1 - \frac{1}{\varepsilon}\right) \frac{q_i^2}{\alpha_i}. \tag{4}$$

An efficient calculation of GB radii uses the following expression that follows from the Coulomb field approximation [29, 30]:

$$\frac{1}{\alpha_i} = \frac{1}{R_i} - \frac{1}{4\pi} \int_{solute, r>R_i} \frac{1}{r^4} dV \tag{5}$$

where R_i is the radius of atom i and the integral is carried out over the solute interior except for a sphere of radius R_i around atom i. Born radii calculated according to Eq. 5 are reasonably accurate for small molecules, but additional corrections improve the accuracy for typical biomolecules [31–34]. The GB formalism is computationally very efficient, maintains reasonable accuracy, and provides analytical

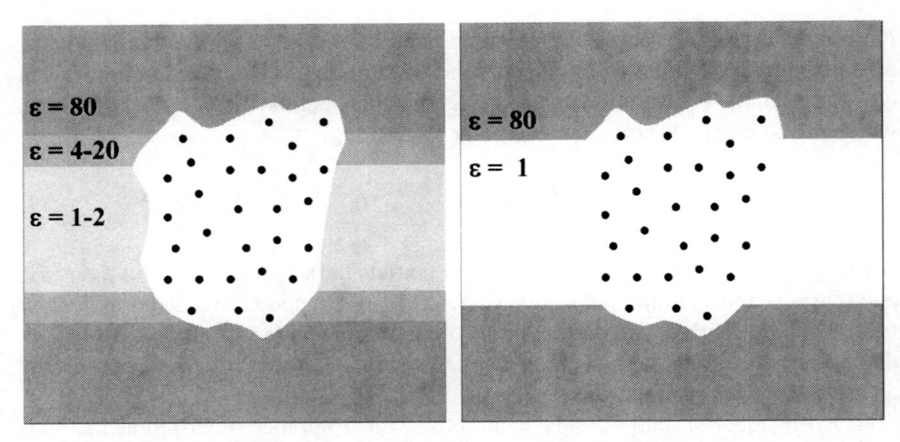

Fig. 1 *Left*: Heterogeneous dielectric model of biological membranes. *Right*: Two-dielectric membrane model. Atomic sites of the solute with explicit partial charges are indicated as black dots

derivatives [25]. As a consequence, GB methods are very attractive for the practical implementation of dielectric continuum implicit solvent models in molecular dynamics simulations of biomolecules [33, 35–40].

Coming back to the implicit modeling of membrane environments, it is immediately clear that a membrane model consisting of multiple dielectric layers, as shown in Fig. 1 and suggested from previous simulation studies [41, 42], is easily handled within the context of PB theory. However, the application of the GB formalism for modeling membrane environments in simulations is less straightforward. Equation 3 only applies to a simple two-dielectric system with $\varepsilon = 1$ inside the solute cavity and a constant dielectric $\varepsilon > 1$ elsewhere. A membrane environment can be modeled with the GB method by making the assumption that the dielectric constant of the hydrophobic membrane interior is the same as the solute interior, i.e., 1, and that the dielectric constant of phospholipid head groups is the same as the surrounding aqueous solvent [43, 44]. Then, a two-dielectric system can be maintained by simply extending the integration in Eq. 5 to an infinite slab representing the membrane interior (see Fig. 1) without the need for further modifications. Two-dielectric models of membrane environments ignore the low, but nonnegligible, polarizability of the lipid tails and a gradual increase of the dielectric constant in the lipid head group region that is in part caused by limited penetration of water molecules. Nevertheless, such models have been implemented and used successfully in folding and aggregation studies of membrane-bound peptides [43, 45, 46].

A truly heterogeneous dielectric environment can also be implemented with the GB formalism by introducing a local dielectric constant, ε_i, for each atomic site and by modifying Eq. 3 slightly [47]:

$$\Delta G_{solvation,elec} = -\frac{1}{2}\sum_{i,j}\left(1-\frac{1}{\varepsilon_{ij}}\right)\frac{q_i q_j}{\sqrt{r_{ij}^2+\alpha_i(\varepsilon)\alpha_j(\varepsilon)e^{-r_{ij}^2/F\alpha_i(\varepsilon)\alpha_j(\varepsilon)}}}, \quad (6)$$

where $\varepsilon_{ij} = (\varepsilon_i + \varepsilon_j)/2$ and the Born radii α_i are calculated as a function of ε [48]. The local dielectric constant could be chosen according to the dielectric layer in which an atom is located. However, solvation energies calculated according to Eq. 6 in this manner would not agree with solutions of the PB equation for layered dielectric environments. The reason is that a charge in the presence of multiple dielectric interfaces leads to polarization at each interface, which is captured by PB theory but not by the GB formalism because the GB model only considers the solute cavity and the immediate dielectric boundary. The consequences of multiple dielectric boundaries can be approximated, however, by introducing an effective dielectric constant that is used in Eq. 6 instead of the actual discrete dielectric layers [47]. The effective dielectric profile can be obtained by solving the PB equation for a probe charge at different locations in a given heterogeneous dielectric system. Practical details on how to obtain such effective dielectric profiles are discussed in Sect. 3.4.

So far, the presence of charged species in the membrane environment has been neglected. Many biological membranes contain a significant fraction of anionic phospholipids that interact strongly with ions from the surrounding electrolyte solution to form an electric double layer. Neutral phospholipids with zwitterionic head groups also interact with ions but to a much lesser degree. Gouy-Chapman theory describes such a system by smearing the phospholipid charge onto a planar surface and solving the PB equation in the presence of an electrolyte solution [49]. The resulting electrostatic potential outside the membrane is found to decay with distance z perpendicular to the membrane, as follows:

$$\phi(z) = \frac{2kT}{e} \ln \frac{1 + \alpha \exp(-\kappa z)}{1 - \alpha \exp(-\kappa z)}, \tag{7}$$

where α is related to the potential at the membrane surface that depends on the surface charge density, κ is the inverse Debye length, k is the Boltzmann constant, T is the temperature, and e is the charge of the electrolyte. Inside the membrane, the electrostatic potential is constant and equal to the value at the surface. The contribution to the solvation free energy from Gouy-Chapman theory is obtained by multiplying the electrostatic potential according to Eq. 7 with the explicit charge distribution of the solute and, thus, readily added to a continuum dielectric implicit membrane model [50].

2.2 Nonpolar Interactions

Implicit models of aqueous solvent sometimes neglect nonpolar contributions to the solvation free energy. However, in implicit models of membranes, the nonpolar component is crucial because the balance between electrostatic and nonpolar components determines whether a given molecule prefers the membrane interior, the membrane–water interface, or aqueous solvent. Electrostatic interactions alone always favor high-dielectric environments where charge–charge interactions are screened most. Nonpolar interactions between proteins and membranes consist of three distinct aspects: cost of cavity formation, solute–solvent van der Waals interactions, and lateral forces near the membrane–water interface caused by membrane surface tension [51]. As a first approximation, solute–solvent van der Waals interactions do not differ much between aqueous solvent and the membrane interior and are, therefore, often neglected in implicit membrane models. In contrast, because the cost of cavity formation is essentially related to the polarity of the solvent, it increases from nearly zero in the membrane interior to the significant cost of cavity formation in aqueous solvent. The cost of cavity formation is well described as a function of the solvent-accessible surface area (SASA):

$$\Delta G_{solvation,cavity} = \gamma \, SASA. \tag{8}$$

If the SASA is expressed as a sum over atomic contributions, $SASA = \sum_i SASA_i$, the prefactor γ can vary as a function of atomic location within the membrane, most conveniently as a function of z, the coordinate perpendicular to the membrane slab. Then, the cost of cavity formation in the membrane environment is obtained as [47]:

$$\Delta G_{solvation,cavity} = \sum_i \gamma(z_i) SASA_i. \tag{9}$$

The profile $\gamma(z)$ can be modeled ad hoc as a sigmoidal function [43] or fitted to free energies of insertion of nonpolar molecules such as O_2 from explicit lipid simulations [52,53]. In the latter case, van der Waals interactions are implicitly included so that Eq. 9 would describe the entire nonpolar contribution to the solvation free energy based on the small molecule data.

Membrane deformation after insertion of a molecule introduces an energetic barrier as a function of membrane surface tension [51]. This effect results in lateral forces that are strongest in the phospholipid head group region because of denser packing than the phospholipid tails. In an implicit model, such forces can be incorporated through an anisotropic contribution to the nonpolar solvation energy given in Eq. 9, where only the projections of the atomic surfaces visible from the $x - y$ directions near the water–membrane interface are considered.

3 Methods

Current practical implementations of implicit membrane models in major simulation packages that can be used in simulations of peptides or proteins are only available in the latest versions of the CHARMM program [54] (c31 and onward). They include four different models, IMM1 [2,50], GBSA/IM [44], GBSW [43], and HDGB [47], that are discussed in more detail in this section.

3.1 IMM1

IMM1 is an extension of the EEF1 model [55] that approximates solute–solvent interactions with a distance-dependent dielectric constant in explicit charge–charge interactions and an empirical self-energy term that results from comparing reference solvation free energies of fully exposed atoms with the actual solvent exposure of a given atom within a protein. Membrane environments are modeled by adjusting the reference solvation free energies according to experimental transfer energies depending on whether an atom is located within the hydrophobic region of the membrane or in aqueous solvent [2]. IMM1 is partially empirical because it is not based directly on PB theory, but it retains a physical force field to describe the protein. IMM1 includes an implementation of Gouy-Chapman theory that allows modeling of membranes that are composed of anionic phospholipids [50].

Compared with the other GB-based implicit membrane models described in Sects. 3.2–3.4, IMM1 is extremely fast, near the cost of vacuum simulations, because the solvation term adds little extra cost. The united-atom force field CHARMM19 [56] and short-cutoff distance-dependent dielectric constant electrostatics that are part of the IMM1 model speed up calculations further. Therefore, IMM1 is suited best for applications that require very extensive sampling and that can tolerate the approximate nature of the IMM1 model (see **Note 1**).

3.2 GBSA/IM

The GBSA/IM model implements a two-dielectric membrane model in which the solute cavity is extended into the hydrophobic region of the membrane [44]. The electrostatic contribution is obtained from a slightly modified version of an older GB model [57] that is fast but less accurate than recent GB versions [25]. The nonpolar contribution is obtained from atomic solvation parameters according to Eq. 9 with $\gamma_i = 0$ inside the membrane and a recommended value of $\gamma_i = 25\,\mathrm{cal/(mol\,Å^2)}$ outside the membrane. The GBSA/IM model can be used in principle with any force field, but it is parameterized only for the united-atom CHARMM19 force field [56] (see **Note 2**).

3.3 GBSW

The GBSW membrane model [43] is similar to the GBSA/IM model but is based on a recent GB implementation in which the volume integral in Eq. 5 is calculated numerically and a higher-order correction to the Coulomb field approximation is included [58]. The dielectric interface in the GBSW model is based on overlapping van der Waals spheres with a smoothing function instead of the more common sharp molecular surface. Because of the smooth interface, the GBSW method is numerically very stable [59]. The membrane model is also a two-dielectric model in which the integration in Eq. 5 is simply extended into the membrane slab. The nonpolar contribution to the solvation energy is calculated also according to Eq. 9, with a recommended value of $30\,\mathrm{cal/(mol\,Å^2)}$ outside the membrane. The effect of the membrane is modulated with a switching function across the interface.

The GBSW membrane model has been used successfully in a number of studies, in particular, peptide–membrane association, peptide folding in the context of a membrane, and peptide–peptide association within the membrane [43, 45, 46] (see **Note 3**).

3.4 HDGB

The HDGB implicit membrane model [40, 47] is based on the GBMV method for homogeneous dielectric environments [31, 32]. The GBMV method is very similar to the GBSW method, with the exception that it approximates the sharp molecular surface instead of a smoothed van der Waals surface. However, the HDGB membrane model implements a multiple dielectric layer model of a biological membrane, as shown in Fig. 1, rather than a two-dielectric approximation. HDGB uses the modified GB expression given in Eq. 6, which requires a dielectric profile for assigning a dielectric constant at a given atom site according to the distance from the membrane center, z.

The dielectric profile is obtained by solving the PB equation for a spherical probe within a layered dielectric system at different values of z. A probe radius of 2 Å is recommended because it matches a typical atom size, and a Δz of 0.5 Å is generally sufficient to sample the dielectric profile. HDGB creates a smooth continuous dielectric profile from the discrete data points through spline interpolation. Multiple dielectric layers offer a greater level of detail in describing the membrane environment; however, it is also more challenging to find appropriate values for the width and dielectric constants of different layers. Although experimental data alone does not provide enough detail, explicit lipid simulations may be used to guide setup of a layered dielectric membrane model. As an example, simulations of DPPC [41] suggest a three-layer system, with $\varepsilon = 2$ from 0 to 10 Å; $\varepsilon = 7$ from 10 to 15 Å; and $\varepsilon = 80$ further than 15 Å from the membrane center [47], although other values are also possible within the uncertainty of the computational data. The corresponding dielectric profile is shown in Fig. 2.

The nonpolar part is parameterized to match the shape of the free energy of O_2 insertion into a DPPC membrane, but scaled so that the value of γ outside the membrane approaches the desired surface tension in water. The following double-switching function is used in HDGB to model the nonpolar profile:

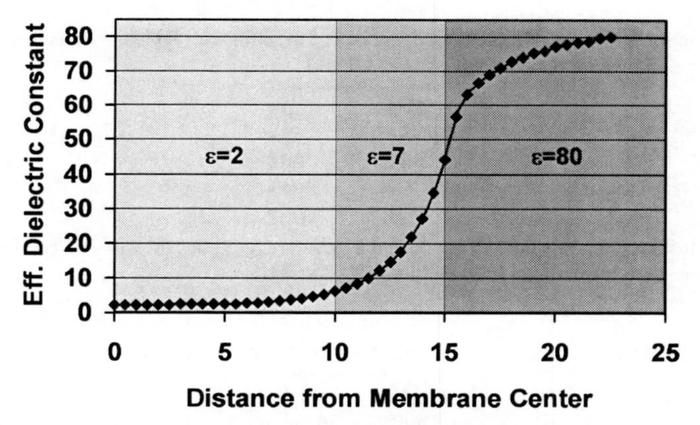

Fig. 2 Effective dielectric profile obtained from PB equation with 2-Å spherical probe charge for a 2/7/80 dielectric layer model of DPPC

$$S(z) = \begin{cases} \dfrac{s(|z| - z_s)^2(3z_m - 2|z| - z_s)}{(z_m - z_s)^3} & 0 \leq |z| < z_m \\ \dfrac{(1-s)(|z|^2 - z_m^2)^2(3z_t^2 - 2|z|^2 - z_m^2)}{(z_t^2 - z_m^2)^3} & z_m \leq |z| < z_t \\ 1 & otherwise \end{cases} \quad (10)$$

with z being the distance from the membrane center and s, z_s, z_m, and z_t the fitted parameters (see the example below for values appropriate for DPPC).

The HDGB is more complicated and computationally more expensive (see Sect. 3.5) than the other implicit membrane models. It requires more careful parameterization for a given membrane type, but the reward is a quantitatively more accurate description of membrane environments. The HDGB method has been able to match amino acid membrane insertion energetics with data from experiments and explicit lipid simulations [47]. Furthermore, the HDGB method has been applied successfully in simulations of integral membrane proteins [40] (see **Note 4**).

3.5 Timing

A major factor in the use of implicit membrane models is the computational efficiency that can be achieved. Table 1 compares simulation times for the models described above. The data shows that all of the GB-based membrane models required much more time than the empirical IMM1 model. Among the GB-based methods, HDGB is most expensive, and the GBSA/IM model with the united-atom CHARMM19 force field is only slightly faster than GBSW with the all-atom CHARMM22 force field [60].

Table 1 Time for 500 steps of molecular dynamics of melittin with typical parameters for each of the implicit membrane models available in CHARMM

IMM1	5 s
GBSA/IM	59 s
GBSW	68 s
HDGB	103 s

The HDGB test used the fast GBMV mode and a radial integration grid with $N_\phi = 4$ [31,32]. Measurements were taken on an Intel Pentium IV (2.8 GHz)

Table 2 Width of hydrophobic layer for selected phospholipid species [64, 65]

DMPC	25.4 Å
DOPC	26.8 Å
DPPC	27.9 Å
POPC	27.1 Å

3.6 Implicit Membrane Geometry

A key parameter in all implicit membrane models is the width of the hydrophobic layer that depends on the type of the phospholipids. The data compiled in Table 2 based on experiments may be used as a reference.

Commonly, implicit membrane models are set up under the assumption that the membrane slab is oriented in x-y direction, whereas z is the direction perpendicular to the membrane, with $z = 0$ corresponding to the center of the hydrophobic region.

3.7 Langevin Dynamics

Implicit solvent simulations with standard molecular dynamics integrators neglect friction and stochastic collisions with the environment. This has no consequences for obtaining correct relative thermodynamics of different conformational states with sufficient sampling, but it does affect kinetic properties and estimates of transition state activation barriers. Friction and stochastic collisions with the environment can be included by carrying out implicit solvent simulations with Langevin dynamics [61, 62]. The use of Langevin dynamics requires the selection of a suitable friction coefficient. In the case of membranes, the friction coefficient is expected to vary depending on the local environment, however, as a first approximation, a uniform friction coefficient may be used. The best choice for the friction coefficient in such simulations remains a subject of current research, but values between 5 and 100 ps^{-1} seem to be reasonable [62].

4 Outlook

The practical application of implicit membrane models in simulations of proteins and peptides is still in its infancy. A number of implicit membrane models have been implemented that offer different advantages and disadvantages from the very fast IMM1 method to the most complex HDGB model. Further improvements of the current models are expected to include Gouy-Chapman theory, implicit solute–solvent van der Waals interactions, and the effect of membrane surface tension. However, all of the discussed models can already be used for simulations of proteins or peptides in membrane environments.

5 Notes

Example CHARMM inputs for running simulations with different implicit membrane models (tested with version c33b1) are given in this section.

1. IMM1 is based on the EEF1 model and requires the same modified CHARMM19 force field with neutralized charged side chains. This force field is part of the CHARMM distribution and loaded at the beginning with:

```
open read unit 10 form name toph19_eef1.1.inp
read rtf card unit 10
close unit 10
open read unit 10 form name param19_eef1.1.inp
read para card unit 10
close unit 10
```

   After the protein solute has been set up, the IMM1 solvation term is turned on with:

```
eef1 setup membrane slvt water slv2 chex width 27.0
temp 300.0 – unit 20 name solvpar.inp
```

   The input file `solvpar.inp` is also supplied with the CHARMM distribution and contains the atomic solvation parameters. The main adjustable parameters are the width of the hydrophobic layer and the temperature. Note that the IMM1 model is used best for temperatures near 300 K.

   If anionic phospholipids are modeled, a Gouy-Chapman term can be added with:

```
eef1 setup membrane slvt water slv2 chex width 27.0
temp 300.0 – gouy anfr 0.2 area 70 offset 3.0 conc
0.1 valence 1 – unit 20 name solvpar.inp
```

where the adjustable parameters are the fraction of anionic lipids (`anfr`), the area per lipid in Å^2 (`area`), the concentration (`conc`) and valence (`valence`) of the electrolyte solution, and the location of the charge plane relative to the hydrophobic layer (`offset`).

Furthermore, the EEF1/IMM1 model requires that distance dependent dielectric is turned on with a short 9 Å group-based cutoff:

```
update ctonnb 7.0 ctofnb 9.0 cutnb 10.0 group rdie
switch vswitch
```

The EEF1/IMM1 implementation in CHARMM is sensitive to the nonbonded list generator that is used during the dynamics run. One option that does work is BYGR.

2. The GBSA/IM membrane model is parameterized for the CHARMM19 force field and, although not strictly required, it is highly recommended to use only this force field. Use of the GBSA/IM model requires that the polar and nonpolar components are set up separately. The nonpolar contribution based on atomic solvation parameters is set up as follows:

```
read saim unit 5
* ASP parameters
*
1.40 27.0 0.   z
ANY  C*   25.0 2.1 0.0
ANY  O*   25.0 1.6 0.0
ANY  N*   25.0 1.6 0.0
ANY  H*   25.0 0.8 0.0
END
```

The relevant parameters are the width of the membrane in angstroms given as the second parameter in the fourth line and the value of γ outside of the membrane that is specified in the subsequent lines for each atom type. It is recommended to use the same value of g for all atom types. Typical values of γ in aqueous solvent may range from 5 to 30 cal/(mol Å^2).

The electrostatic contribution is setup as follows:

```
gbim p1 0.415 p2 0.239 p3 1.756 p4 10.51 p5 1.1
lambda 0.730 tmemb 27.0
```

The parameters `p1-5` and `lambda` have to be specified but should not be altered unless a different force field is used. The thickness of the membrane is given with `tmemb`.

3. The GBSW membrane model can be used with any force field. It is activated with:

```
scalar wmain = radius
gbsw sgamma 0.03 tmemb 27.0
msw 2.5
```

The first command copies the van der Waals radii to be used by GBSW for defining the dielectric interface. It is possible to use other atomic radii, for example, the set proposed by Nina et al. [63].

Options of the `gbsw` command are the membrane width (`tmemb`), the half-width of the membrane switching function (`msw`) and the value of γ for the nonpolar contribution outside the membrane (`sgamma`) in kcal/(mol Å^2).

The GBSW implementation requires the use of an electrostatic switching function.

4. The HDGB method is called from within the GBMV command, as in the following example:

```
open read unit 10 form name eps.profile
gbmv corr 3 uneps 10 - a1 0.3255 a3 1.085 a4 -0.14
a5 -0.1 - zs 0.5 zm 9.2 zt 25 st0 0.32 sa 0.015
```

HDGB is turned on with `corr 3` and requires the dielectric profile as input from an external file (the unit is given with `uneps`). The parameters a1 to a5 determine how Born radii are calculated in GBMV for different dielectric environments. The remaining parameters describe the nonpolar contribution to the solvation free energy, with $\gamma(z)$ calculated according to Eq. 10 (`st0` is "s" in Eq. 10). The maximum value of γ in aqueous solvent is given with `sa`.

The membrane width is not explicitly given in HDGB. For each membrane type, a new dielectric profile has to be generated and the nonpolar parameters have to be adjusted accordingly. The values given in the example correspond to DPPC with a membrane width of approximately 28 Å. As a first guess, one can compress or stretch the dielectric and nonpolar profiles to match the width of different membrane types.

HDGB also requires the use of an electrostatic switching function.

References

1. Lin, J.-H., Baker, N. A. and McCammon, J. A. (2002). Bridging implicit and explicit solvent approaches for membrane electrostatics. *Biophysical Journal* **83**, 1374–1379.
2. Lazaridis, T. (2003). Effective energy function for proteins in lipid membranes. *Proteins* **52**, 176–192.
3. Lomize, A. L., Pogozheva, I. D., Lomize, M. A. and Mosberg, H. I. (2006). Positioning of proteins in membranes: A computational approach. *Protein Science* **15**, 1318–1333.
4. Ducarme, P., Rahman, M. and Brasseur, R. (1998). IMPALA: A simple restraint field to simulate the biological membrane in molecular structure studies. *Proteins-Structure Function and Bioinformatics* **30**, 357–371.

5. Jahnig, F. and Edholm, O. (1992). Modeling of the structure of bacteriorhodopsin—a molecular-dynamics study. *Journal of Molecular Biology* **226**, 837–850.
6. Sanders, C. R. and Schwonek, J. P. (1993). An approximate model and empirical energy function for solute interactions with a water-phosphatidylcholine interface. *Biophysical Journal* **65**, 1207–1218.
7. Nolde, D. E., Arseniev, A. S., Vergoten, G. and Efremov, R. G. (1997). Atomic solvation parameters for proteins in a membrane environment. Application to transmembrane alpha-helices. *J Biomol Struct Dyn* **15**, 1–18.
8. Efremov, R. G., Nolde, D. E., Vergoten, G. and Arseniev, A. S. (1999). A solvent model for simulations of peptides in bilayers. I. Membrane-promoting alpha-helix formation. *Biophysical Journal* **76**, 2448–2459.
9. Efremov, R. G., Nolde, D. E., Vergoten, G. and Arseniev, A. S. (1999). A solvent model for simulations of peptides in bilayers. II. Membrane-spanning alpha-helices. *Biophysical Journal* **76**, 2460–2471.
10. BenTal, N., BenShaul, A., Nicholls, A. and Honig, B. (1996). Free-energy determinants of alpha-helix insertion into lipid bilayers. *Biophysical Journal* **70**, 1803–1812.
11. Kessel, A., Cafiso, D. S. and Ben-Tal, N. (2000). Continuum solvent model calculations of alamethicin-membrane interactions: Thermodynamic aspects. *Biophysical Journal* **78**, 571–583.
12. Murray, D., BenTal, N., Honig, B. and McLaughlin, S. (1997). Electrostatic interaction of myristoylated proteins with membranes: simple physics, complicated biology. *Structure* **5**, 985–989.
13. Roux, B. and MacKinnon, R. (1999). The cavity and pore helices the KcsA K+ channel: Electrostatic stabilization of monovalent cations. *Science* **285**, 100–102.
14. Roux, B., Berneche, S. and Im, W. (2000). Ion channels, permeation, and electrostatics: Insight into the function of KcsA. *Biochemistry* **39**, 13295–13306.
15. Im, W. and Roux, B. (2002). Ion permeation and selectivity of OmpF porin: A theoretical study based on molecular dynamics, brownian dynamics, and continuum electrodiffusion theory. *Journal of Molecular Biology* **322**, 851–869.
16. Grossfield, A., Sachs, J. and Woolf, T. B. (2000). Dipole lattice membrane model for protein calculations. *Proteins-Structure Function and Genetics* **41**, 211–223.
17. Sharp, K. A. and Honig, B. (1990). Electrostatic interactions in macromolecules—theory and applications. *Annual Review of Biophysics and Biophysical Chemistry* **19**, 301–332.
18. Warwicker, J. and Watson, H. C. (1982). Calculation of the electric potential in the active site cleft due to α-helix dipoles. *Journal of Molecular Biology* **157**, 671–679.
19. Gilson, M. K., Sharp, K. A. and Honig, B. H. (1987). Calculating the electrostatic potential of molecules in solution: method and error assessment. *Journal of Computational Chemistry* **9**, 327–335.
20. Holst, M., Baker, N. and Wang, F. (2000). Adaptive multilevel finite element solution of the Poisson-Boltzmann equation I. Algorithms and examples. *Journal of Computational Chemistry* **21**, 1319–1342.
21. Baker, N., Holst, M. and Wang, F. (2000). Adaptive multilevel finite element solution of the Poisson-Boltzmann equation II. Refinement at solvent-accessible surfaces in biomolecular systems. *Journal of Computational Chemistry* **21**, 1343–1352.
22. Baker, N. A. (2005). Improving implicit solvent simulations: a Poisson-centric view. *Current Opinion in Structural Biology* **15**, 137–143.
23. Luo, R., David, L. and Gilson, M. K. (2002). Accelerated Poisson-Boltzmann calculations for static and dynamic systems. *Journal of Computational Chemistry* **23**, 1244–1253.
24. Lu, B. Z., Chen, W. Z., Wang, C. X. and Xu, X.-j. (2002). Protein molecular dynamics with electrostatic force entirely determined by a single Poisson-Boltzmann calculation. *Proteins* **48**, 497–504.
25. Feig, M., Onufriev, A., Lee, M. S., Im, W., Case, D. A. and Brooks III, C. L. (2004). Performance comparison of generalized Born and Poisson Methods in the calculation of electrostatic solvation energies for protein structures. *Journal of Computational Chemistry* **25**, 265–284.

26. Prabhu, N. V., Zhu, P. J. and Sharp, K. A. (2004). Implementation and testing of stable, fast implicit solvation in molecular dynamics using the smooth-permittivity finite difference Poisson-Boltzmann method. *Journal of Computational Chemistry* **25**, 2049–2064.
27. Im, W., Beglov, D. and Roux, B. (1998). Continuum solvation model: computation of electrostatic forces from numerical solutions to the Poisson-Boltzmann equation. *Computer Physics Communications* **111**, 59–75.
28. Friedrichs, M., Zhou, R. H., Edinger, S. R. and Friesner, R. A. (1999). Poisson-Boltzmann analytical gradients for molecular modeling calculations. *Journal of Physical Chemistry B* **103**, 3057–3061.
29. Still, W. C., Tempczyk, A., Hawley, R. C. and Hendrickson, T. (1990). Semianalytical Treatment of solvation for molecular mechanics and dynamics. *Journal of the American Chemical Society* **112**, 6127–6129.
30. Qiu, D., Shenkin, P. S., Hollinger, F. P. and Still, W. C. (1997). The GB/SA continuum model for solvation. A fast analytical method for the calculation of approximate Born radii. *J Phys Chem A* **101**, 3005–3014.
31. Lee, M. S., Salsbury, F. R., Jr. and Brooks, C. L., III. (2002). Novel generalized Born methods. *Journal of Chemical Physics* **116**, 10606–10614.
32. Lee, M. S., Feig, M., Salsbury, F. R., Jr. and Brooks, C. L., III. (2003). New analytical approximation to the standard molecular volume definition and its application to generalized Born calculations. *Journal of Computational Chemistry* **24**, 1348–1356.
33. Onufriev, A., Bashford, D. and Case, D. A. (2004). Exploring protein native states and large-scale conformational changes with a modified generalized born model. *Proteins-Structure Function and Bioinformatics* **55**, 383–394.
34. Grycuk, T. (2003). Deficiency of the Coulomb-field approximation in the generalized Born model: An improved formula for Born radii evaluation. *Journal of Chemical Physics* **119**, 4817–4826.
35. Calimet, N., Schaefer, M. and Simonson, T. (2001). Protein molecular dynamics with the generalized Born/ACE solvent model. *Proteins* **45**, 144–158.
36. Tsui, V. and Case, D. A. (2000). Molecular dynamics simulations of nucleic acids with a generalized Born solvation model. *Journal of the American Chemical Society* **122**, 2489–2498.
37. Fan, H., Mark, A. E., Zhu, J. and Honig, B. (2005). Comparative study of generalized Born models: Protein dynamics. *Proceedings of the National Academy of Sciences of the United States of America* **102**, 6760–6764.
38. Geney, R., Layten, M., Gomperts, R., Hornak, V. and Simmerling, C. (2006). Investigation of salt bridge stability in a generalized born solvent model. *Journal of Chemical Theory and Computation* **2**, 115–127.
39. Chocholousova, J. and Feig, M. (2006). Implicit solvent simulations of DNA and DNA-protein complexes: Agreement with explicit solvent vs experiment. *Journal of Physical Chemistry B* **110**, 17240–17251.
40. Tanizaki, S. and Feig, M. (2006). Molecular dynamics simulations of large integral membrane proteins with an implicit membrane model. *Journal of Physical Chemistry B* **110**, 548–556.
41. Stern, H. A. and Feller, S. E. (2003). Calculation of the dielectric permittivity profile for a nonuniform system: Application to a lipid bilayer simulation. *Journal of Chemical Physics* **118**, 3401–3412.
42. Zhou, F. and Schulten, K. (1995). Molecular dynamics study of a membrane-water interface. *Journal of Physical Chemistry* **99**, 2194–2207.
43. Im, W., Feig, M. and Brooks III, C. L. (2003). An implicit membrane generalized Born theory for the study of structure, stability, and interactions of membrane proteins. *Biophysical Journal* **85**, 2900–2918.
44. Spassov, V. Z., Yan, L. and Szalma, S. (2002). Introducing an implicit membrane in generalized Born/solvent accessibility continuum solvent models. *Journal of Physical Chemistry B* **106**, 8726–8738.
45. Im, W. and Brooks, C. L. (2004). De novo folding of membrane proteins: An exploration of the structure and NMR properties of the fd coat protein. *Journal of Molecular Biology* **337**, 513–519.

46. Im, W. and Brooks, C. L. (2005). Interfacial folding and membrane insertion of designed peptides studied by molecular dynamics simulations. *Proceedings of the National Academy of Sciences of the United States of America* **102**, 6771–6776.
47. Tanizaki, S. and Feig, M. (2005). A generalized Born formalism for heterogeneous dielectric environments: Application to the implicit modeling of biological membranes. *Journal of Chemical Physics* **122**, 124706.
48. Feig, M., Im, W. and Brooks III, C. L. (2004). Implicit solvation based on generalized Born theory in different dielectric environments. *Journal of Chemical Physics* **120**, 903–911.
49. Mclaughlin, S. (1989). The Electrostatic Properties of Membranes. *Annual Review of Biophysics and Biophysical Chemistry* **18**, 113–136.
50. Lazaridis, T. (2005). Implicit solvent simulations of peptide interactions with anionic lipid membranes. *Proteins-Structure Function and Bioinformatics* **58**, 518–527.
51. Gullingsrud, J. and Schulten, K. (2003). Gating of MscL studied by steered molecular dynamics. *Biophysical Journal* **85**, 2087–2099.
52. Marrink, S. J. and Berendsen, H. J. C. (1996). Permeation process of small molecules across lipid membranes studied by molecular dynamics simulations. *Journal of Physical Chemistry* **100**, 16729–16738.
53. Marrink, S.-J. and Berendsen, H. J. C. (1994). Simulation of water transport through a lipid membrane. *Journal of Physical Chemistry* **98**, 4155–4168.
54. Brooks, B. R., Bruccoleri, R. E., Olafson, B. D., States, D. J., Swaminathan, S. and Karplus, M. (1983). CHARMM: A program for macromolecular energy, minimization, and dynamics calculations. *Journal of Computational Chemistry* **4**, 187–217.
55. Lazaridis, T. and Karplus, M. (2000). Effective energy functions for protein structure prediction. *Current Opinion in Structural Biology* **10**, 139–145.
56. Neria, E., Fischer, S. and Karplus, M. (1996). Simulation of activation free energies in molecular systems. *Journal of Chemical Physics* **105**, 1902–1921.
57. Dominy, B. N. and Brooks III, C. L. (1999). Development of a generalized Born model parametrization for proteins and nucleic acids. *Journal of Physical Chemistry B* **103**, 3765–3773.
58. Im, W., Lee, M. S. and Brooks, C. L., III. (2003). Generalized Born model with a simple smoothing function. *Journal of Computational Chemistry* **24**, 1691–1702.
59. Chocholousova, J. and Feig, M. (2006). Balancing an accurate representation of the molecular surface in generalized Born formalisms with integrator stability in molecular dynamics simulations. *Journal of Computational Chemistry* **27**, 719–729.
60. MacKerell, A. D., Jr., Bashford, D., Bellott, M., Dunbrack, J. D., Evanseck, M. J., Field, M. J., Fischer, S., Gao, J., Guo, H., Ha, S., Joseph-McCarthy, D., Kuchnir, L., Kuczera, K., Lau, F. T. K., Mattos, C., Michnick, S., Ngo, T., Nguyen, D. T., Prodhom, B., Reiher, W. E., Roux, B., Schlenkrich, M., Smith, J. C., Stote, R., Straub, J., Watanabe, M., Wiorkiewicz-Kuczera, J., Yin, D. and Karplus, M. (1998). All-atom empirical potential for molecular modeling and dynamics studies of proteins. *Journal of Physical Chemistry B* **102**, 3586–3616.
61. Brooks, C. L., Berkowitz, M. and Adelman, S. A. (1980). Generalized Langevin theory for many-body problems in chemical-dynamics—Gas-surface collisions, vibrational-energy relaxation in solids, and recombination reactions in liquids. *Journal of Chemical Physics* **73**, 4353–4364.
62. Zagrovic, B. and Pande, V. (2003). Solvent viscosity dependence of the folding rate of a small protein: distributed computing study. *Journal of Computational Chemistry* **24**, 1432–1436.
63. Nina, M., Beglov, D. and Roux, B. (1997). Atomic radii for continuum electrostatics calculations based on molecular dynamics free energy simulations. *Journal of Physical Chemistry* **101**, 5239–5248.
64. Kucerka, N., Tristram-Nagle, S. and Nagle, J. F. (2006). Structure of fully hydrated fluid phase lipid bilayers with monounsaturated chains. *Journal of Membrane Biology* **208**, 193–202.
65. Kucerka, N., Tristram-Nagle, S. and Nagle, J. F. (2006). Closer look at structure of fully hydrated fluid phase DPPC bilayers. *Biophysical Journal* **90**, L83–L85.

Part IV
Protein Structure Determination

Chapter 11
Comparative Modeling of Proteins

Gerald H. Lushington

Summary Three-dimensional analysis of protein structures is proving to be one of the most fruitful modes of biological and medical discovery in the early 21st century, providing fundamental insight into many (perhaps most) biochemical functions of relevance to the cause and treatment of diseases. Fully realizing such insight, however, would require analysis of too many distinct proteins for thorough laboratory analysis of all proteins to be feasible, thus, any method capable of accurate, efficient *in silico* structure prediction should prove highly expeditious. The technique generally acknowledged to provide the most accurate protein structure predictions, called comparative modeling, has, thus, attracted substantial attention and is the focus of this chapter. Although other reviews have reported on the method development and research history of comparative modeling, our discussion herein focuses on the general philosophy of the method and specific strategies for successfully achieving reliable and accurate models. The chapter, thus, relates aspects of template selection, sequence alignment, spatial alignment, loop and gap modeling, side chain modeling, structural refinement, and validation.

Keywords: Comparative modeling · Homology · Loop modeling · Proteins · Sequence alignment · Structure alignment · Structure refinement · Structure validation · Threading

1 Introduction

From the first atomic-level resolution of a protein structure (whale myoglobin by Kendrew et al. [1]) onward, three-dimensional (3D) protein models have provided a wealth of insight into biomolecular properties and processes, inspiring a growing thirst for structural detail. For example, the stated mandate of the Structural Genomics Initiative (http://www.structuralgenomics.org/) is to solve 3D structures for every unique protein in nature. As for the human genome project, such objectives

From: Methods in Molecular Biology, vol. 443, Molecular Modeling of Proteins
Edited by Andreas Kukol © Humana Press, Totowa, NJ

are of a scope (~100,000 proteins, not counting posttranslational modifications and conformational variants) that requires not only determination but innovation, in that conventional resolution methods, such as x-ray crystallography and nuclear magnetic resonance (NMR) are too time consuming and demanding to address such goals. An increasingly viable alternative is efficient and analytically rigorous computational modeling. This chapter, thus, offers a brief discussion of *in silico* protein structure prediction, focusing mainly on comparative modeling. Whereas other papers (e.g., Martí-Renom et al. [2], Baker and Šali [3]) provide comprehensive reviews of the underlying method development and research achievements in the field, this chapter discusses the motivation for using comparative modeling and outlines the practical considerations to be made in assembling such a model.

A key inspiration for projects such as the Structural Genomics Initiative was the realization that the human genome sequencing project was not the panacea for biological understanding that some had hoped for, but, rather, exposed how many factors other than genetic coding portions of our DNA help to define interspecies and intraspecies diversity, medical causality, and other key issues of organism-scale biology. A more apt currency for characterizing life may lie in proteomics, wherein one can identify specific units responsible for a given biological function or dysfunction, and, thus, surmise the underlying molecular mechanisms. A key source of insight for the latter is protein structure. Combined with simple rules of physics (e.g., thermodynamics, electrostatics, etc.) and chemistry (bond formation and breaking), 3D molecular structure information can be readily extrapolated toward understanding of intermolecular association, enzyme function, structural response, etc., collectively covering most of the basic molecular causality behind the functional or dysfunctional processes experienced by an organism.

Pharmaceutical research frequently follows the paradigm of discovering a target for a given disease, characterizing the target, and finding ways of beneficially modulating its behavior. Because most targets are proteins, a key tool in the first step is comparison of protein expression signatures for diseased and healthy tissue samples. Proteins whose presence is noticeably amplified or suppressed in dysfunctional tissue are flagged as potential diagnostic biomarkers and as possible therapeutic targets. Therapeutics are then sought either to restore normal healthy balance in target function or to compensate for imbalances. 3D structural knowledge of a prospective target is invaluable for the latter, providing key insight into potential receptors that might be pharmacologically targeted. Insight into nontarget protein structures is also very useful for intuiting potential drug side effects, because proteins with receptors similar to the target are at pronounced risk of inadvertent inhibition by chemicals designed for the latter. Broad understanding of protein structures should further aid in expression-based target identification by helping to pinpoint protein–protein interactions that may obscure whether a deviant expression profile in a protein is a primary cause of a given dysfunction or a secondary symptom. Specifically, knowing 3D structures can inform us of which proteins should colocalize in a given cellular environment, and which are likely to engage in direct (i.e., actual physical association) or indirect (e.g., exchange of metabolite or transmitter) interactions.

Protein 3D structures may be obtained in a number of ways. Most high accuracy structures in the protein databank (PDB; http://www.rcsb.org/pdb) were resolved through crystallographic techniques, with the rest mainly arising from NMR analysis. Although some proteins have been resolved via both techniques, crystallography is best suited to high-resolution determination of structures for soluble, predominantly globular, proteins most amenable to crystallization, whereas NMR affords more a dynamic interpretation of the structure (i.e., multiple conformers are usually resolved from the same data) and is viable for many proteins that resist crystallization, albeit generally at a lower level of resolution. Theoretically, the entire proteome should be accessible to either NMR or crystallography. Although some of the most massive proteins still pose major challenges of data deconvolution, it is reasonable to expect that analytical enhancements will eventually conquer those holdouts as well. It is more uncertain, however, whether experimental resolution of the entire known proteome will ever be practical, because both crystallography and NMR remain technically very demanding, time consuming, and expensive. The underlying assumption of the Structural Genomics Initiative is that computational modeling may yet accomplish what is spectroscopically impractical.

The main schemes for computational protein structure prediction include: 1) self-assembly simulations, 2) associative models based on sequence pattern recognition, and 3) comparative modeling. All of these methods were inspired in part by seminal findings of Christian Anfinsen et al. that an unraveled ribonuclease amino acid (AA) chain could, in plain solution, coalesce within a reasonable (minutes to hours) time frame to form a protein functionally indistinguishable from native in vivo ribonuclease [4]. This implied that: 1) a protein can be uniquely identified by its AA sequence, 2) this sequence uniquely encodes the in vivo protein function, and 3) the AA sequence is capable of consistent self-assembly into functional form, based exclusively on intramolecular interactions among sequence AAs plus interatomic interactions with surrounding solvent. The third observation, thus, directly suggested that assembly simulations could be a reliable mode for protein structure prediction, whereas the first two observations were key to the eventual formulation of pattern recognition and comparative modeling techniques.

Practical formulation of associative and comparative modeling methods required a substantial basis in empirical understanding of protein structure. One key development was the accumulation of a reasonable volume of protein sequence data beginning in the early 1950s, leading to gradual elucidation of strong correlative patterns between protein sequence similarity on one hand and analogous function on the other. This permitted the classification of proteins into families and superfamilies [5], with an underlying assumption being that members of the same protein superfamily are all evolutionarily related (a.k.a., homologous), and members of the same family are closely related.

A second important precursor to protein structure prediction was the acquisition of crystal structures from the late 1950s onward. These structures generally validated the earlier family and superfamily classifications in that proteins with similar sequences and function were usually found to have very similar 3D structures. Analysis of structural data also revealed that all proteins tended to assemble as a sum of a limited number of unique substructure forms. At a fine-grained level, it was noted that all proteins tended to adopt similar hydrogen bond-stabilized features such as α-helices, β-sheets, and a number of distinct turn and hairpin structures, collectively referred to as *secondary structure*. At coarser levels, it also became apparent that the full manifold of protein structures could be classified into a relative small number (~500 [6]) of unique *folds*: characteristic self-stabilizing collections of secondary structure elements. Not surprisingly, proteins within the same family (as classified by sequence) almost invariably possessed similar folds. Such correspondence between sequence similarity trends versus structural similarity completes the basis for associative and comparative modeling techniques. Associative models are, thus, based on probabilistic tendencies of certain AA combinations to adopt a given secondary structure and tendencies of certain secondary structure elements to adopt a specific fold, whereas the more general tendency of proteins with similar sequences to adopt similar 3D structures is the foundation for comparative modeling.

Although each of the different structure prediction methods is potentially applicable to modeling a given target protein, each has limitations to heed. Self-assembly simulations, commonly known as protein-folding models, are especially computationally demanding because of huge numbers of unique conformers that must be considered for even modest-sized proteins. With such large conformer manifolds, one may also have to empirically choose between multiple unique but comparably favorable structures. A more severe problem exists for the fraction of protein population whose thermodynamically optimal conformer is not the one observed in vivo, the most famous example of which is the prion protein cellular (PrPC), whose native form seems to be higher in energy than a rogue form, prion protein scrapie (PrPSc). Presence of the latter rogue form of the protein in living tissue is thought to catalyze the exothermic refolding of healthy PrPC to the biologically inutile PrPSc, with the result being prion diseases such as scrapie, Creutzfeldt-Jakob Disease, or bovine spongiform encephalopathy (BSE) [7]. Because of the greater thermodynamic stability of the latter (PrPSc), a good folding algorithm based on the underlying free energy effects assumed by Anfinsen to drive the assembly process [4] should converge to the rogue PrPSc form rather than the biologically healthy PrPC. In truth, energy is often not the sole driving force behind assembly, in that life forms have developed sophisticated tools called chaperonins that perform protein assembly quality control, correcting misfolds, and sometimes preferentially effecting assembly of higher energy structural forms. Relevant interactions between chaperonins and their client polypeptides are very complex and entail much mechanistic detail that is not well understood [8]. Folding simulations that explicitly take chaperonin contributions into account have recently been undertaken [8, 9]; however, it is not yet clear how effective and practical they will prove to be in de novo structure prediction.

Accurate associative modeling is also fairly challenging because of prospects for compounding errors. Even for the most sophisticated prediction algorithms, the first step of secondary structure prediction from raw sequence is generally thought to be at most approximately 80% successful on a per-residue basis, thus, for a typical protein (hundreds of AAs), leading to dozens of local errors at the outset. Because local errors in backbone torsion can easily extrapolate to large errors in the global structure (e.g., mislocation of entire lobes), many proteins are poorly predicted. Furthermore, the stated 80% success rate applies primarily to globular proteins, and indications are that similar predictions for transmembrane proteins are significantly worse [10].

Results from the 6th Critical Assessment of Techniques for Protein Structure Prediction competition (CASP6) (Gaeta, Italy, Dec. 4–8, 2004) reveal that comparative models attain better protein structure predictions on average than nontemplate methods. In the full set of systems for which both comparative and folding models were generated, with fully automated comparative models achieving a mean root mean-squared deviation (RMSD) of 2.17 Å relative to experimental crystal structures, and further improvement to 2.09 Å RMSD given human intervention in template selection, alignment modification, etc. By contrast, folding methods (with and without intervention) attained mean RMSD values of 2.41 Å and 2.51 Å, respectively. Comparative models are also relatively computationally expedient; however, their dependence on quality 3D templates structurally and functionally similar to the target precludes many targets from consideration. Fortunately, the proteome has a relatively small number of unique folds, thus, coordinated pursuit of novel folds by experimentalists might yield at least one reasonable template structure for any given protein. The Structural Genomics Initiative, thus, aims to identify and target a key subset of currently structurally uncharacterized proteins that, according to sequence family classification, should provide a set of viable templates for all prospective targets.

2 Methods

In practice, comparative modeling is best viewed not as one technique but rather as a strategy for assembling information from various component methods (including assembly and associative techniques) toward a unified 3D structure prediction. In general, these component steps can be approximately summarized as follows:

1. Identify template proteins with structural similarity to the target as gauged (optimally) from sequence-based homology, or from physicochemical similarity.
2. Align the target sequence with all relevant template sequences.
3. Spatially align all of the template structures into a single framework, and use the sequence alignment to project the target protein backbone onto this framework.
4. Estimate structures for target protein fragments that are ill represented by the template manifold, or else omit them from the predicted structure.
5. Align target side chains with analogs within the template structures, or intelligently guess their disposition according to known spatial and torsional preferences.
6. Ameliorate unphysical contacts and strains via conformational searches.
7. Evaluate the final relaxed model for physical tenability.

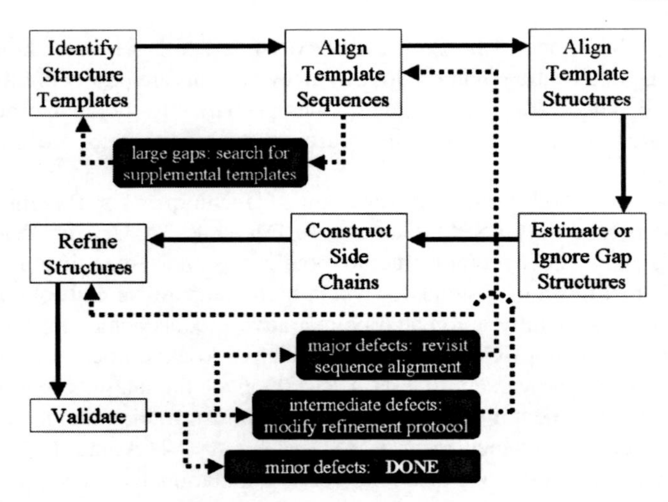

Fig. 1 Flow diagram for comparative modeling of proteins showing standard process (*solid arrows*) and feedback/refinement mechanisms (*dashed arrows*)

Each step above entails various methodological and strategic considerations, some of which provide opportunities for iterative feedback to previous steps, as is shown graphically in Fig. 1. These considerations will be elaborated on in the remainder of the chapter.

2.1 Template Identification

When identifying prospective templates via sequence homology, a reasonable template–target sequence identity is critical for minimizing the risk of errors in sequence alignment and for maximizing the likelihood that the entire template fold is to the target. The widely used minimum sequence identity criterion is 30% sequence identity over the extent of the mutual target–template alignment. The precise origin of this number is difficult to trace, in that the seminal paper typically cited for template identification, by Chothia and Lesk, offered a more conservative criterion of 50% identity, although their objective was to find highly similar structures with less than 1.0-Å RMSD in the position of backbone atoms [11]. Conversely, others have claimed that identity levels as low as 20 to 22% are frequently still viable [12]. Nonetheless, the empirical 30% criterion is broadly accepted, has stood the test of recent careful analyses [13], and, given a careful, well-scrutinized sequence alignment, seems to offer good odds for reliable structure prediction.

Although many targets exist for which no templates with at least 30% sequence identity are available, it has been estimated via fold statistics that approximately 70% of all possible targets of interest should have a template of reasonable structural similarity already present in the PDB. This wealth of templates may be explained from the fact that structural conservation can often be largely retained even in cases of very distant ancestral commonality. Furthermore, shared structure is also possible courtesy of evolutionary convergence, such that unrelated proteins may gradually adapt similar structures because of an inherently strong functional value of having such a structure. Thus, if no template meeting the 30% sequence identity is identified through homology searches, a reasonable model may still be achieved via a technique known as "threading," which evaluates target sequences against a library of unique fold representatives (a collection of structurally unique proteins such as those collected in the class, architecture, topology, and homologous superfamily [CATH] database [14]) according to residue by residue similarity in terms of spatial size, hydrophilicity/hydrophobicity, helix- or sheet-forming propensity, etc. Functionally special residues such as prolines (whose closed-ring structure induces a kink in AA chains), cysteines (well known for their fold-stabilizing disulfide bonds), and, sometimes, ionic residues (aspartate, glutamate, lysine, and arginine—all capable of forming salt bridges) are frequently given additional weight in any assessment. Because sequence identity is generally less useful for assessing threading templates, a consensus Z-score is instead used to gauge statistical significance for the quality of one template candidate relative to others being considered. Z-score scales vary across the manifold of different threading programs, but typically provide an indication of how likely it is that a given target–template pair are actually members of a common protein family or super family, whether they merely share a common fold, or whether they have no obvious relationship. An example of such ranges, as reported for the widely used PROSPECT-II program [15], is provided in Fig. 2.

The Z-score provides a reasonable scheme for identifying the plausible templates, but has a margin of error. A study contrasting performance of PROSPECT-II with other threading programs found that the top-scoring PROSPECT-II match was a valid fold-conserving template 84.1% of the time when the template manifold contained species within the same family as the target, 52.6% of the time when superfamilial (but no familial) templates were available, and 27.7% if the best templates merely shared a common fold [15]. These numbers improved to 88.2%, 64.8%, and 50.3% when examining the top five matches for possible valid templates. These are reasonably successful ratios relative to competing threading models, but do highlight the possibility that invalid templates may achieve high scores, and that one should scrutinize multiple top-scoring candidates, looking for templates known to prefer similar regions of cells as the target species, known to perform functions at least vaguely similar to the target, etc.

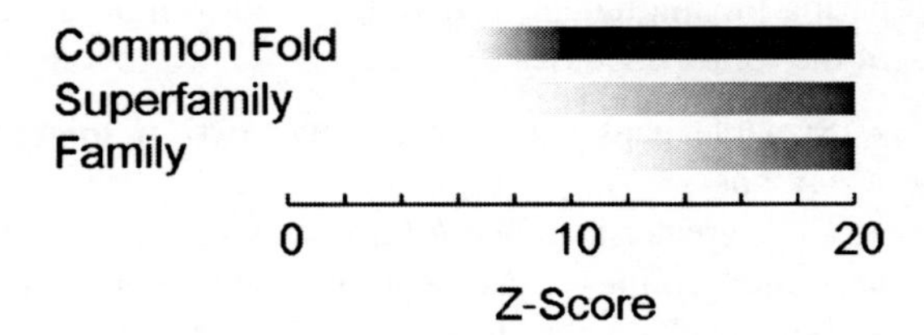

Fig. 2 Observed correspondence between PROSPECT Z-scores computed for pairs of proteins and the likelihood of their sharing a common fold or having a familial relationship

Note that in cases in which templates have been identified by homology, but the sequence identity is somewhat below the safe range (i.e., less than 50%, and especially if less than 30%), threading can provide an excellent point of validation. If threading analysis on the target template does not rank the target at a level that seems to assure a conserved fold, the template should be scrutinized with care and skepticism.

Also note that regardless of whether homology or threading is used for template identification, there is often strength in numbers. If multiple plausible templates that exhibit some compositional diversity but have similar 3D folding patterns are used, the quality of the resulting sequence alignment and comparative model typically benefits, because multiple sequence alignments are typically more accurate than pairwise alignments, and simultaneous accounting for multiple structures may tend to smoothen out local structural aberrations. One should avoid using template relatives that were resolved in substantially different conformations, however, because this may result in a final model corresponding to an unphysical hybrid. For metalloproteins, where the presence or absence of the metal ions can substantially impact the resulting protein conformation, one should identify the desired metal ion state of the target and choose templates with analogous ion populations.

2.2 Sequence Alignment

The various programs commonly used to template identification generally also yield a tentative sequence alignment relative to the target. In homologous cases with greater than 50% target–template sequence conservation over the mutually aligned portion of the structure, it is generally assumed that the alignment prediction algorithm will produce a qualitatively reliable alignment with only modest local misalignments (no positional errors more than several residues). Over a data set of broadly varying protein similarity, the PROSPECT-II assessment of threading reliability was that the program could achieve approximately a 60% average accuracy in prediction the alignment position of any given residue, and typically located each residue within four AA positions of the correct spot approximately 80% of the time [15]. In fairly strong threading models (e.g., PROSPECT Z-scores >10) one can likely assume better performance, perhaps on par with good homology alignments. However, in all cases, and especially in those with poorer identity or Z-scores, careful manual validation is a good policy. This can be achieved by comparing the alignment relative to the known 3D (template) structure to identify any of the following cases:

- Sizeable gaps (i.e., greater than two or three AAs) present in template core regions
- Gaps greater than one AA in known template sheets or helices
- Target prolines located within known template helices
- Positional displacement of more than one or two AAs for template cysteine residues known to engage in disulfide bonds

- Displacement of more than two to three AAs for template ionic residues known to form salt bridges

Alignments containing instances such as those above can yield unphysical comparative models for which fold conservation may be less favored or even no longer feasible. Manual adjustments may, thus, be made to alleviate such errors, at the expense of opening or extending gaps in exterior loops, as long as relatively modest penalties are incurred in the overall alignment score. Any templates that have a significant number of alignment problems that cannot be alleviated through minor manual adjustment should be viewed with skepticism. Furthermore, the number of irreconcilable alignment problems observed for a given target–template pair is a good source of evaluating and prioritizing that template's suitability relative to other candidates.

The presence of template gaps in regions that should correspond to solvent-exposed loops in the target species is rarely a key criterion in evaluating a template or a target–template alignment. Such loops have a relatively modest impact on the basic fold of a protein, thus, from an evolutionary perspective, they tend to have the largest frequency of noncritical point mutations, insertions, and deletions among homologs. If the gap in such a loop is relative small (less than 5 AAs), and is not thought to play an important role in properties of interest (i.e., not being part of a known receptor or protein–protein interaction site) it might be justifiably omitted, although it is generally more satisfying to piece it together according to empirical loop libraries (e.g., [16, 17]) or as an arbitrary structure such as a beta turn, with the expectation that its relatively large mobility will be amenable to subsequent refinement steps. For gaps that are significantly longer than 10 AAs, however, the structural integrity of your predicted model is best served by finding a legitimate template for the unrepresented loop itself. This can entail addition of a lower-ranked template that may be less suitable for global alignment, but that does afford a reasonable alignment to the gapped regions. If such a template is thought to have regions other than the gap in question that are inappropriate for modeling the rest of the target (i.e., very poor alignment, or known to have an inappropriate conformation), one may discard all of the template except that specifically corresponding to the gap region.

2.3 Spatial Alignment of the Target

Once templates have been selected and a sequence alignment is in place, construction of a preliminary structural model of the target is straightforward and is typically available in black-box form in most comparative modeling programs. The assembly methods typically construct a backbone model first, and then incorporate side chains into the resulting framework. Some methods may assemble the core region of a protein (solvent-inaccessible portions of the structure plus conserved secondary structure elements) first, then treat exposed loops.

Most backbone modeling methods begin by superimposing all templates onto a common framework and computing a consensus framework defined by mean positions of corresponding C_αs. One of three different schemes is then typically used: 1) rigid body assembly of target fragments corresponding to nongapped portions of the target–template alignment, as is implemented in the COMPOSER program [18]; 2) segment matching whereby target protein fragments are projected onto the consensus backbone with torsional angles being established through reference to fragment polypeptide libraries (e.g., SEGMOD [19]); or 3) construction of a compromise model that minimizes steric and torsional restraints of the target backbone as it is projected onto the template framework (e.g., MODELLER [20]). All of the above methods have proponents, and it is unclear whether the ultimate model accuracy varies much as a function of method choice, in that subsequent refinement is usually required regardless of technique. The restraints-based formalism is probably the most widely used by comparative modelers at this point.

One major complication to the assembly process may arise when the target protein contains a terminal domain that is represented by a different template than is used for the main core of the protein. In such cases, it may be unclear from the relevant templates how the terminal domain should pack onto the core framework. One method that we have applied for such a scenario [21] is to assemble the main core and the terminal domain as separate models, and to estimate the preferred core–terminus packing via protein–protein docking analysis as performed by GRAMM [22], ZDock [23], and other programs. Protein–protein docking methods typically offer a suitability score for each predicted complex, thus, helping to guide the selection of a complex to serve as the packing model. Another practical and critical consideration is whether a predicted complex places the two units in a position where it is possible to chemically rejoin the broken AA chain without unphysical strain on the backbone. Given a very low probability of predicting a complex that perfectly places the final AA of one unit in covalent binding distance to the initial AA of the next unit, our strategy has been to omit a portion of the target sequence within the core–terminus boundary region during initial model construction for these two domains, with the intention of re-integrating this portion as a flexibly modeled gap (see the next section). Specifically, our recommended protocol is to omit a number of residues (N_R) that are not predicted to be part of a defined secondary structure element (i.e., unstructured coil), and then constrain the choice of docked complexes to those that place the final C_α of the first domain within a distance of less than $N_R \times 3.0$ Å from the first C_α of the next domain (i.e., somewhat less than the maximum unstrained C_α to C_α distance of approximately $N_R \times 3.8$ Å). This method has not yet been exhaustively validated, but is based on reasonable logic and provides a potentially workable solution to an otherwise very challenging problem.

2.4 Loop and Gap Modeling

Although the previous backbone assembly step should yield structure predictions for many (hopefully most) of the loops in your target, it is common for some to be represented by poor sequence conservation or to appear in gapped regions of the alignment. Gaps covering 5 AA positions or less pose minimal concerns because they can typically be patched in with reasonable accuracy by referring to polypeptide structure libraries (e.g., [16, 17, 19]), but gaps of greater length are difficult to reliably model via methods other than template comparison. In cases in which no template for the gap can be found from either homology searches or threading, it is still possible to attempt a prediction based either on the same peptide libraries available for short loops or an algorithmic scheme akin to protein folding strategies (e.g., for molecular dynamics [MD], see Chap. 1; for Monte Carlo conformational searches, see Chap. 5, etc.). Neither of these strategies is guaranteed to produce a model with close correspondence to its real optimal structure, especially for gap lengths larger than 10 AAs. In general, however, many loops without a suitable template remain unresolved precisely because they inherently have high conformational mobility. In such cases, nonoptimal conformers may well still correspond to in vivo accessible structures.

2.5 Side Chain Modeling

Conserved disulfide bonds and salt bridges are typically incorporated into the target model directly during the backbone assembly process. Beyond this, side chain positions for highly conserved residues may also be inferred directly from the template, although their conformations are known to vary significantly from one protein to another when sequence identity is less than 50%, and are considered to be poorly conserved for identities less than 30% [24]. Fortunately, a number of effective side chain packing algorithms have been developed and validated (see Huang et al. [25] for an excellent review), and are often implemented as black-box features in comparative modeling programs.

2.6 Refinement

The nature of comparative modeling, whereby protein structures are predicted by analogy to different but related proteins, invariably leads to some degree of structural error. If the template (or template manifold) contains no sequence gaps relative to the target and only minor variations in sequence, the resulting error may be less than the resolution accuracy of the template(s), thus, further correction to the predicted target structure would be unnecessary. However if the target–template alignment contains gaps or has sequence identity less than 70%, some structural refinement is advisable. In cases in which some templates have been selected to cover regions of the sequence poorly represented by primary templates, boundary effects arise in which one template may significantly perturb the projected target backbone derived from other templates and vice versa. Library models used for loop patching lack specific environment information necessary to adapt the loop to the target of interest and, thus, yield imperfect backbones. It is finally important to note that most backbone errors are amplified in the side chain prediction. Fortunately, most models that have been assembled with care and without unrealistic assumptions will possess only modest errors that may be corrected in a physically natural manner. In vivo, a protein that has been perturbed from its native conformation because of some minor disturbance (e.g., intermolecular interaction, change in ionic state, etc.) stands a reasonable chance of reverting to normal structure when a stable normal environment is restored. By analogy, one may consider the deviations arising from a careful but imperfect protein assembly to be comparable to an environmental perturbation, thus, any simulation that subjects the protein model to conditions akin to a normal in vivo environment should encourage reversion to a reasonable structure. MD methods are exceptionally well suited to this task: protein simulation is one of the most common MD applications (see Chap. 1), and excellent force fields (Chap. 4) have, thus, been developed for simulating the interactions of proteins (and their constituent cofactors, ions, etc.) with solvent environments comparable to in vivo conditions. Various reviews on the MD strategies, techniques, and resources applicable to protein structure prediction and refinement are available (e.g., refs. [26,27]). The main drawback to MD simulations is computational expense, however; thus, when seeking to mitigate unphysical structural aspects arising from model assembly, one may find that constant-temperature simulations may not accomplish all necessary refinements in a viable time frame. An alternative is to perform simulated annealing calculations wherein the protein is gradually warmed up to a fairly hot temperature (1000 K is a reasonable limit; beyond this one risks inducing potentially irreversible unphysical structural changes) and then slowly cooled back to ambient temperature. After multiple thermal cycles of this sort (typically 5–10, with each cycle lasting from 10–100 ps), most errors in the original structure are corrected, generally leaving a fairly plausible conformer. Some comparative modeling programs, such as MODELLER [20], contain an embedded MD code and perform annealing as an option of the structure prediction process.

Special caution is needed when modeling transmembrane proteins. Many MD methods and parameters have been tailored for the specific case of soluble proteins immersed in a polar (generally aqueous) media, and may yield unphysical conformations for inherently hydrophobic membrane-binding portions of a protein. Special methods for simulating transmembrane proteins are discussed in Chaps. 8 to 10 of this volume and, e.g., in a review by Im and Brooks [28].

2.7 Validation

Any protein structure determination, from a rough comparative model to a high-resolution synchrotron-based crystal structure, will differ somewhat from the real in vivo conformation. A number of validation tools have, thus, arisen to evaluate structural models and detect aspects that seem to differ conformationally from standard bond distance, angle, torsion, or contact ranges derived from extensive assessment of known structures. Conveniently, many of the best validation tools are available online, e.g.:

- Anolea: http://www.swissmodel.unibas.ch/anolea/
- Biotech Validation Suite for Protein Structures: http://biotech.ebi.ac.uk:8400/
- EVA: http://maple.bioc.columbia.edu/eva/
- PROCHECK: http://www.biochem.ucl.ac.uk/~ roman/procheck/procheck.html

In comparing the above tools, one finds some redundant checks, but each of the above four utilities has unique features, thus, a rigorous structure evaluation should consider all of them. Among the various warnings that are likely to be issued for a comparative model, some minor problems may be alleviated by more refinement, however, care should be exercised in that overrefinement is often itself a source of error. Errors that are more serious may be indicative of poor (hopefully correctable) choices in the original sequence alignment or loop assembly in the region highlighted. Other issues may arise not from the modeling process but rather from using templates that themselves contain errors. To reduce instances of the latter, one may perform similar validation analysis on candidate templates before use and, thus, screen out inferior structures.

Judgments regarding whether a validation warning warrants countermeasures hinge on how severe the error seems to be, whether it lies in a particular interesting region of the molecule, and on planned analysis to be performed on the model. For small molecule docking studies, accurate structures are required in the receptor region, and side chain conformations can be critical, thus, one would likely try to alleviate any appreciable error within reasonable nonbonding radius (typically ~8 Å) of the putative binding site. Protein–protein docking studies and MD analyses are generally much less sensitive.

A well-constructed and validated structure model can open many doors for subsequent analysis. In addition to valuable insight derived from simple visual inspection, the model can form a reliable basis for many other modeling analyses, as is discussed extensively in other chapters of this book.

References

1. Kendrew, J.C., Bodo, G., Dintzis, H.M., Parrish, R.G., Wyckoff, H., and Phillips, D.C. (1958) A three-dimensional model of the myoglobin molecule obtained by x-ray analysis. *Nature.* **181,** 662–666.
2. Baker, D. and Šali, A. (2001) Protein structure prediction and structural genomics. *Science.* **294,** 93–96.
3. Martí-Renom, M.A., Stuart, A.C., Fiser, A., Sánchez, R., Melo, R., and Sali A. (2000) Comparative protein structure modeling of genes and genomes.*Biomol. Struct.* **29,** 291–325.
4. Anfinsen, C.B., Redfield, R.R., Choate, W.I., Page, J., and Carroll, W.R. (1965) Studies on the gross structure, cross-linkages, and terminal sequences in ribonuclease. *J. Biol. Chem.* **207**, 201–210.
5. Dayhoff, M.O. (1972) *Atlas of Protein Sequence and Structure.* National Biomedical Research Foundation, Georgetown University, Washington DC.

6. Chothia, C. (1992) Proteins. One thousand families for the molecular biologist. *Nature*. **357,** 543–544.
7. Prusiner, S.B. (1991) Molecular biology of prion diseases. *Science*. **252,** 1515–1522.
8. Takagi, F., Koga, N., and Takada, S. (2003) How protein thermodynamics and folding mechanisms are altered by the chaperonin cage: *Mol. Sim. Proc. Natl. Acad. Sci. USA*. **100,** 11367–11372.
9. Baumketner, A., Jewett, A., and Shea, J.E. (2003) Effects of confinement in chaperonin assisted protein folding: rate enhancement by decreasing the roughness of the folding energy landscape. *J. Mol. Biol.* **332,** 701–713.
10. Rost, B. (2001) Protein secondary structure prediction continues to rise. *J. Struct. Biol.* **134,** 204–218.
11. Chothia, C. and Lesk, A.M. (1986) The relation between the divergence of sequence and structure in proteins. *EMBO J.* **5,** 823–826.
12. Chung, S.Y. and Subbiah, S. (1996) How similar must a template protein be for homology modeling by side-chain packing methods? *Pac. Symp. Biocomput.* 126–141.
13. Forrest, L.R., Tang, C.L., and Honig, B. (2006) On the accuracy of homology modeling and sequence alignment methods applied to membrane proteins. *Biophys. J.* **91,** 508–517.
14. Pearl, F., Todd, A., Sillitoe, I., Dibley, M., Redfern, O., Lewis, T., Bennett, C., Marsden, R., Grant, A., Lee, D., Akpor, A., Maibaum, M., Harrison, A., Dallman, T., Reeves, G., Diboun, I., Addou, S., Lise, S., Johnston, C., Sillero, A., Thornton, J., and Orengo, C. (2005) The CATH Domain Structure Database and related resources Gene3D and DHS provide comprehensive domain family information for genome analysis. *Nucl. Acids Res.* **33,** D247–D251.
15. Dongsup, K., Xu, D., Guo, J.T., Elrott, K., and Xi, Y. (2003) PROSPECT II: protein structure prediction program for genome-scale applications. *Prot. Eng.* **16,** 641–650.
16. Fernandez-Fuentes, N., Oliva, B., and Fiser, A. (2006) A supersecondary structure library and search algorithm for modeling loops in protein structures. *Nucl. Acids Res.* **34,** 2085–2097.
17. Kolodny, R., Koehl, P., Guibas, L., and Levitt, M. (2002) Small libraries of protein fragments model native protein structures accurately. *J. Mol. Biol.* **323,** 297–307.
18. Sutcliffe, M.J., Haneef, I., Carney, D., and Blundell, T.L. (1987) Knowledge-based modelling of homologous proteins. Part I. Three dimensional frameworks derived from the simultaneous superposition of multiple structures. *Protein Eng.* **1,** 377–384.
19. Levitt, M. (1992) Accurate modeling of protein conformation by automatic segment matching. *J. Mol. Biol.* **226,** 507–533.
20. Šali, A. and Blundell, T.L. (1993) Comparative protein modeling by satisfaction of spatial restraints. *J. Mol. Biol.* **234,** 779–815.
21. Lushington, G.H., Zaidi, A., and Michaelis, M.L. (2005) Theoretically predicted structures of plasma membrane Ca^{2+}-ATPase and their susceptibilities to oxidation. *J. Mol. Graph. Modeling* **24,** 175–185.
22. Tovchigrechko, A. and Vakser, I.A. (2005) Development and testing of an automated approach to protein docking. *Proteins*. **60,** 296–301.
23. Wiehe, K., Pierce, B., Mintseris, J., Tong, W., Anderson, R., Chen, R., and Weng, Z. (2005) ZDOCK and RDOCK performance in CAPRI rounds 3, 4, and 5. *Proteins*. **60,** 207–221.
24. Chung, S.Y. and Subbiah, S. (1996) A structural explanation for the twilight zone of protein sequence homology. *Structure*. **4,** 1123–1127.
25. Huang, E.S., Koehl, P., Leavitt, M., Pappu, R.V., and Ponder, J.W. (1998) Accuracy of side-chain prediction upon the near-native protein backbones developed by ab initio folding methods. *Proteins*. **33**, 204–217.
26. Caflisch, A. and Paci, E. (2005) Molecular dynamics simulations to study protein folding and unfolding. *Protein Folding Handbook*. **2,** 1143–1169.
27. Karplus, M. and Kuriyan, J. (2005) Molecular dynamics and protein function. *Proc. Natl. Acad. Sci. USA*. **102,** 6679–6685.
28. Im, W. and Brooks, C.L., III. (2005) Interfacial folding and membrane insertion of designed peptides studied by molecular dynamics simulations. *Proc. Natl. Acad. Sci. USA*. **102,** 6771–6776.

Chapter 12

Transmembrane Protein Models Based on High-Throughput Molecular Dynamics Simulations with Experimental Constraints

Andrew J. Beevers and Andreas Kukol

Summary Elucidating the structure of transmembrane proteins domains with high-resolution methods is a difficult and sometimes impossible task. Here, we explain the method of combining a limited amount of experimental data with automated high-throughput molecular dynamics (MD) simulations of α-helical transmembrane bundles in an explicit lipid bilayer/water environment. The procedure uses a systematic conformational search of the helix rotation with experimentally constrained MDs simulations. The experimentally determined helix tilt and rotational angle of a labeled residue with site-specific infrared dichroism allows us to select a unique high-resolution model from a number of possible energy minima encountered in the systematic conformational search.

Keywords: Alpha helix · Experimental constraints · Infrared spectroscopy · Lipid bilayer · Membrane proteins · Molecular dynamics

1 Introduction

Membrane protein structures are underrepresented in the Protein Data Bank [1], currently constituting less than 0.5% of available high-resolution protein structures, yet they are encoded by approximately 30% of the human genome [2] and are estimated to form more than half of available drug targets [3]. The most common methods used, x-ray crystallography and solution-state nuclear magnetic resonance (NMR) spectroscopy, face enormous obstacles with membrane proteins because of the difficulties in obtaining crystals diffracting to sufficient resolution and the insolubility of membrane proteins in aqueous solvents, requiring the use of proteoliposomes or detergent–protein aggregates, which exceed the size limit of NMR spectroscopy. Attenuated total reflection Fourier-transform infrared spectroscopy (ATR-FTIR) is the method of choice for the structural analysis of proteins in the lipid membrane environment [4], yielding secondary structure information and information

From: Methods in Molecular Biology, vol. 443, Molecular Modeling of Proteins
Edited by Andreas Kukol © Humana Press, Totowa, NJ

regarding the maximum helix tilt angle with respect to the membrane normal of an α-helical transmembrane protein. Absorption of infrared radiation is caused by the excitation of vibrational (and rotational) energy levels, and through specific labeling of atoms in amino acid residues with rare isotopes (e.g., $^{13}C{=}^{18}O$ carbonyl group), vibrations of individual groups can be analyzed, thus, achieving atomic resolution [5, 6]. This is the basis of the technique of site-specific infrared dichroism (SSID) allowing determination of the orientation of transmembrane α-helical bundles [5,7], which has been applied in combination with molecular modeling to the structural elucidation of the transmembrane domains of various proteins [8–11]. This chapter describes the methodology for a conformational search procedure of α-helical bundles in an explicit lipid bilayer based on molecular dynamics (MD) simulations incorporating experimental constraints from SSID and conventional ATR-FTIR spectroscopy. In principle, the methods discussed are applicable to any sort of experimental constraints, e.g., from solid-state NMR spectroscopy, which is another emerging technique for structural investigation of transmembrane proteins [12].

The methods are exemplified with the oncogenic mutant ErbB-2 transmembrane domain. ErbB-2 is an epidermal growth factor receptor involved in the mediation of cell growth and differentiation [13, 14]. ErbB receptors are composed of an extracellular ligand-binding region, a transmembrane region, and a cytoplasmic tyrosine kinase. Activation of the tyrosine kinase occurs by ligand-induced dimerization, a generally accepted activation pathway for all receptor tyrosine kinases [15]. The rat oncogene neu encodes a mutant ErbB-2 receptor, which contains a single Val664Glu mutation in the transmembrane domain [16–18]. This mutation causes permanent association and activation of ErbB-2 that leads, finally, to tumor formation.

2 Theory

MD simulations of biomolecules (discussed in Chap. 1 of this volume) calculate the thermal fluctuation of a molecule applying Newton's laws of motion. The force field (discussed in Chap. 4) for biomolecules is often based on a description of the molecule as hard spheres and harmonic springs and incorporates the interactions between all particles in the simulation system, yielding the potential energy in dependence of the coordinates. Compared with the complexity of physical interactions between molecules in a test tube, computational force fields are oversimplified. Furthermore, one MD simulation represents the trajectory of one molecule, whereas an experiment in a test tube using 1 mL of a 1 μM solution is the statistical average over 6×10^{14} molecules. Additionally, the modeler has to ensure that the molecule subjected to MD simulations can overcome energy barriers to find the global energy minimum.

Shortcomings of the force field can be addressed by including experimental data as constraints, which add to the potential energy of the system. Additionally, MD simulations of the same molecule should be performed several times, starting with different initial random atom velocities. The problem of overcoming energy barriers

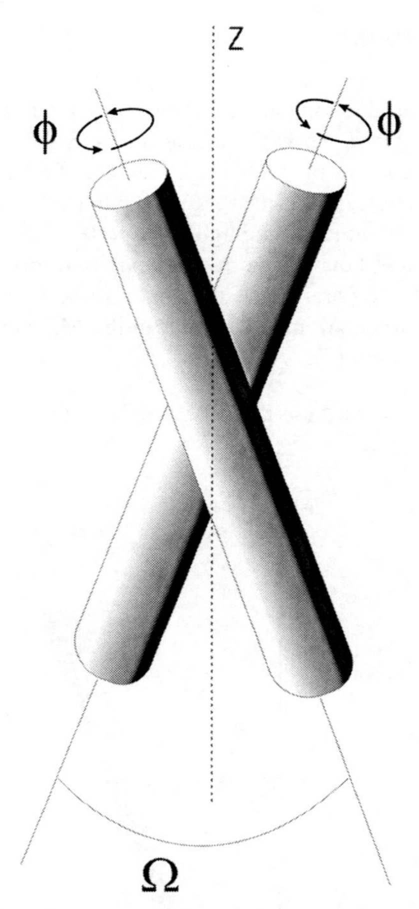

Fig. 1 The method of rotating the peptides by increments of the angle ϕ, resulting in the array of starting structures used for MD simulation. The angle Ω represents the helical crossing angle

is sometimes addressed by increasing the temperature of the simulation. It should, however, be noted that the parameters of the force field are usually optimized for the thermodynamic standard temperature of 298 K. In this chapter, the approach of using a systematic array of starting structures is adopted. For the common structural arrangement of transmembrane α-helical bundles, the structural parameters to vary are the helix rotation, ϕ (Fig. 1), and the helix crossing angle, Ω.

Another important point for the MD simulation is the accurate representation of the complex system composed of water molecules, protein, and lipid molecules. With increasing computer power, an explicit atomistic representation of water and lipid molecules is adopted in current the state-of-the-art simulations, as discussed in Chap. 8 of this volume.

2.1 Defining Constraints

As mentioned in the introduction, any experimental method may be used to derive constraints in proteins. Here, we will concentrate on SSID as an example applied to the transmembrane domain of the ErbB-2 receptor. This technique allows us to determine the helix tilt angle, β, and the rotational pitch angle, ω, of specific residues in an α-helix, while a rotational pitch angle of $\omega = 0°$ is defined as the position at which the C=O carbonyl bond of the residue points in the direction of the helix tilt [5] (Fig. 2). From these parameters, the angle of the C=O bond to the z-axis, θ, which is used as an orientational constraint in the MD simulation, is calculated using the following equation [7]:

$$\cos\theta = \cos\alpha\cos\beta - \sin\alpha\sin\beta\cos(\omega + 17°),$$

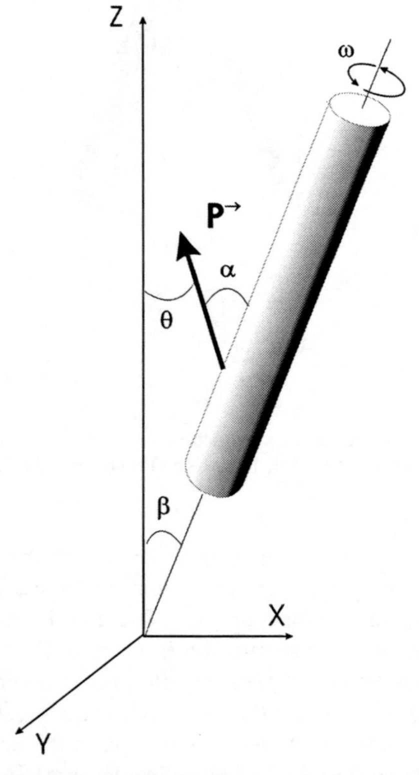

Fig. 2 Definition of the helix tilt β, the rotational pitch angle ω, and the angle θ representing the angle between a particular labeled amide I transition dipole moment P and the z-axis. ω is defined as 0° when the transition dipole moment points in the direction of the helix tilt. The angle α is defined as 38° for all α-helices [19]

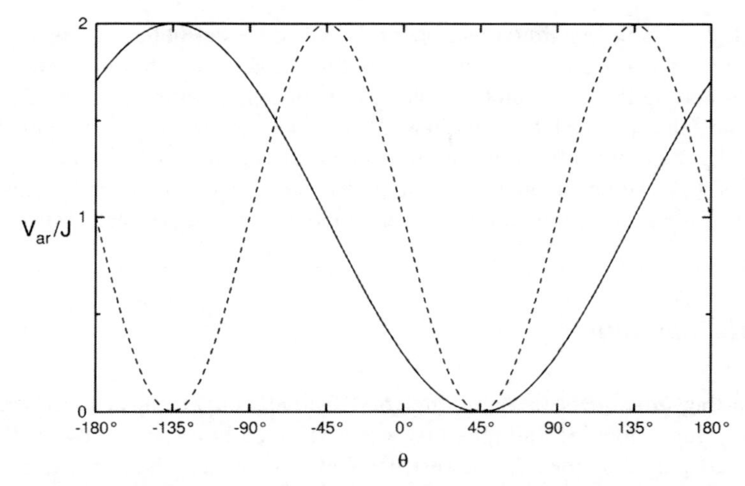

Fig. 3 Plot of the potential energy $V_{ar} = \mathrm{k_{ar}}\{1 - \cos[n(\theta - \theta_0)]\}$ function in J (joule) for orientational constraints with $\mathrm{k_{ar}} = 1$, $\theta_0 = 45°$, and n = 1 (*solid curve*) or n = 2 (*dashed curve*)

where α represents the angle between the transition dipole moment and the molecular director (given as 38° in all α-helices [19]).

The angle θ of a specified C=O bond to the z-axis is constrained using the potential energy term implemented in the GROMACS MD simulation package [20, 21] using the equation:

$$V_{ar} = \mathrm{k_{ar}}\{1 - \cos[n(\theta - \theta_0)]\},$$

where $\mathrm{k_{ar}}$ is the force constant and θ_0 represents the experimentally defined angle. The multiplicity n = 2 is usually chosen to treat parallel and antiparallel vectors equally (see Fig. 3). The same equation is used to constrain tilt angles by setting the angle between a vector connecting the Cα atoms of residue i and i + 7 (two full helical turns) and the z-axis to the experimentally defined angle.

Additional experimental information from FTIR spectroscopy is the secondary structure of the transmembrane peptide, which has been found to be α-helical in the case of the ErbB-2 transmembrane domain. This information is added to the force field by constraining the distance between the carbonyl oxygen of residue i and the amide hydrogen of residue i + 4; it is these groups that give rise to the well-defined intramolecular hydrogen bonding of α-helices.

2.2 Creation of Starting States

The first part of the computation is to create the stating states by systematic rotation of the α-helices around the helical axis (Fig. 1) by defined increments of ϕ, such as 10° (36 possible rotations) and adopting two different helical crossing angles $\Omega = \pm 25°$, which leads to 72 different starting structures. It has been shown that α-helical homo-oligomers can be treated by symmetric rotation, particularly when, during the subsequent MD simulations, no symmetry restrictions are applied [22]. Hetero-oligomers must be subjected to asymmetric rotation, which, unfortunately, generates a far larger amount of starting structures, e.g., 41,572 structures for a tetramer, if the same search parameters are applied as in the symmetric search.

2.3 MD Simulation

Each starting structure is then subjected to MD simulation, ideally at various random initial starting velocities. Initially, this was performed in a vacuum [8–10, 22, 23], but this chapter describes the newest development of including an explicit lipid bilayer–water environment [24,25]. The current approach inserts each starting structure into a hole created in a pre-equilibrated bilayer using a modified version of the GROMACS software, which uses an outward-directed force on the lipid molecules in the central section of the bilayer to create a hole [26]. This process is shown in more detail in Chap. 8.

2.4 Clustering

Each starting state will adopt an orientation that minimizes its potential energy during the MD simulation. As previously mentioned, it is highly unlikely that the MD simulation of one α-helical bundle will explore all of its possible rotations. Therefore, each helical complex will adopt an energy minimum close to its starting point. However, various structures will converge to a local energy minimum, thus, forming several clusters of structures, which can be identified by root mean square deviation (RMSD) comparisons between each pair of structures. Once a cluster is identified, an average structure is created by averaging the coordinates of each structure within the cluster. This may generate a slightly deformed structure, which needs to be subjected to energy minimization followed by MD simulations. Alternatively, a representative structure of the cluster could be chosen that is most similar to all other cluster members.

2.5 Analysis

As a result, a number of clusters emerge, from which averages are calculated. The cluster averages have a variety of rotational pitch angles, because the angle between the C=O bond and the z-axis is used as orientational constraint. Through slight deformation of the ideal α-helical geometry, this constraint can be fulfilled at various rotational pitch angles. Furthermore, the structures are the result of the interplay between the experimental constraints and the tendency of the structure to adopt an energy minimum in the "imperfect" computational force field, which also takes into account side chain packing and protein–lipid interaction for which there is no experimental data from SSID. Therefore, it is important that the rotational pitch angles of the labeled sites are calculated for each model structure and compared with the experiment. The correct model is not necessarily the lowest energy structure, but the structure that fits the experimental data with the lowest deviation, as was the case for the structure of the oncogenic neu/erbb-2 dimer [25] and the transmembrane domain structure of the HIV-1 vpu pentamer [9].

3 Method

The procedure uses the CHI suite of programs [22] for the crystallography and NMR system [27] (see **Note 1**), and the GROMACS [21] suite for MD simulations. CHI programs have been modified for use in the lipid search procedure, and the automation of the lipid MD simulation is carried out with software developed by the authors of this chapter (see **Note 2**). Unmodified CHI programs are identified below by the prefix "chi," e.g., chi_create, whereas modified CHI programs have the prefix "lchi," e.g., lchi_search.

3.1 Constraints

The first step of the procedure is to specify the experimental data; these are the angles between the C=O bond and the z-axis from SSID and distance restraints to specify the experimentally known helix geometry (see **Note 3**). These form part of a GROMACS topology file (Fig. 4).

3.2 Starting Structure Creation

chi_create generates an α-helical bundle structure according to sequence and dimerization specification in the parameter file chi_param. An array of starting structures is created with *lchi_search*, e.g., for the ErbB-2 example, 36 different rotations at 2 different crossing angles are used for simulation at 4 different initial random velocities, leading to $36 \times 2 \times 4 = 288$ starting structures.

```
[ angle_restraints_z ]
;     ai     aj     funct    th0      fc      mult
      148    149    1        -64.2    5000    2
      83     84     1        -55.6    5000    2
      225    226    1        -59.3    5000    2
      6      78     1        -28.1    5000    2
      78     135    1        -28.1    5000    2
      87     143    1        -25.2    5000    2
      143    213    1        -25.2    5000    2
      152    221    1        -21.9    5000    2
      221    282    1        -21.9    5000    2

[ distance_restraints ]
; ai    aj     type   index   type'   low    up1    up2    fac
49      86     1      0       2       0.16   0.21   0.23   5000
58      95     1      0       2       0.16   0.21   0.23   5000
75      101    1      0       2       0.16   0.21   0.23   5000
84      110    1      0       2       0.16   0.21   0.23   5000
93      118    1      0       2       0.16   0.21   0.23   5000
99      129    1      0       2       0.16   0.21   0.23   5000
108     134    1      0       2       0.16   0.21   0.23   5000
116     142    1      0       2       0.16   0.21   0.23   5000
127     151    1      0       2       0.16   0.21   0.23   5000
132     160    1      0       2       0.16   0.21   0.23   5000
140     177    1      0       2       0.16   0.21   0.23   5000
149     186    1      0       2       0.16   0.21   0.23   5000
158     195    1      0       2       0.16   0.21   0.23   5000
175     204    1      0       2       0.16   0.21   0.23   5000
184     212    1      0       2       0.16   0.21   0.23   5000
193     220    1      0       2       0.16   0.21   0.23   5000
202     228    1      0       2       0.16   0.21   0.23   5000
210     236    1      0       2       0.16   0.21   0.23   5000
218     241    1      0       2       0.16   0.21   0.23   5000
226     250    1      0       2       0.16   0.21   0.23   5000
234     259    1      0       2       0.16   0.21   0.23   5000
```

Fig. 4 An example of an itp file containing constraints for the angle to the z-axis for the CO= bond, the local helix tilt (between Cα atoms of residues i and i + 7 either side), and distances between amine hydrogens and carbonyl oxygens of residues i and i + 4 in the transmembrane region. ai, aj refer to the atom numbers while the other parameters (funct, th0, fc, etc.) specify the parameters of the potential energy function (see the GROMAC manual for a detailed explanation)

3.3 MD Simulation

A pre-equilibrated lipid bilayer of 128 DMPC molecules is used with the bilayer normal oriented parallel to the z-axis. The following procedure is performed automatically for each starting structure using GROMACS and software developed by the authors:

1. A hole is created in the bilayer by applying an outward-directed force on water and lipid molecules in its center [26].
2. A starting structure is inserted into this hole.

3. Counterions are added to neutralize overall charge.
4. The resulting lipid/water/ions/peptide system is subjected to energy minimization and MD simulation for 100 ps with the peptide atoms restrained in their current positions.
5. Position restraints are removed. The system is energy minimized and then subjected to a 200-ps MD simulation. Examples of input files for minimization, restrained MD, and unrestrained MD are shown in Fig. 5.
6. Finally, the system is subjected to energy minimization and analyzed. Energy, helix rotation, crossing angle, and helix shift is recorded.

Energy Minimisation

```
; VARIOUS PREPROCESSING OPTIONS
cpp                         = /lib/cpp
include                     = -Ilib
define = -DFLEX_SPC
; RUN CONTROL PARAMETERS
integrator                  = steep
; start
time and timestep in ps
tinit                       = 0 dt = 0.002
nsteps                      = 250
; number of steps for center of mass motion removal
nstcomm                     = 1

; Energy MINIMIZATION OPTIONS
; Force tolerance and initial step-size
emtol                       = 100
emstep                      = 0.01
; Max number of iterations in relax_shells
niter                       = 0
; Frequency of steepest descents steps when doing CG
nstcgsteep                  = 1000

; OPTIONS FOR ELECTROSTATICS AND VDW
; Method for doing electrostatics
coulombtype                 = Cut-off
rcoulomb-switch             = 0
rcoulomb                    = 1.8
```

Fig. 5 Parameter files for used for energy minimization, position restraint MD, and unrestrained MD. The parameters specify the algorithm used for simulation, the choice of temperature and pressure coupling, the treatment of electrostatic and van der Waals interactions, the use of extra input files for position restraints, and whether random velocities should be generated at the start up

```
; Dielectric constant (DC) for cut-off or DC of reaction
field
epsilon-r                = 1

; Method for doing Van der Waals
vdw-type                 = Cut-off
; cut-off lengths        =
rvdw-switch              = 0
rvdw                     = 1.0

; OPTIONS FOR BONDS
constraints              = none
; Type of constraint algorithm
constraint-algorithm     = Lincs
; Do not constrain the start configuration
unconstrained-start      = no
; Relative tolerance of shake =
shake-tol                = 0.0001
; Highest order in the expansion of the constraint coupling matrix
lincs-order              = 4
; Lincs will write a warning to the stderr if in one step a bond
; rotates over more degrees than =
lincs-warnangle          = 30
; Convert harmonic bonds to morse potentials
morse                    = no
```

Position Restraint MD

```
; VARIOUS PREPROCESSING OPTIONS
cpp                      = /lib/cpp
include                  = -Ilib
define                   = -DPOSRES

(everything else like Unrestrained MD)
```

Unrestrained MD

```
; VARIOUS PREPROCESSING OPTIONS
title                    =
cpp                      = /lib/cpp
include                  = -Ilib
```

Fig. 5 continued

```
; RUN CONTROL PARAMETERS
integrator               = md
; start time and timestep in ps
tinit                    = 0
dt                       = 0.002
nsteps                   = 1000000 ; 2 ns
; number of steps for center of mass motion removal
nstcomm                  = 1

; OPTIONS FOR ELECTROSTATICS AND VDW
; Method for doing electrostatics
coulombtype              = Cut-off
rcoulomb-switch          = 0


rcoulomb                 = 1.8
; Dielectric constant (DC) for cut-off or DC of reaction field
epsilon-r                = 1
; Method for doing Van der Waals
vdw-type                 = Cut-off
; cut-off lengths        =
rvdw-switch              = 0
rvdw                     = 1.0

; OPTIONS FOR WEAK COUPLING ALGORITHMS
; Temperature coupling   =
tcoupl                   = Berendsen
; Groups to couple separately
tc-grps                  = DMPC Protein SOL_Cl
; Time constant (ps) and reference temperature (K)
tau-t                    = 0.1 0.1 0.1
ref-t                    = 300 300 300
; Pressure coupling      =
Pcoupl                   = Berendsen
Pcoupltype               = Isotropic
; Time constant (ps), compressibility (1/bar) and reference P (bar)

   tau-p                    = 1
   compressibility          = 4.5E-5
   ref-p                    = 1

   ; SIMULATED ANNEALING CONTROL
   annealing                = no

   ; GENERATE VELOCITIES FOR STARTUP RUN
   gen-vel                  = yes
   gen-temp                 = 300
   gen-seed                 = 173529
```

Fig. 5 continued

```
; OPTIONS FOR BONDS
constraints              = all-bonds
; Type of constraint algorithm
constraint-algorithm     = Lincs
; Do not constrain the start configuration
unconstrained-start      = no
; Relative tolerance of shake
shake-tol                = 0.0001
; Highest order in the expansion of the constraint coupling matrix

lincs-order              = 4
; Lincs will write a warning to the stderr if in one step a bond
; rotates over more degrees than
lincs-warnangle          = 30
; Convert harmonic bonds to morse potentials
morse                    = no
```

Fig. 5 continued

3.4 Clustering

1. The RMSD between each pair of structures is calculated using chi_rmsd.
2. Clusters are then identified by RMSD comparisons between each structure. Structures are allocated to a cluster if the RMSD between each member of the cluster is lower than a set limit (2.0 Å for the ErbB-2 example) and a minimum number of structures (five for ErbB-2) contribute to that cluster.
3. Cluster average structures are calculated by averaging the coordinates, energy minimization, and experimentally constraint MD simulation of 2 ns after insertion in the lipid bilayer, following the same protocol as detailed in Sect. 3.3.

An overview of the resulting structures and clusters following the steps in Sects. 3.3 and 3.4 is given in Fig. 6.

3.5 Analysis

The cluster average structure is subjected to energy minimization and analyzed. Energy, helix rotation, crossing angle, and helix shift is recorded.

The resulting structures are manually compared with experimentally obtained rotational pitch angles, and the structure with the lowest deviation is the preferred model, as shown in Fig. 7 for the ErbB-2 example [25] (see **Note 4**).

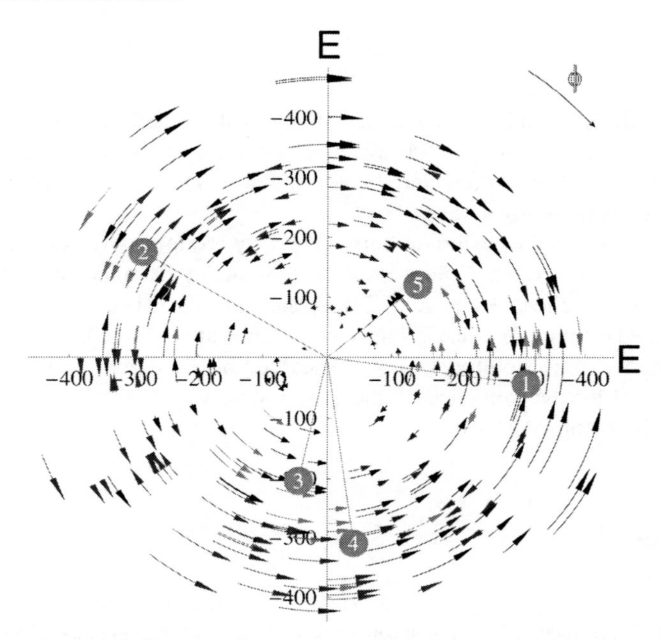

Fig. 6 Polar plot of the energies of structures obtained from an MD simulation of mutant ErbB-2 in dependence of the helix rotation angle, ϕ. The distance from the center indicates negative energy (E) in kJ/mol. Each individual structure is indicated by a *triangle*, whereas the clustered averages are shown as *numbered circles*. The arcs represent the movement of structures from their starting positions with respect to the helix rotation angle, ϕ, during the MD simulation

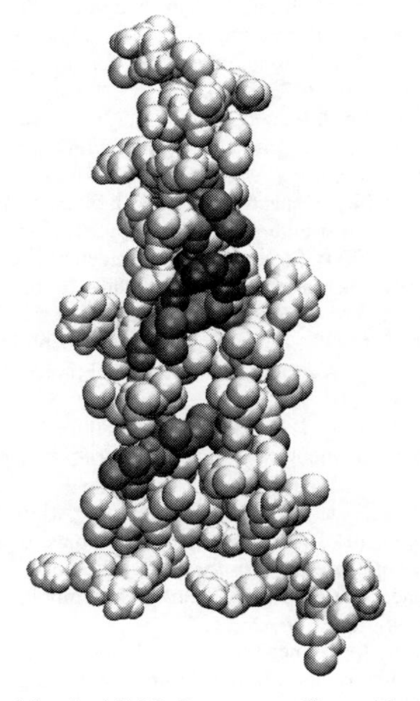

Fig. 7 Space-fill rendering of the final ErbB-2 structure. Glu residues are rendered in *black* and labeled residues in *dark grey* [25]

4 Notes

1. CHI is available from Paul Adams, http://www.csb.yale.edu/userguides/ datamanip/chi/html/chi.html and CNS is available from http://cns.csb.yale.edu.
2. The modified CHI programs and the additional software is available on request from the corresponding author.
3. Distance constraints to maintain helix geometry are normally not necessary for simulations in an explicit lipid bilayer because helices remain stable if they naturally occur as helices. However, FTIR spectroscopy has been used to determine the secondary structure, thus, distance constraints are included.
4. At this stage, some structures can be discarded because of significant deformations caused by the influence of the constraints, which are incompatible with some helix rotations.

References

1. Berman, H. M., Westbrook, J., Feng, Z., Gilliland, G., Bhat, T. N., Wessig, H., Shindyalov, I. N. and P.E., B. (2002). The protein data bank. *Nucleic Acids Research* **28**, 235–242.
2. Wallin, E. and von Heijne, G. (1998). Genome-wide analysis of integral membrane proteins from eubacterial, archaean and eukaryotic organisms. *Protein Science* **7**, 1029–1038.
3. Terstappen, G. C. and Reggiani, A. (2001). In silico research in drug discovery. *Trends Pharmacol Sci* **22**, 23–26.
4. Tatulian, S. A. (2003). Attenuated total reflection Fourier transform infrared spectroscopy: A method of choice for studying membrane proteins and lipids. *Biochemistry* **42**, 11898–11907.
5. Arkin, I. T., MacKenzie, K. R. and Brünger, A. T. (1997). Site-directed dichroism as a method for obtaining rotational and orientational constraints for orientated polymers. *Journal of the American Chemical Society* **119**, 8973–8980.
6. Torres, J., Kukol, A., Goodman, J. M. and Arkin, I. T. (2001). Site-specific examination of secondary structure and orientation determination in membrane proteins: the peptidic $^{13}C{=}^{18}O$ group as a novel infrared probe. *Biopolymers* **59**, 396–401.
7. Kukol, A. (2005). Site-specific IR spectroscopy and molecular modelling combined towards solving transmembrane protein structure. *Spectroscopy* **19**, 1–16.
8. Kukol, A. and Arkin, I. T. (2000). Structure of the Influenza C virus CM2 protein transmembrane domain obtained by site-specific infrared dichroism and global molecular dynamics searching. *Journal of Biological Chemistry* **275**, 4225–4229.
9. Kukol, A. and Arkin, I. T. (1999). Vpu transmembrane peptide structure obtained by site-specific Fourier transform infrared dichroism and global molecular dynamics searching. *Biophysical Journal* **77**, 1594–1601.
10. Kukol, A., Torres, J. and Arkin, I. T. (2002). A structure for the trimeric MHC Class II-associated invariant chain transmembrane domain. *Journal of Molecular Biology* **320**, 1109–1117.
11. Kukol, A., Adams, P. D., Rice, L. M., Brunger, A. T. and Arkin, I. T. (1999). Experimentally based orientational refinement of membrane protein models: a structure for the Influenza A M2 H+ channel. *Journal of Molecular Biology* **286**, 951–962.
12. Lindblom, G. and Grobner, G. (2006). NMR on lipid membranes and their proteins. *Current Opinion in Colloid and Interface Science* **11**, 24–29.
13. Olayioye, M. A., Neve, R. M., Lane, H. A. and Hynes, N. E. (2000). The ErbB signalling network: receptor heterodimerization in development and cancer. *EMBO Journal* **19**, 3159–3167.

14. Leahy, D. J. (2004). Structure and function of the epidermal growth factor (EGF/ErbB) family of receptors. *Advances in Protein Chemistry* **68**, 1–27.
15. Schlessinger, J. (2000). Cell signalling by receptor tyrosine kinases. *Cell* **103**, 211–225.
16. Bargmann, C. I. and Weinberg, R. A. (1988). Increased tyrosine kinase activity associated with the protein encoded by the activated neu oncogene. *Proceedings of the National Academy of Sciences USA* **85**, 5394–5398.
17. Bargmann, C. I., Hung, M.-C. and Weinberg, R. A. (1986). Multiple independent activations of the neu oncogene by a point mutation altering the transmembrane domain of p185. *Cell* **45**, 649–657.
18. Bargmann, C. I. and Weinberg, R. A. (1988). Oncogenic activation of the neu-encoded receptor protein by point mutation and deletion. *EMBO Journal* **7**, 2043–2052.
19. Tsuboi, M. (1962). Infrared dichroism and molecular conformation of α-form poly-gamma-benzyl-L-glutamate. *Journal of Polymer Science* **59**, 139–153.
20. Berendsen, H. J. C., van der Spoel, D. and van Drunen, R. (1995). GROMACS: A message passing parallel molecular dynamics implementation. *Comp. Phys. Comm.* **91**, 43–56.
21. Lindahl, E., Hess, B. and van der Spoel, D. (2001). Gromacs 3.0: A package for molecular dynamics simulation and trajectory analysis. *J. Mol. Mod.* **7**, 306–317.
22. Adams, P. D., Arkin, I. T., Engelman, D. M. and Brünger, A. T. (1995). Computational searching and mutagenesis suggest a structure for the pentameric transmembrane domain of phospholamban. *Structural Biology* **2**, 154–162.
23. Adams, P. D., Engelman, D. M. and Brunger, A. T. (1996). Improved prediction for the structure of the dimeric transmembrane domain of glycophorin a obtained through global searching. *Proteins: Structure, Function and Genetics* **26**, 257–261.
24. Beevers, A. J. and Kukol, A. (2006). Systematic molecular dynamics searching in a lipid bilayer: Application to the glycophorin A and oncogenic ErbB-2 transmembrane domains. *J Mol Graph Model* **25**, 226–233.
25. Beevers, A. J. and Kukol, A. (2006). The transmembrane domain of the mutant ErbB-2 receptor. A structure obtained from site-specific infrared dichroism and molecular dynamics simulations. *Journal of Molecular Biology* **361**, 945–953.
26. Faraldo-Gomez, J. D., Smith, G. R. and Sansom, M. S. P. (2002). Setting up and optimisation of membrane protein simulations. *European Biophysical Journal* **31**, 217–227.
27. Brunger, A. T., Adams, P. D., Clore, G., Gros, W., Grosse-Kunstleve, R., Jiang, J., Kuszewski, J., Nilges, M., Pannu, N., Read, R., Rice, L. M., Simonson, T. and Warren, G. (1998). Crystallography and NMR system: A new software suite for macromolecular structure determination. *Acta Crystallographica D Biological Crystallography* **54**, 905–921.

Chapter 13
Nuclear Magnetic Resonance-Based Modeling and Refinement of Protein Three-Dimensional Structures and Their Complexes

Gloria Fuentes, Aalt D.J. van Dijk, and Alexandre M.J.J. Bonvin

Summary Nuclear magnetic resonance (NMR) has become a well-established method to characterize the structures of biomolecules in solution. High-quality structures are now produced, thanks to both experimental and computational developments, allowing the use of new NMR parameters and improved protocols and force fields in structure calculation and refinement. In this chapter, we give a short overview of the various types of NMR data that can provide structural information, and then focus on the structure calculation methodology itself. We discuss and illustrate with tutorial examples both "classical" structure calculation and refinement approaches as well as more recently developed protocols for modeling biomolecular complexes.

Keywords: Docking · NMR · Refinement · Structure calculation · Validation of structures

1 Introduction

The first step of a structure determination by nuclear magnetic resonance (NMR) spectroscopy consists in the acquisition of NMR data, typically using heteronuclear multidimensional experiments that allow the assignment of the chemical shifts of all atoms/spins of a molecule (^{1}H, ^{15}N, ^{13}C). Once the signals have been assigned, ^{13}C- and ^{15}N-edited three-dimensional (3D) nuclear Overhauser enhancement spectroscopy (NOESY) spectra are generally used to obtain interatomic distances from nuclear Overhauser effects (NOE); these provide the required structural information to define the 3D structure of the protein [1, 2]. In addition to distance restraints, other parameters, such as J-couplings [3] and residual dipolar couplings (RDCs) [4] can be measured, providing additional structural information to define the structure of a protein. The experimental NMR parameters are then typically used in restrained molecular dynamics simulations following some kind of simulated

From: Methods in Molecular Biology, vol. 443, Molecular Modeling of Proteins
Edited by Andreas Kukol © Humana Press, Totowa, NJ

annealing scheme (MD/SA) to generate 3D structures [5]. These are nowadays usually refined in explicit solvent (water), which has been shown to significantly improve the quality of the structures [6,7]. The resulting ensembles of structures should satisfy as many restraints as possible, together with general chemical properties of proteins (such as bond lengths and angles). The whole approach will converge, provided enough data of sufficient quality are available, allowing the determination of an ensemble of structures with a given fold.

In the last few years, a lot of attention is directed toward understanding biomolecular interactions, and NMR is playing an important role here [8], especially in its ability to detect weak and transient interactions [9]. When dealing with complexes, NMR suffers, however, because of the size limitation problem, and, therefore, complementary computational methods, such as docking, are becoming increasingly popular. Docking is defined as the modeling of the 3D structure of a complex from its known constituents, and its combination with a limited amount of (NMR-) data (so called *data-driven docking*) is extremely powerful and has found a wide range of applications [10].

In this chapter, we discuss first "classical" NMR structure calculation and refinement methods and then address the modeling of protein–protein complexes. These will be illustrated with tutorial examples making use of the program CNS [11] with ARIA-derived [12] scripts from the RECOORD [7] webpage and of the HADDOCK package [13].

2 Theory

2.1 NMR Structural Information Sources

Several NMR parameters providing structural information can be measured for use in structure calculations and refinement. These will be briefly reviewed in the following subsections.

2.1.1 Nuclear Overhauser Effects

Classical protein structure determination by NMR relies on a dense network of distance restraints derived from NOEs between nearby hydrogen atoms in a protein [1,2].

The NOE originates from cross-relaxation between dipolar-coupled spins that involve a transfer of magnetization from one spin to another. The NOE approximately scales with the distance r between the two spins as $1/r^6$. Because of this $1/r^6$ dependency, NOEs are only detected between protons less than 5 to 6 Å away in space. They provide essential information for defining the tertiary structure of a protein.

2.1.2 Chemical Shifts

Although chemical shifts are very sensitive probes of the chemical environment of a spin, their dependency on the 3D structure is complex. Although several software packages exist that allow prediction of chemical shifts, such as ShiftX [14], SHIFTS [15], SHIFTCALC [16], and PROSHIFT [17], both computational and accuracy limits have prevented their common use as restraints in structure calculations, although direct refinement against chemical shifts has been described [18]. Their deviations from random coil values provide, however, valuable information regarding secondary structure preferences; they can be used to restrict the local conformation of a residue to a given region of the Ramachandran plot, either through torsion angle restraints [19] or by special database potential functions [20].

2.1.3 J-Couplings

Scalar or J-couplings are mediated through chemical bonds connecting two spins. The energy levels of each spin are slightly altered depending on the spin state of scalar coupled spins (α or β), resulting in splitting of the resonance lines. Particularly informative are the vicinal, three bonds scalar coupling constants, ^{3}J, between atoms separated by three covalent bonds from each other, which are correlated to the enclosed torsion angle, Θ, by an empirical correlation, the Karplus curve [21]. In particular, ^{3}J(HN-Hα) and ^{3}J(Hα-Hβ) give information regarding the φ-angle and the χ_1 angle in an amino acid, respectively. The use of ^{3}J(Hα-Hβ) coupling does require stereospecific assignments of diastereotopic proton Hβ2/Hβ3 pairs.

The main difference with NOEs is that scalar coupling constants only provide information regarding the local conformation of a polypeptide chain. J-couplings have been used as dihedral angle restraints [1] or direct J-coupling restraints [22,23] in NMR structure calculations.

2.1.4 Hydrogen Bonds

Slow hydrogen exchange indicates that an amide proton is protected from the solvent, which is usually interpreted as involvement in a hydrogen bond [24]. The acceptor atom cannot, however, be identified directly, and one has to rely on NOEs around the postulated hydrogen bond or assumptions regarding regular secondary structures to define it. Hydrogen bond restraints should be used with caution, although they can be very useful in the case of large proteins when not enough NOE data is available yet. Note that hydrogen bonds can now also directly be detected from cross-hydrogen bond scalar coupling measured from constant time HNCO spectra [25, 26]. These can provide useful restraints for structure calculations [27]. Hydrogen bond restraints are introduced into the structure calculation as distance restraints, typically by confining the donor hydrogen/acceptor distance to a given range.

2.1.5 Residual Dipolar Couplings

During the past years, RDCs have become an increasingly important source of structural information [28, 29]. They can be measured in solution by weakly aligning the molecule using a variety of methods [30]. RDCs provide angular information between the internuclear vector for which they are measured and a set of globally defined axes in the molecule, namely those of the alignment tensor. The measured RDCs are given by:

$$D^i(\beta^i\alpha^i) = 0.5D_0[A_a(3\cos^2\beta^i - 1) + \tfrac{3}{2}A_r(\cos 2\alpha^i \sin^2\beta^i)].$$

Here, A_a is the axially symmetric part of the alignment tensor, equal to $[A_{zz} - 1/2(A_{xx} + A_{yy})]$ and A_r is the rhombic component of the alignment tensor, equal to $(A_{xx} - A_{yy})$, where A_{xx}, A_{yy}, and A_{zz} are the x, y, and z-components of the alignment tensor, respectively; α^i and β^i are the azimuthal and polar angles of the vector for which the RDC is reported, in the frame of the alignment tensor. D_0 is the strength of the (static) dipolar coupling defined as:

$$D_0 = -\left(\frac{\mu_0}{4\pi}\right)\frac{\gamma_i\gamma_j h}{2\pi^2 r_{NH}^3},$$

which, in the case of N-NH RDCs is equal to 21.7 kHz. r_{NH} is the length of the NH vector, μ_o is the magnetic permeability of vacuum, γ_i is the gyromagnetic ratio of spin i, and h is Planck's constant. The structural information is contained in the angles α and β; note that if an RDC is measured between two atoms that are not at a fixed distance from each other, there is also a distance dependence (via the r term in D_0). RDCs can be added as orientational restraints to the target function of the structure calculation algorithm [31]. Usually, only RDCs measured for internuclear vectors with a fixed distance are used.

2.1.6 Diffusion Anisotropy

Diffusion anisotropy (relaxation) data contain orientational information comparable to RDCs [32]. NMR relaxation is characterized by relaxation times T_1 and T_2, and the ratio T_1/T_2 can be used to define diffusion anisotropy restraints in NMR structure calculations [33]. Again, the orientation information comes from the angles of internuclear vectors in an external frame, which, in the case of diffusion anisotropy data, corresponds to the orientational diffusion tensor frame.

2.1.7 Paramagnetic Restraints

If a paramagnetic metal ion is present in a protein, the NMR signals of the nuclei in a shell around it will be affected [34] by several effects, including contact and pseudocontact shifts, relaxation rate enhancements, and cross-correlation effects.

In principle, these can provide both distance and orientation information. They have been implemented as restraints in various structure calculation software packages [35, 36].

2.2 Structure Calculation Software

The experimental information sources discussed above (Sect. 2.1) can be used as restraints in the calculation process. They have been implemented in several computer programs, among which are CNS [11], Xplor-NIH [37], CYANA [38], SCULPTOR [39], the SANDER module of AMBER [40], and even GROMACS [41]. The most commonly used are CYANA and Xplor/CNS.

Structure calculations are usually based on some molecular dynamic simulated annealing (SA) protocol performed in torsion angle and/or Cartesian space, followed by a final refinement phase in explicit solvent (water). A general feature of these protocols is that they use a "target function" that measures how well the calculated structure fits the experimental data and the chemical information; the lower this function, the better the agreement. The chemical information is defined in the force field that contains terms such as bond length, bond angles, van der Waals interactions, etc. Often, the description of long-range nonbonded interactions is simplified to increase the speed of the calculations by considering only repulsions between atoms and neglecting electrostatic interactions. A full nonbonded representation, including van der Waals (Lennard-Jones) and electrostatic (Coulomb) interactions, is typically reintroduced for final refinement in explicit solvent [6].

2.3 Structural Statistics and Structural Quality

The first step in structure validation is the selection of NMR structures from a large ensemble of calculated structures. The most widely used structure selection procedure is based on the agreement with the experimental data (small number of violations) and a low energy of the structures; typically ensembles of approximately 20 lowest energy models are selected, although this number is arbitrary. Ideally, the selected ensemble should represent the available conformational space accessible to the structure while satisfying the experimental restraints. From this ensemble, a representative structure is usually defined; no real consensus exists, however, on how it should be selected. We recommend selection of the structure that differs the least from all other structures within the ensemble, i.e., the closest to the average structure.

The final ensemble is subsequently subjected to structural validation to obtain an indication of the quality and structural statistics. In practice, several quality indicators are often used to assess the quality of the NMR ensembles, such as:

- The goodness of fit to the experimental data, by analyzing restraint violations
- The precision of the ensemble, measured by positional root mean square deviation (RMSD)
- Several chemical and stereochemical quality indicators that are generally used to assess the local and overall quality of protein structures, many of them based on knowledge from high resolution x-ray structures [42]

Table 1 lists the most commonly used validation programs. The use of some of these programs will be described later in Sect. 3.

Table 1 Internet resources of NMR-related programs and databases mentioned in this chapter

Software	Internet address	Purpose
CNS	http://cns.csb.yale.edu/v1.1	Multilevel hierarchical approach for the most commonly used algorithms in macromolecular structure determination (NMR, crystallography)
RECOORD	http://www.ebi.ac.uk/msd-srv/docs/NMR/recoord/scripts.html	Database of recalculated NMR structures with the CNS scripts used in the tutorial example
HADDOCK	www.nmr.chem.uu.nl/haddock/ (installation notes: http://www.nmr.chem.uu.nl/haddock/installation.html)	High ambiguity driven protein–protein docking based on biochemical and/or biophysical information
PDB	http://www.rcsb.org/pdb/Welcome.do	An information portal to biological macromolecules structures
BMRB	http://www.bmrb.wisc.edu	Biological Magnetic Resonance Data Bank
CCPN	http://www.ccpn.ac.uk	A collaborative computing project for NMR
PROCHECK	http://www.biochem.ucl.ac.uk/~roman/procheck/procheck.html	Checks the stereochemical quality of a protein structure, producing a number of PostScript plots analyzing its overall and residue-by-residue geometry
PROCHECK_NMR	http://www.biochem.ucl.ac.uk/~roman/procheck_nmr/manual/manprochint.html	PROCHECK-NMR is a suite of programs that have been derived from the PROCHECK programs to analyze ensembles of protein structures solved by NMR
PROCHECK_COMP	http://www.biochem.ucl.ac.uk/~roman/procheck_comp/procheck_comp.html	Compares residue-by-residue geometry of a set of closely related protein structures
WHATIF	http://swift.cmbi.kun.nl/whatif/ web server: http://swift.cmbi.kun.nl/WIWWWI	Versatile molecular modeling package that is specialized on working with proteins and the molecules in their environment such as water, ligands, nucleic acids, etc.

Table 1 Continued

Software	Internet address	Purpose
WHATCHECK	http://swift.cmbi.ru.nl/gv/whatcheck	The protein verification tools from the WHAT IF program
MOLPROBITY	http://molprobity.biochem.duke.edu	Structure validation and all-atom contact analysis for proteins, nucleic acids, and their complexes
QUEEN	http://www.cmbi.kun.nl/software/queen/index.spy?site=queen&action=Home	Quantitative evaluation of experimental NMR restraints
TALOS	http://spin.niddk.nih.gov/bax/software/TALOS/info.html	Protein backbone angle restraints from searching a database for chemical shift and sequence homology
profit	http://www.bioinf.org.uk/software/profit/	Least square-fitting program that performs the basic function of fitting one protein structure to another
NACCESS	http://wolf.bms.umist.ac.uk/naccess/nacwelcome.html	Stand-alone program that calculates the accessible area of a molecule from a PDB format file
Xmgrace	http://plasma-gate.weizmann.ac.il/Grace	Plotting tool
molmol	http://hugin.ethz.ch/wuthrich/software/molmol	Molecular graphics program
Rasmol	http://www.umass.edu/microbio/rasmol/index2.htm	Molecular graphics program

2.4 NMR-Based Modeling of Complexes

In principle, the structural information sources discussed above (Sect. 2.1) apply as well for structure calculation of protein–protein complexes. Again, NOEs are the most important information source, and, when available, RDCs are very useful as well [8, 43]. However, it is often difficult to obtain intermolecular NOEs, especially in the case of weakly interacting and transient complexes. For those cases, however, NMR remains a powerful method that provides several ways of mapping the interface between the components of a complex.

In one approach, the so-called chemical shift perturbation (CSP) experiment, $^{1}H^{15}N$-HSQC spectra of one ^{15}N-labeled component of the complex are recorded in the absence and presence of increasing amounts of its partner [9]. Changes in chemical shift after addition of the partner reveal residues that are possibly involved in the interaction. In cross-saturation or saturation transfer (SAT) experiments [44], the observed protein is ^{15}N labeled and perdeuterated with its amide deuterons exchanged back to protons, whereas the "donating" partner protein is unlabeled. Saturation of the unlabeled protein leads, by cross-relaxation mechanisms, to signal attenuation (typically monitored by $^{1}H^{15}N$-HSQC spectra) of those residues in the labeled protein that are in close proximity. Finally, in the case of paramagnetic systems, several of the above-mentioned paramagnetic effects can also be used to map interfaces [45].

To make use of those interface mapping data, NMR-based docking approaches have been developed [13,46–49]. One of these is HADDOCK [13], which combines a limited amount of (NMR-)data with docking in so called *data-driven docking*. Data-driven docking in HADDOCK follows a three-stage procedure:

1. Rigid body energy minimization.
2. Semiflexible refinement following an SA protocol during which increasing amounts of flexibility are allowed:
   (a) High temperature rigid-body search
   (b) Rigid body simulated annealing (SA)
   (c) Semiflexible SA with flexible side chains at the interface
   (d) Semiflexible SA with fully flexible interface (both backbone and side chains)
3. Final refinement in explicit solvent (water or DMSO).

During the docking, the (NMR-)data are introduced as "ambiguous interaction restraints" (AIRS). These are defined between active and passive residues, active residues being the residues that, based on the experimental data, have been identified to be involved in the interaction, and passive residues being their surface neighbors. An AIR is defined between each active residue and all active and passive residues of the partner protein. This restraint is only fulfilled when the active residue will make contact with at least one of the active or passive residues of the partner protein, which means that the restraint will indeed drive the docking.

The ranking of the docking solutions is performed using a "HADDOCK-score," which is a combination of several terms including restraint energies, intermolecular energies (van der Waals and electrostatic), desolvation energy, buried surface area, etc.

3 Methods

In this tutorial section, we describe the procedures to generate various type of NMR restraints and their use in structure calculation using CNS with the RECOORD scripts. This will be followed by a description of the steps to be followed for NMR-based modeling of biomolecular complexes using HADDOCK. As a convention, the commands to be executed are highlighted in grey using a Courier font. Information regarding the various programs and web pages used in this tutorial can be found in Table 1.

3.1 Restraint Generation

The keystone of NMR structure determination consists of three different types of experimental structure restraints: distance restraints, dihedral angle restraints, and orientational restraints.

3.1.1 Distance Restraints

A cross-peak in a NOESY spectrum indicates special proximity between two nuclei. Thus, each peak can be converted into a maximum distance between the nuclei, usually between 1.8 and 6.0 Å. This distance can be obtained according to the intensity of the NOESY peak (proportional to the distance to the minus 6th power, $1/r^6$). This intensity–distance relationship is only approximate, thus, usually a distance range is assumed. The assignment of the NOESY peaks to the correct nuclei based on the chemical shifts is of crucial importance. The manual detection of NOEs is an intensive and time-consuming job. Some programs, such as CANDID [50] and ARIA [12], can perform this task in an automated fashion coupled to the structure calculation protocol.

A common problem in NMR has been the limited availability of software allowing easy conversion between different data formats, which makes data exchange and use of different programs a tedious process. The CCPN Data Model for macromolecular NMR [51] is intended to cover all data needed for macromolecular NMR spectroscopy from the initial experimental data to the final validation. The ccpNmr FormatConverter application allows the import and export of data from and to a large variety of formats (Fig. 1).

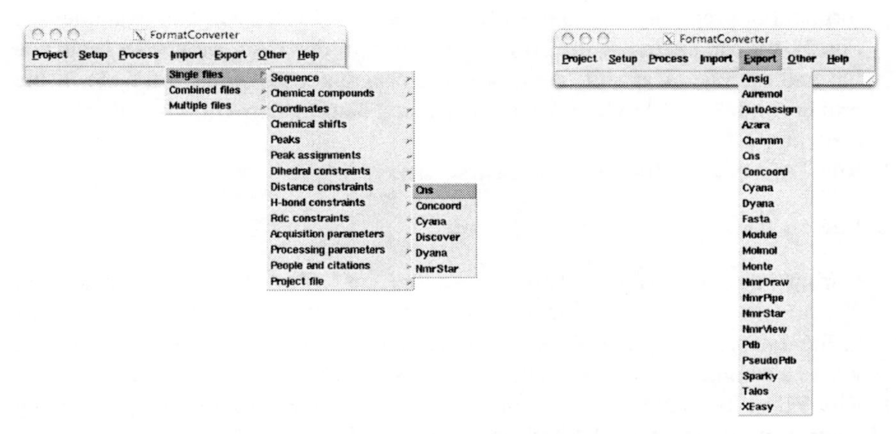

Fig. 1 Graphical user interface (GUI) layer of the CCPN FormatConverter (http://www.ccpn.ac.uk)

```
REMARK Ubiquitin input for TALOS, HA2/HA3 assignments arbitrary.

DATA SEQUENCE MQIFVKTLTG KTITLEVEPS DTIENVKAKI QDKEGIPPDQ QRLIFAGKQL
DATA SEQUENCE EDGRTLSDYN IQKESTLHLV LRLRGG

VARS   RESID RESNAME ATOMNAME SHIFT
FORMAT %4d   %1s     %4s      %8.3f

   1 M       HA            4.23
   1 M       C           170.54
   1 M       CA           54.45
   1 M       CB           33.27
   2 Q       N           123.22
   2 Q       HA            5.25
   2 Q       C           175.92
   2 Q       CA           55.08
   2 Q       CB           30.76
...
  10 G       N           108.89
  10 G       HA2           4.35
  10 G       HA3           3.61
  10 G       C           174.07
  10 G       CA           45.46
...
```

Fig. 2 Example of the input shift table required in TALOS

3.1.2 Dihedral Angle Restraints

J-coupling and secondary chemical shifts can be used to define restraints on the torsion angles of the chemical bonds, typically the ϕ, ψ, and χ_1 angles can be generated and included in the protocol. They can be calculated applying the Karplus equation [21] to the measured J-couplings, or by using chemical shifts in programs such as TALOS [52] or CSI [53]. We will describe here the use of TALOS.

TALOS is a database system for empirical prediction of ϕ and ψ backbone torsion angles using a combination of available chemical shifts (Hα, Cα, Cβ, CO, N) for a protein sequence. To use TALOS, the following steps should be followed:

1. Create a directory for the predictions from where all the following commands will be executed.
2. Prepare the input table with the sequence and shift assignments in the required format. For preparing the input shift table required by TALOS (for example, see Fig. 2), we can again use the FormatConverter. In this case, we need a sequence file and the chemical shift table, and the program will export the table in the proper format for TALOS, taking into account naming conventions and shift referencing.
3. Run TALOS to perform the database search:

   ```
   talos.tcl -in myshifts.tab
   ```

   During the searching phase, a series of files will be created in "`pred/res.*.tab`." Each of these files contains the 10 best matches in the database for a given residue. In addition, a file "`pred.tab`" is created, where a summary of the prediction results is stored.
4. Run VINA to summarize the results. This can be done with one of the following commands, depending whether a structure template is available or not:

   ```
   vina.tcl -in myshifts.tab
   vina.tcl -in myshifts.tab -ref mystructure.pdb
   ```

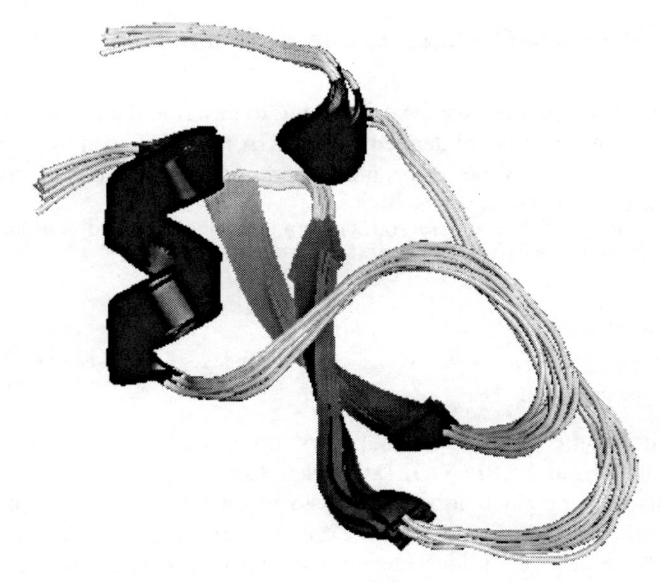

Fig. 3 Ensemble of final water-refined structures for PDB entry 1bf0 recalculated with the RECOORD scripts

This step will adjust the individual prediction files to identify outliers in the detected matches and it will prepare a new summary file. This step is optional, because, in the previous step, a "`pred.tab`" was already created.

5. Run RAMA to inspect and adjust the predictions made by the program:

```
rama.tcl –in myshifts.tab
rama.tcl –in myshifts.tab –ref mystructure.pdb
```

During the manual prediction, you will classify the results for a given residue as "Good," "Ambiguous," or "Bad." For that purpose, you have to examine the ϕ/ψ distributions of the detected matches and decide which ones should be included in the prediction and which ones are outliers. The prediction files will be overwritten to reflect any changes made interactively, and a final "`pred.tab`" will be created containing the classification and predictions (average and standard deviations) for the ϕ and ψ angles for each residue.

To convert the TALOS predictions into CNS/Xplor restraints, we can use the perl script `talos2xplor.pl`, which can be obtained from the Biomolecular NMR laboratory at UAH (http://daffy.uah.edu/nmr/analysis.html). The script will ask you for an input a TALOS prediction file, the minimum $\pm$ error (e.g., 20°) you want to include in the restraints, and an output CNS/Xplor restraint file.

3.2 NMR Structure Calculation and Refinement

For the structure calculation part, we are going to describe the use of the program CNS [11] with an SA protocol derived from ARIA [54], followed by refinement in explicit solvent [7]. All of the scripts mentioned in this section can be downloaded from the RECOORD webpage (see Table 1).

Start by creating a folder where you will run the calculations, download there the tar file containing the RECOORD scripts, and decompress it:

```
mkdir struct-calc
cd struct-calc/
wget http://www.ebi.ac.uk/msd- srv/docs/NMR/recoord/files/RECOORDscripts.tgz
tar xzfv RECOORDscripts.tgz
```

In case the wget command does not work, use a web browser to download the scripts manually from the RECOORD webpage (see Table 1).

Before starting the calculations, you need to set up your current path for the scripts to work. To do this, you need to edit `changeScriptsDir.sh` found in the `RECOORDscripts/` folder, change the path for `newDir` in line 8 by your current path (you can find it by typing "`pwd`" in the shell) and execute it:

```
cd RECOORDscripts
nedit changeScritpsDir.sh #(change line 8 for pwd)
./changeScritpsDir.sh
cd ..
```

(`nedit` is a text editor; if not installed, use your preferred editor instead).

Most of the scripts use the CNS executable, so check that CNS is properly installed.

The last step is to setup a working directory, assigning a project name for the protein on which you are going to work. This project name will be used to generate the file names at the different stages of the protocol. We will use the 1bf0 structure [55] as an example, with the corresponding NMR restraints available for this entry from the BioMagResBank (BMRB) [56].

```
mkdir 1bf0
cd 1bf0
wget http://www.pdb.org/pdb/files/1bf0.pdb.gz
gunzip 1bf0.pdb.gz
```

The easiest way of obtaining the restraints in a format ready to be used in this protocol is to go to the BMRB from the PDB entry, select `4-filtered-FRED` in the stage window, select the distance restraints in XPLOR/CNS format by clicking on it, and then click on "`170823`" in `mrblock_id` and copy and paste these restraints in a text file called unambig.tbl (see **Note 1**).

3.2.1 Generation of Molecular Topology Files

We can generate the molecular topology of the protein using the RECOORD script `generate.sh` (a modified version of the CNS script `generate_easy.inp`), either from the primary sequence or from a PDB coordinate file, depending on availability. Here, we will use the downloaded PDB file (see **Note 2**):

```
../RECOORDscripts/generate.sh 1bf0.pdb
```

If you give a name such as `1bf0.pdb`, a topology file called `1bf0_cns.mtf` will be generated. You should check the `ERRORS_generate` file created inside the `1bf0/` folder for possible errors. In this particular case, you can see that the script reported many errors but they are basically nomenclature errors and they can be ignored. A new pdb file called `1bf0_cns.pdb` is also generated with the proper CNS nomenclature (display the structure in your favorite molecule viewer to make sure that it looks reasonable and that the script worked properly).

3.2.2 Generation of Extended Starting Structure

The next step is the generation of an extended starting conformation, which will be used as input in the SA protocol. For this, use the RECOORD script, `generate_extended.sh`.

```
../RECOORDscripts/generate_extended.sh 1bf0_cns.mtf
```

Keeping the name given before, an extended structure called `1bf0_cns_extended.pdb` will be generated. In addition, it is advisable to check the `ERRORS_generate_extended` file in the same working directory for errors.

3.2.3 SA Stage

Use the script `annealing.sh` to start the structure calculation run. This script will generate a CNS parameter file (`run.cns`) with all details and specifications of the protocol that is used. The NMR restraints are contained in table files. We can use three different types of restraints, depending on their availability: `unambig.tbl` (NOE distance restraints), `hbonds.tbl`, and `dihedrals.tbl`. Note that the `annealing.sh` script should be run from a higher level than the previous two scripts.

```
cd ..
../RECOORDscripts/annealing.sh 1bf0
```

Individual job files will be generated and executed for each model you want to calculate. By default, two models will be generated in the created `str/` folder, with names similar to 1bof_cns_[1-2].pdb. The CNS input and output files can be found in the directory cnsRef/, together with possible error files (see **Note 3**). The header of every PDB file generated contains information regarding violations and energy values.

3.2.4 Water Refinement Stage

Once the SA phase is finished and all resulting structures have been written into the str/ directory, we can proceed to water refinement. For this purpose, we are going to use the script re_h2o.sh.

```
../RECOORDscripts/re_h2o.sh 1bf0
```

In the str/ directory, a new directory called wt/ will be created, the best energy structures will be copied there and subsequently refined (see **Note 4**). The final ensemble of structures obtained following this protocol is shown in Fig. 3.

3.3 Structure Validation and Quality Assessment

3.3.1 Restraint Violations

To obtain statistics regarding distance and dihedral angle violations for the water refined ensemble, use the scripts calcViol.sh and analysViol.sh, which analyze and summarize violations, respectively.

```
cd 1bf0
../RECOORDscripts/calcViol.sh 1bf0_cns str/wt/violations
0.3 convertOff 1bf0_cns.mtf unambig.tbl
```

where the input parameters correspond to the entry name (in this specific case, 1bf0_cns), the directory in which the coordinate files can be found (in this case the directory containing the water refined structures 1bf0_cns_w_[1-25].pdb), the violation distance cut-off (a frequently used value is 0.3 Å), the conversion switch to CNS format (it is optional and by default set to convertOff), the topology and restraint files (these are also optional, with, as the default, 'entryname'_cns.mtf and unambig.tbl). Files with violations statistics will be created in the new violations/ folder, with names as viol_1bf0_cns_w_0.3. Once the violations have been calculated, they can be analyzed with analysViol.sh:

```
../RECOORDscripts/analysViol.sh 1bf0_cns violations
```

where violations/ is the directory created previously with the calcViol.sh script. The results are summarized in the violations folder in the viol_results file.

3.3.2 Structural Validation

Various software tools are available to assess the stereochemical quality of the generated protein structures. Some of the most widely used packages are PROCHECK [57] and WHATIF [58]. Procheck provides a detailed graphical indication of the quality of a protein structure, giving an assessment of both the overall quality of the structure, as compared with well-defined structures of the same resolution, and of highlight regions that may need further investigation.

The command to run PROCHECK, once the program is properly installed, is:

```
procheck filename [chain] resolution
```

where filename indicates the coordinates file in Brookhaven format, `chain` is an optional one-letter chain-ID, in case several chains are included in the model, and `resolution` is a real number giving the resolution of the structure, to select the structures from the database to compare with our model.

Because PROCHECK only allows the analysis of a single structure at a time, it is worthwhile to also use PROCHECK_COMP or PROCHECK_NMR [59], a suite of programs that have been derived from the original PROCHECK programs, to compare, residue by residue, the geometry of a set of closely related protein structures, such as those in an NMR ensemble. To run PROCHECK_COMP, you need to create a file, e.g., `1bf0.list`, containing the names of the structures you want to analyze, and then type the command:

```
procheck_comp 1bf0.list
```

Both programs produce easily interpreted color postscript files that can be viewed using `ghostview` or similar programs. Type, for example, "`gs 1bf0_01.ps`" to display the Ramachandran plot showing the ϕ/ψ torsion angles for all residues in the structure. The coloring/shading on the plot represents the different regions: the darkest areas correspond to the "core" regions representing the most favorable combinations of ϕ/ψ values. `1bf0_06.ps` shows various graphs and diagrams of protein geometrical properties as a function of the amino acid sequence, allowing you to possibly distinguish regions with normal geometry from those that might be poorly defined and present unusual geometry (see **Note 5**).

Another very useful protein validation tool is WHATCHECK, based on WHATIF, which also uses reference values from an x-ray database for most of the checks carried out. A great advantage of WHATCHECK is that the reference database of high-resolution protein structures is larger than in PROCHECK and continuously updated. Further, it provides many more checks and is more critical. WHATIF is also available as a web server (see Table 1), where a variety of quality parameters can be obtained by uploading a PDB coordinates file to the server [42].

3.3.3 Precision of the Ensemble

The precision of a structure can be estimated by measuring the conformational variance over an ensemble of models. Usually, this variance has been expressed as the positional RMSD of the individual models from the mean structure. This parameter is useful for estimating the precision of the calculation, but does not report on the accuracy. The later can only be calculated if a standard reference is available.

The positional RMSD from the mean and the prediction of secondary structure elements can be obtained using the molecular graphical program Molmol [60]. The least-square fitting program Profit can also perform the basic function of fitting a protein structure to another and allows for much more flexibility. It can be used to calculate the accuracy of structures, provided a reference structure is known. This program can be used in a direct interactive fashion in a terminal window or using scripts. A very simple script called here `profit.in` could be written as follows:

```
reference a.pdb
mobile b.pdb
! specifies the residues to fit on
! in this case: 10-20 in the reference with 30-40 in
! the mobile zone 10-20:30-40
! specifies the atom subsets for both
! fitting and RMS calculation.
atom CA,C,N
fit
! writes the fitted coordinates to a file
write b_fiton_a.pdb
quit
```

To execute it, simply type:

```
profit < profit.in
```

3.4 Modeling of Complexes by Data-Driven Docking Using HADDOCK

We describe here the use of the HADDOCK2.0 package (see Table 1) for the modeling of a protein–protein complex. In the following, we will use data from the `haddock2.0/examples/e2a-hpr` directory. You should first copy this directory to the directory in which you are working (see **Note 6**):

```
cp -r $HADDOCK/examples/e2a-hpr  .
```

3.4.1 Preparation of PDB Files and Input Data

If you are using an ensemble of structures, split the file such that each individual PDB file contains only one structure (see **Note 7**). As input data, you should combine CSP data (or other data indicating residues at the interface) and solvent accessibility data calculated with NACCESS; use only those residues that have both a high enough CSP and a high enough relative accessibility. In the example, the (average) per residue solvent accessibilities calculated with NACCESS are already provided in the files `e2a_1F3G.rsa` and `hpr/hpr_rsa_ave.lis` (the latter containing the average for the 10 starting structures for hpr). From these files, you can select the residues with high enough (e.g., >40–50%) accessibility (see **Note 8**). You could calculate the accessibility values yourself using the following command:

```
naccess e2a_1F3G.pdb
```

3.4.2 Definition of Active and Passive Residues

Passive residues are defined as the solvent-accessible surface neighbors of active residues. To define them you can display your molecule in a space-filling model using, for example, rasmol:

```
rasmol e2a_1F3G.pdb
```

and color the active residues, for example, in red. Then, filter out the residues having a low solvent accessibility and select all surface neighbors to define the passive residues (color them, for example, in green), which, again, you should filter with the solvent accessibility criterion. In the `e2a-hpr` example, several rasmol scripts are provided with the respective residues already colored according to this scheme:

```
e2a_rasmol_active.script, e2a_rasmol_active_passive.script
```

and similar for hpr.

You will use the active and passive residues for both molecules to generate AIRs; for this, go to the HADDOCK project setup section on http://www.nmr.chem.uu.nl, click on “generate AIR restraint file” and follow the instructions. You should save the resulting file as `ambig.tbl` in the working directory; note that, in the `e2a-hpr` example directory, `ambig.tbl` is already present (see **Note 9**).

3.4.3 Setup of a New Run: new.html

To set up a new run, return to the project setup page on http://www.nmr.chem.uu.nl, click on “start a new project” and follow the instructions. Depending on the experimental data you have available, you will input various data files, such as ambiguous restraints, unambiguous restraints, RDCs, etc. After saving the `new.html` file to disk, type "haddock2.0" in the same directory. This will generate a run directory containing all of the necessary information to run haddock. An example of a new.html file can be found in the e2a-hpr directory as new.html-example (see **Note 10**).

3.4.4 Run.cns

The next step is to define all parameters to perform the docking run. For this, enter the newly created directory:

```
cd run1
```

You will find a file called run.cns containing all the parameters to run the docking. You need to edit this file and define a number of project-specific parameters, such as the semiflexible segments at the interface or fully flexible segments and other parameters governing the structure docking (see **Note 11**). You can edit your run.cns file via "project setup" on http://www.nmr.chem.uu.nl. More information is available via the "run.cns" option in the manual section on http://www.nmr.chem.uu.nl.

3.4.5 Docking Run

To actually start the docking run with HADDOCK, in the directory containing the run.cns file (see **Note 12**) type:

```
haddock2.0 >& haddock.out &
```

As more extensively explained in "The Docking" section in the HADDOCK manual, the entire protocol consists of four stages:

1. *Topologies and structures generation*: The resulting topologies (*.psf) and coordinates (*.pdb) files are written into the begin/ directory (see **Note 13**).
2. *Randomization and rigid body energy minimization:* The generated docked structures are written into structures/it0/. When all structures have been generated, HADDOCK will write the PDB filenames sorted according to the criterion defined in the run.cns into file.cns, file.list, and file.nam in the structures/it0 directory.
3. *Semiflexible SA*: The best 200 structures after rigid body docking (this number is defined in run.cns and can be modified) will be subjected to a semiflexible SA in torsion angle space. The temperatures and number of steps for the various stages are defined again in the run.cns parameter file. The resulting refined structures are written into structures/it1. At the end of the calculation, HADDOCK generates the file.cns, file.list, and file.nam files containing the filenames of the generated structures sorted accordingly to the criterion defined in the run.cns parameter file (see **Note 14**). At the end

of this stage, the structures are analyzed and the results can be found in the `structures/it1/analysis` directory (see Sects. 3.4.6 and 3.4.7).

4. *Flexible explicit solvent refinement.* The `re_h2o.inp` (or `re_dmso.inp`, if the chosen solvent is DMSO) CNS script is used for this step. The resulting structures are written in the `structures/it1/water` directory. At the end of the explicit solvent refinement, HADDOCK generates the `file.cns`, `file.list`, and `file.nam` files containing the filenames of the generated structures sorted accordingly to the criterion defined in the run.cns parameter file. Finally, the structures are analyzed and the results can be found in the `structures/it1/water/analysis` directory (see Sects. 3.4.6 and 3.4.7).

3.4.6 Automatic Analysis

A number of analysis scripts are automatically run after the semiflexible and explicit solvent refinement stages, with the results placed into `structures/it1/analysis` and `structures/it1/water/analysis`, respectively. Here we discuss a few of the most relevant output files.

- `e2a-hpr_rmsd.disp`: Contains the pairwise RMSD matrix; this file is used as input for RMSD clustering.
- `noe.disp`: Contains the number of distance restraints violations per structure and averaged over the ensemble over all distance restraint classes and for each class (unambiguous, ambiguous, hbonds) separately. Comparable files are generated when you have RDC restraints (`sani.disp`) or relaxation data restraints `(dani.disp)`.
- `energies.disp`: Contains the various energy terms per structure and averaged over the ensemble.
- `ana_*.lis`: There is a set of files called `ana*.lis` where * can be `dihed_viol`, `dist_viol_all`,`hbond_viol`, `hbonds`, `nbcontacts`, `noe_viol_all`, `noe_viol_ambig`, or `noe_viol_unambig`. The "viol" refers to violations, and those files contain listings of violations, including the number of times a restraint is violated and the average distance and violation per restraint. In addition, `ana_hbonds.lis` gives a listing of hydrogen bonds, and `ana_nbcontacts.lis` gives a listing of nonbonded contacts.
- `ene-residue.disp`: Contains intermolecular energies for all interface residues.
- `nbcontacts.disp`: Contains nonbonded contacts.

3.4.7 Manual Analysis

An important part of the analysis needs to be performed manually. A number of scripts and programs are provided for this purpose in the `tools` directory. These allow collection of various statistics on the generated models and, more importantly, clustering of solutions and their analysis on a per-cluster basis.

- *Collecting statistics of the models with* `ana_structure.csh`: Copy this script from the `tools` directory into `structures/it1` or `structures/it1/water`. This script should be run once the `file.list` file has been created. It extracts from the various PDB files various energy terms, violation statistics, and the buried surface area, and calculates the RMSD of each structure compared with the lowest energy structure (if the location of ProFit is defined [see installation and software links on http://www.nmr.chem.uu.nl/haddock]). Several files called "`structures * .stat`" are created, which contain the same information but sorted in different ways. The most important file is `structures_haddock-sorted.stat`, which is sorted based on the HADDOCK-score. You can generate a plot of the HADDOCK-score as a function of the RMSD (using Xmgrace, for example). A script called `make_ene-rmsd_graph.csh` is provided in `$HADDOCKTOOLS` for this purpose. Specify two columns to extract data from and a filename:

  ```
  $HADDOCKTOOLS/make_ene-rmsd_graph.csh 3 2
  structures_haddock-sorted.stat
  ```

  This will generate a file called `ene_rmsd.xmgr`, which you can display with xmgrace:

  ```
  xmgrace ene_rmsd.xmgr
  ```

- *Clustering of solutions using* `cluster_struc`: The clustering is run automatically in `it1/analysis` and `it1/water/analysis` based on the criteria defined in the `run.cns` file. However, try using different cut-offs for the clustering because it is difficult to know *a priori* the best RMSD cut-off. This will depend on the system under study and the number of experimental restraints used to drive the docking (see **Note 15**).
  `cluster_struc` reads the `e2a-hpr_rmsd.disp` file containing the pairwise RMSD matrix and generates clusters. The usage is (in the `analysis` directory):

  ```
  cluster_struc [-f] e2a-hpr_rmsd.disp cut-off
  min_cluster_size > cluster.out
  ```

  Here, cut-off indicates the RMSD cut-off and `min_cluster_size` is the minimum number of structures in a cluster (typically a number like 4 or 5) (−f is optional, see **Note 16**).

The output looks like:

```
cluster 1 → 2 3 4 5 6 7 8 9 10 11 12 13 14 15 16 17
18 19 23 24 27 28 43
cluster 2 → 25 26 29 32 34 35 57 71 73 20 21 44 39 46
...
```

The numbers correspond to the structure number in the analysis file. For example, 2 corresponds to the second structure in analysis, i.e, the second structure in `file.list in it1 or it1/water.`

- *Analysis of the clusters with `ana_cluster.csh`*: This script takes the output of `cluster_rmsd` to perform an analysis of the various clusters, calculating average energies, RMSDs, and buried surface area per cluster. To run it, type with as argument the output file of the clustering, e.g.:

```
$HADDOCKTOOLS/ana_cluster.csh [-best #]
analysis/cluster.out
```

The `-best` # is an optional argument to generate additional files with cluster averages calculated only on the best # structures of a cluster. The best structures are selected based on the criteria defined in `run.cns`, i.e., the sorting found in `file.list`. This allows removal of the dependency of the cluster averages based on the size of the respective clusters (see **Note 17**). The `ana_cluster.csh` script analyzes the clusters in a similar way as the `ana_structure.csh` script, but, in addition, generates average values over the structures belonging to one cluster. It creates a number of files for each cluster containing the cluster number `clustX` in the name (see **Note 18**). In addition, files containing various averages over clusters are created, `cluster_xxx.txt`; these contains the average and standard deviation of various terms such as intermolecular energy (xxx = ene) etc. In addition, files combining all of the above information and sorted based on various criteria are provided: `clusters.stat` that contains the various cluster averages, unsorted, and `clusters_xxx-sorted.stat`, where xxx is the energy term according to which the values are sorted (e.g., xxx = ene for intermolecular energy, etc.). The most relevant is `clusters_haddock-sorted.stat`.

- *Rerunning the HADDOCK analysis on a cluster basis*: Having performed the cluster analysis, you can now rerun the HADDOCK analysis for the best structures of each cluster to obtain various statistics on a "cluster bests" basis. For this, one needs the cluster-specific `file.nam_clust#`, `file.list_clust#`, and `file.cns_clust#` files. A script called `make_links.csh` is provided that will move the original `file.nam`, `file.list`, and `file.cns` files to `file.nam_all`, `file.list_all`, `file.cns_all`, and the same with the analysis directory. It will then create links to the appropriate files (`file.nam_clust#,...`) and to a new `analysis_clust#` directory.
For example, to rerun the analysis for the best 10 structures of the first cluster type in the water directory:

```
$HADDOCKTOOLS/make_links.csh clust1_best10
cd ../../..
haddock2.0
```

The cd command brings you back into the main run directory, from where you again start HADDOCK. Only the analysis of the best 10 structures of the first cluster in the water will be run. Once finished, go to the respective analysis directory and inspect the various files. The RMSD from the average structures should now be low (check `rmsave.disp`).

Having run the HADDOCK analysis on a cluster basis for each cluster, you should now have new directories in the water directory, called `analysis_clustX_best10`. Each analysis directory now contains cluster-specific statistics. You can also visualize the clusters. For Rasmol, first use the `joinpdb` perl script to concatenate the various PDB files into one singe file:

```
$HADDOCKTOOLS/joinpdb —o e2a-hpr_clust1.pdb e2a-hprfit_*.pdb
rasmol - nmrpdb e2a-hpr_clust1.pdb
```

In general, the cluster with the lowest HADDOCK score will be considered the best model. Scoring in docking is, however, a difficult problem and we recommend, if possible, the use of additional information for validation, such as, for example, mutagenesis data, if available. The selected model should explain as much as possible what is known about the system.

4 Notes

1. If dihedrals or any other types of restraints are available, they can be obtained in a similar way. The names assigned will be `dihedrals.tbl` and `hbonds.tbl`.
2. This only works if a PDB coordinates file is available. Otherwise, use `generate_seq.inp` and `generate_template.inp` from CNS to create such a PDB.
3. Once you have everything set up in a proper way to work, you can edit the script and make some changes for some protocol parameters. You can, for example, change the number of models to generate. It is set to 2 by default, but more common numbers would be 100 or 200. For systems that are more complex, you can switch to a longer annealing protocol, by doubling the number of steps to be carried out. Depending on whether you are going to use a cluster or your own computer, you should change the submit command. Remember also to change the sleeping time between submitting jobs, especially if you are not using a cluster and you do not want to have 100 jobs running on your computer at the same time! In such a case, choose a sleep time that matches the time needed for one structure calculation.

4. You should also edit this script and change the number of structures to refine because, by default, it is set to only 1. Increase this number to 25 models. They will be assigned names such as `1bf0_cns_w_[1-25].pdb`. The CNS input and output will be directed to the directory `cnsWtRef/`.
5. For visualizing these plots after running `procheck_comp`, the number tag is kept for the Ramachandran plot, however, for the residue properties plot, the number tag is now `_07.ps`.
6. The `$HADDOCK` environment variable should be defined if HADDOCK was properly installed.
7. Make sure that the format of the PDB files containing your starting structures is correct. There should be an END statement at the end, and there should be no SEGID (the SEGID is a four character long string at columns 73–76 in the PDB format) or ChainID (the ChainID is a chain identifier following the residue name in column 22). If you use a crystal structure, make sure that there are no missing residues.
   Another point concerns ions; if proper care is not taken, they can give problems in torsion angle dynamics. To deal with this, the script `covalions.cns` defines artificial bonds to connect the ion to the protein. If you have another ion than is defined in the first line of the script, add it there. In addition, make sure that their name in the PDB file matches the ion name defined in the `ion.top` file in the toppar directory. To avoid having a N- or C-terminal patch applied to them, they should also be defined in the `topallhdg5.3.pep` file (look for the "`first IONS`" and "`last IONS`" statements).
8. The cut-off is not a hard limit; check the accessibilities and possibly include residues with lower accessibilities but functionally important groups.
9. Distance restraints can be used in HADDOCK in `ambig.tbl` or `unambig.tbl`. These are treated in the same way, except that the random removal option (`noecv=true`) only is applied to `ambig.tbl`. By default, one would use `ambig.tbl; unambig.tbl` could be used, for example, to provide extra NOEs or other data for which one wants to use different force constants.
10. An important setting in `new.html` is the value of `N_COMP`. This should be set to be equal to the number of components of the complex (two in case of a dimer, three for a trimer, etc.). Note that it can also be set to one, in which case, HADDOCK could be used for refinement instead of docking.
11. HADDOCK allows the definition of fully flexible regions: these are treated as fully flexible throughout all stages, except the initial rigid-body docking. This should be useful for cases in which part of a structure is disordered or unstructured or when docking small flexible molecules onto a protein. This option also allows the use of HADDOCK for structure calculations of complexes when classical NMR restraints are available to drive the folding.
12. This causes the HADDOCK program to run in the background. If, at some stage, HADDOCK stops producing new structures and the run is not yet finished, search for error messages in the output files: `gunzip xxx.out.gz` where `xxx.out.gz` is a particular output file, and look for ERR in this file.

Also, kill the current HADDOCK process:

```
ps -ef | grep haddock
kill -9 id
```

Here, id is the process id that is returned by the `ps -ef` command.

13. The OPLS force field used by HADDOCK is a mixed united/all-atom force field; all atoms, including protons, are described; the later, however, do not have vdw parameters but are accounted for in the carbon parameters to which they are attached. From version 2.0 of HADDOCK, nonpolar hydrogen atoms are deleted by default to speed up the calculation; this does not really affect the resulting structures because the missing hydrogens are actually accounted for in the united atoms parameters. You can change this behavior by setting `delenph=true` in `run.cns`. This should be performed if classical NOE distance restraints are used.
14. A typical error would be that only one or two structures in `it1` are not successfully calculated. Often, you can cope with this by changing the random seed in `run.cns (iniseed)` and restart HADDOCK. Otherwise, try to decrease the `timestep` (e.g., 0.001 instead of 0.002). If none of this works, simply copy the missing structures from the `it0` directory so that the run can proceed.
15. For the RMSD calculation, the structures are superimposed on the interface backbone atoms of molecule A and the RMSD is calculated on the interface backbone atoms of molecule B; this might be called ligand interface RMSD. The resulting RMSD values are larger than would be obtained by fitting the whole molecule, which explains the large cutoff value that is used by default (7.5 Å). If only a small fraction of the structures do fall into clusters, try increasing the cut-off.
16. The `-f` option stands for full linkage, a method that generates larger clusters in which the structures within a cluster can, thus, differ more.
17. It is better to use a small number of structures (e.g., five) for comparison of the clusters than to use all structures of each cluster, because, in this way, the comparison will not depend on the cluster size.
18. The ordering of the structures in the `file.nam_clustXX` files comes from the clustering. The PDB files might, therefore, no longer be sorted accordingly to a defined criterion.

References

1. Wüthrich, K., *NMR of proteins and nucleic acids*. Wiley: New York, 1986.
2. Neuhaus, D. and Williamson, M. P., *The nuclear Overhauser effect in structural and conformational analysis*. John Wiley & Sons: 2000.
3. Altona, C., Vicinal coupling constants & conformation of biomolecules. In *Encyclopedia of Nuclear Magnetic Resonance*, Harris, D. M. G. a. K. R., Ed. John Wiley, London: 1996; pp 4909–4922.

4. Bax, A., Kontaxis, G. and Tjandra, N. (2001) Dipolar couplings in macromolecular structure determination. *Methods in Enzymology* **339**, 127–174.
5. Guntert, P. (1998) Structure calculation of biological macromolecules from NMR data. *Quarterly Reviews of Biophysics* **31**, 145–237.
6. Linge, J. P., Williams, M. A., Spronk, C. A. E. M., Bonvin, A. M. J. J. and Nilges, M. (2003) Refinement of protein structures in explicit solvent. *Proteins* **50**, 496–506.
7. Nederveen, A. J., Doreleijers, J.F., Vranken, W.F., Miller, Z., Spronk, C.A.E.M, Nabuurs, S.B., Güntert, P., Livny, M., Markley, J.L., Nilges, M., Ulrich, E.L., Kaptein, R., and Bonvin, A.M.J.J. (2005) Recoord: A recalculated coordinates database of 500+ proteins from the pdb using restraint data from the biomagresbank. *Proteins: Struc. Funct. & Bioinformatics* **59**, 662–672.
8. Bonvin, A. M. J. J., Boelens, R. and Kaptein, R. (2005) NMR analysis of protein interactions. *Current Opinion in Chemical Biology* **9**, 501–508.
9. Zuiderweg, E. R. (2002) Mapping protein-protein interactions in solution by NMR spectroscopy. *Biochemistry* **41**, 1–7.
10. van Dijk, A. D. J., Boelens, R., and Bonvin, A. M. J. J. (2005) Data-driven docking for the study of biomolecular complexes. *Febs Journal* **272**, 293–312.
11. Brunger, A. T., Adams, P. D., Clore, G. M., DeLano, W. L., Gros, P., Grosse-Kunstleve, R. W., Jiang, J.-S., Kuszewski, J., Nilges, N., Pannu, N.S., Read, R.J., Rice, L.M., Simonson, T., and Warren, G.L. (1998) Crystallography and NMR system (CNS): A new software system for macromolecular structure determination. *Acta Crystallogr. D Biol.* **54**, 905–921.
12. Linge, J. P., Habeck, M., Rieping, W. and Nilges, M. (2003) Aria: Automated NOE assignment and NMR structure calculation. *Bioinformatics* **19**, 315–316.
13. Dominguez, C., Boelens, R. and Bonvin, A. M. J. J. (2003) Haddock: A protein-protein docking approach based on biochemical or biophysical information. *Journal of the American Chemical Society* **125**, 1731–1737.
14. Neal, S., Nip, A. M., Zhang, H. Y. and Wishart, D. S. (2003) Rapid and accurate calculation of protein h-1, c-13 and n-15 chemical shifts. *Journal of Biomolecular NMR* **26**, 215–240.
15. Xu, X. P. and Case, D. A. (2001) Automated prediction of ^{15}n, ^{13}cα, ^{13}cβ and ^{13}c$'$ chemical shifts in proteins using a density functional database. *Journal of Biomolecular NMR* **21**, 321–333.
16. Williamson, M. P., Kikuchi, J. and Asakura, T. (1995) Application of h-1-NMR chemical-shifts to measure the quality of protein structures. *Journal of Molecular Biology* **247**, 541–546.
17. Meiler, J. (2003) Proshift: Protein chemical shift prediction using artificial neural networks. *Journal of Biomolecular NMR* **26**, 25–37.
18. Clore, G. M. and Gronenborn, A. M. (1998) New methods of structure refinement for macromolecular structure determination by NMR. *Proceedings-National Academy of Sciences USA* **95**, 5891–5898.
19. Luginbuehl, P., Szyperski, T. and Wuethrich, K. (1995) Statistical basis for the use of ^{13}cα chemical shifts in protein structure determination. *Journal of Magnetic Resonance Series B* **109**, 229.
20. Kuszewski, J., Qin, J., Gronenborn, A. M. and Clore, M. G. (1995) The impact of direct refinement against ^{13}cα and ^{13}cβ chemical shifts on protein structure determination by NMR. *Journal of Magnetic Resonance Series B* **106**, 92.
21. Karplus, M. (1963) Vicinal proton coupling in nuclear magnetic resonance. *Journal American Chemistry Society* **85**, 2870–2871.
22. Kim, Y. P., J. H. Prestegard (1990) Refinement of the NMR structures for acyl carrier protein with scalar coupling data. *Proteins* **8**, 377–385.
23. Torda, A. E., Brunne, R. M., Huber, T. and Kessler, H. (1993) Structure refinement using time-averaged j-coupling constant restraints. *Journal of Biomolecular NMR* **3**, 55.
24. Wagner, G. and Wüthrich, K (1982) Amide proton exchange and surface conformation of the basic pancreatic trypsin inhibitor in solution. *Journal Molecular Biology* **160**, 343–361.
25. Pervushin, K., Ono, A., Fernandez, C., Szyperski, T., Kainosho, M. and Wuthrich, K. (1998) NMR scalar couplings across Watson-Crick base pair hydrogen bonds in DNA observed by transverse relaxation-optimized spectroscopy. *Proceedings of the National Academy of Sciences of the United States of America* **95**, 14147–14151.

26. Cordier, F., Rogowski, M., Grzesiek, S. and Bax, A. (1999) Observation of through-hydrogen-bond 2hjhc' in a perdeuterated protein. *Journal of Magnetic Resonance* **140**, 510–512.
27. Bonvin, A. M. J. J., Houben, K., Guenneugues, M., Kaptein, R. and Boelens, R. (2001) Rapid protein fold determination using secondary chemical shifts and cross-hydrogen bond ^{15}N-$^{13}C'$ scalar couplings ($^{3hb}j_{nc'}$). *Journal of Biomolecular NMR* **21**, 221–233.
28. Bax, A. (2003) Weak alignment offers new NMR opportunities to study protein structure and dynamics. *Protein Science* **12**, 1–16.
29. Bax, A. and Grishaev, A. (2005) Weak alignment NMR: A hawk-eyed view of biomolecular structure. *Current Opinion in Structural Biology* **15**, 563–570.
30. Prestegard, J. H., Bougault, C. M. and Kishore, A. I. (2004) Residual dipolar couplings in structure determination of biomolecules. *Chemical Reviews* **104**, 3519–3540.
31. Tjandra, N., Omichinski, J. G., Gronenborn, A. M., Clore, G. M. and Bax, A. (1997) Use of dipolar ^{1}h-^{15}N and ^{1}H-^{13}C couplings in the structure determination of magnetically oriented macromolecules in solution. *Nature Structural Biology* **4**, 732–738.
32. Fushman, D., Varadan, R., Assfalg, M. and Walker, O. (2004) Determining domain orientation in macromolecules by using spin-relaxation and residual dipolar coupling measurements. *Progress in Nuclear Magnetic Resonance Spectroscopy* **44**, 189–214.
33. Tjandra, N., Garrett, D. S., Gronenborn, A. M., Bax, A. and Clore, G. M. (1997) Defining long range order in NMR structure determination from the dependence of heteronuclear relaxation times on rotational diffusion anisotropy. *Nature Structural Biology* **4**, 443–449.
34. Bertini, I., Luchinat, C., Parigi, G. and Pierattelli, R. (2005) NMR spectroscopy of paramagnetic metalloproteins. *Chembiochem* **6**, 1536–1549.
35. Banci, L., Bertini, I., Cavallaro, G., Giachetti, A., Luchinat, C. and Parigi, G. (2004) Paramagnetism-based restraints for xplor-nih. *Journal of Biomolecular NMR* **28**, 249–261.
36. Bertini, I., Luchinat, C. and Parigi, G. (2002) Paramagnetic constraints: An aid for quick solution structure determination of paramagnetic metalloproteins. *Concepts in Magnetic Resonance* **14**, 259–286.
37. Schwieters, C. D., Kuszewski, J. J. and Clore, G. M. (2006) Using xplor-nih for NMR molecular structure determination. *Progress in Nuclear Magnetic Resonance Spectroscopy* **48**, 47–62.
38. Guntert, P., Mumenthaler, C. and Wuthrich, K. (1997) Torsion angle dynamics for NMR structure calculation with the new program dyana. *Journal of Molecular Biology* **273**, 283–298.
39. Hus, J. C., Marion, D. and Blackledge, M. (2000) De novo determination of protein structure by NMR using orientational and long-range order restraints. *Journal of Molecular Biology* **298**, 927–936.
40. Case, D. A., Cheatham, T. E., Darden, T., Gohlke, H., Luo, R., Merz, K. M., Onufriev, A., Simmerling, C., Wang, B. and Woods, R. J. (2005) The amber biomolecular simulation programs. *Journal of Computational Chemistry* **26**, 1668–1688.
41. Van der Spoel, D., Lindahl, E., Hess, B., Groenhof, G., Mark, A. E. and Berendsen, H. J. C. (2005) Gromacs: Fast, flexible, and free. *Journal of Computational Chemistry* **26**, 1701–1718.
42. Spronk, C. A. E. M., Nabuurs, S. B., Krieger, E., Vriend, G. and Vuister, G. W. (2004) Validation of protein structures derived by NMR spectroscopy. *Progress in Nuclear Magnetic Resonance Spectroscopy* **45**, 315–337.
43. Clore, G. M. (2000) Accurate and rapid docking of protein-protein complexes on the basis of intermolecular nuclear Overhauser enhancement data and dipolar couplings by rigid body minimization. *Proc Natl Acad Sci U S A* **97**, 9021–9025.
44. Takahashi, H., Nakanishi, T., Kami, K., Arata, Y. and Shimada, I. (2000) A novel NMR method for determining the interfaces of large protein-protein complexes. *Nature Structural Biology* **7**, 220–223.
45. Sakakura, M., Noba, S., Luchette, P. A., Shimada, I. and Prosser, R. S. (2005) An NMR method for the determination of protein-binding interfaces using dioxygen-induced spin-lattice relaxation enhancement. *Journal of the American Chemical Society* **127**, 5826–5832.
46. Clore, G. M. and Schwieters, C. D. (2003) Docking of protein-protein complexes on the basis of highly ambiguous intermolecular distance restraints derived from 1H/15N chemical shift mapping and backbone 15N-1H residual dipolar couplings using conjoined rigid body/torsion angle dynamics. *J Am Chem Soc* **125**, 2902–2912.

47. Dobrodumov, A. and Gronenborn, A. M. (2003) Filtering and selection of structural models: Combining docking and NMR. *Proteins* **53**, 18–32.
48. Fahmy, A. and Wagner, G. (2002) Treedock: A tool for protein docking based on minimizing Van der Waals energies. *J Am Chem Soc* **124**, 1241–1250.
49. McCoy, M. A. and Wyss, D. F. (2002) Structures of protein-protein complexes are docked using only NMR restraints from residual dipolar coupling and chemical shift perturbations. *J Am Chem Soc* **124**, 2104–2105.
50. Herrmann, T., Guntert, P. and Wuthrich, K. (2002) Protein NMR structure determination with automated NOE assignment using the new software candid and the torsion angle dynamics algorithm dyana. *Journal of Molecular Biology* **319**, 209–227.
51. Vranken, W. F., Boucher, W., Stevens, T. J., Fogh, R. H., Pajon, A., Llinas, M., Ulrich, E. L., Markley, J. L., Ionides, J. and Laue, E. D. (2005) The ccpn data model for NMR spectroscopy: Development of a software pipeline. *Proteins* **59**, 687–696.
52. Cornilescu, G., Delaglio, F. and Bax, A. (1999) Protein backbone angle restraints from searching a database for chemical shift and sequence homology. *Journal of Biomolecular NMR* **13**, 289–302.
53. Wishart, D. S. and Sykes, B. D. (1994) Chemical shifts as a tool for structure determination. *Methods In Enzymology*, 363.
54. Linge, J. P., O'Donoghue, S. I. and Nilges, M. (2001) Automated assignment of ambiguous nuclear Overhauser effects with aria. *Methods in Enzymology* **339**, 71–90.
55. Gilquin, B., Lecoq, A., Desne, F., Guenneugues, M., Zinn-Justin, S. and Menez, A. (1999) Conformational and functional variability supported by the bpti fold: Solution structure of the ca2+ channel blocker calcicludine. *Proteins* **34**, 520–532.
56. Seavey, B. R., Farr, E. A., Westler, W. M. and Markley, J. L. (1991) A relational database for sequence-specific protein NMR data. *Journal Of Biomolecular NMR* **1**, 217–236.
57. Laskowski, R. A., Macarthur, M. W., Moss, D. S. and Thornton, J. M. (1993) Procheck—a program to check the stereochemical quality of protein structures. *Journal of Applied Crystallography* **26**, 283–291.
58. Vriend, G. (1990) What if—a molecular modeling and drug design program. *Journal of Molecular Graphics* **8**, 52–56.
59. Laskowski, R. A., Rullmann, J. A. C., MacArthur, M. W., Kaptein, R. and Thornton, J. M. (1996) Aqua and procheck-NMR: Programs for checking the quality of protein structures solved by NMR. *Journal of Biomolecular NMR* **8**, 477–486.
60. Koradi, R., Billeter, M. and Wüthrich, K. (1996) Molmol: A program for display and analysis of macromolecular structures. *Journal Molecular Graphics* **14**, 51–55.

Part V
Conformational Change

Chapter 14
Conformational Changes in Protein Function

Haiguang Liu, Shubhra Ghosh Dastidar, Hongxing Lei, Wei Zhang, Matthew C. Lee, and Yong Duan

Summary Conformational changes are the hallmarks of protein dynamics and are often intimately related to protein functions. Molecular dynamics (MD) simulation is a powerful tool to study the time-resolved properties of protein structure in atomic details. In this chapter, we discuss the various applications of MD simulation to the study of protein conformational changes, and introduce several selected advanced techniques that may significantly increase the sampling efficiencies, including locally enhanced sampling (LES), and grow-to-fit molecular dynamics (G2FMD).

Keywords: AMBER · G2FMD · LES · Molecular dynamics · Protein conformation

1 Introduction

Protein functions are mostly dependent on their three-dimensional structures. Therefore, conformational changes are often intimately linked to their functions. After protein molecules are first synthesized from the ribosomal machinery, large-scale conformational changes take place to fold the nascent proteins to their native or functionally active conformations. Subsequently, conformational changes at smaller scale are often necessary for many proteins to carry out their functions. Proteins can switch among different conformations in response to their environment to meet their functional roles. In many cases, such conformational changes are the results of binding by other molecules that regulate protein activity. Changes can also be induced by modifying amino acids (e.g., posttranslational modifications on SH2 domains) [1]. In all of these cases, molecular dynamics (MD) simulations can be applied to study the conformational changes.

Application of MD to the study of biomolecular dynamics is now a well-established approach [2]. Over the years, a wide array of MD methods has been developed to investigate an ever-increasing range of biological phenomena [3]. Some notable examples include protein folding [4], induced-fit protein–ligand interactions

From: Methods in Molecular Biology, vol. 443, Molecular Modeling of Proteins,
Edited by Andreas Kukol © Humana Press, Totowa, NJ

[5], and ion-channel flexibility [6]. In recent years, the explosive increase in the number of available high-quality x-ray crystal structures covering many representative families of functionally important proteins has presented an unprecedented opportunity to carry out comparative studies of protein dynamics. Information amassed from such comparative studies will, no doubt, help us decipher the principles of structure–function relationship of proteins. However, studies of protein conformational dynamics were often constrained by technical limitations of experimental methods and information regarding functional motions were limited to simple comparisons of different crystal conformations of a protein and extrapolating the path between the various states. In this respect, the methods of MD offer a powerful alternative for molecular biologists to explore the conformational landscapes of proteins in a more rigorous manner.

Conformational changes of proteins can be categorized into four main classes depending on the time scales of the dynamics and the extent of global structural change. The major types of changes are as follows: 1) large scale conformational change of the backbone leading to folding or unfolding, 2) small scale structural change after change of the environment (e.g., solvent) or after binding with ligand, 3) conformational change of a specific region (e.g., loop) of a large protein without significant global structural change, and 4) conformational changes of the side chains for fine tuning of the stability of the protein and their complexes. The major categories of dynamics will be discussed in Sect. 3 of this chapter, whereas folding and unfolding will be covered in Chap. 15.

Two major bottlenecks limit the applications of MD simulations to the exploration of protein dynamics. The first bottleneck is technological. Despite recent advances in computer technologies, the overwhelming demand of MD on computational resources still sets a limit on the practically achievable simulation time scale. The second bottleneck is thermodynamics in nature. Even if the first bottleneck may be overcome, proteins may be trapped in their local (free) energy minima, thus, simulation trajectories may be confined to around specific regions in the conformational space without ever sampling the more interesting and biologically relevant transitions. Fortunately, many techniques have been developed to overcome these difficulties. For example, techniques such as replica-exchange molecular dynamics (REMD), locally enhanced sampling (LES), grow-to-fit molecular dynamics (G2FMD), and other methods could be brought to bear, depending on the level of conformational search required. These methods can greatly enhance sampling of the conformational space.

However, difficulties still exist when dealing with very large conformational changes. For instance, domain movement may require milliseconds or longer. For such cases, coarse-grained models provide an alternative to atomic level simulations for studying the collective motions. One example of such coarse-grained simulation is elastic network model, which we discuss in **Note 1**.

In this chapter, we focus on two special techniques that have been successfully applied to enhance sampling of simulations during the study of conformational changes. Because REMD is mainly used in protein-folding studies, it will be discussed in Chap. 15 together with folding and unfolding studies.

Using the MD simulation techniques mentioned above, we have recently explored the conformational dynamics associated with the functions of HIV-1 integrase [7] and of HPPK [5]. In the case of HIV-1 integrase, the dynamics of the catalytic domain surface loop was found to be closely related to its function. In the case of HPPK, conformational changes associated with three flexible surface loops not anticipated by x-ray crystal structures were identified. We have also developed methods to sample protein side chain conformations [8]. The results of these studies can be found from the cited publications. Here we describe the methods.

2 Theory

The MD simulation is based on Newton's equation of motion and the details are discussed in Chap. 1.

2.1 Locally Enhanced Sampling

The LES method was first introduced by Elber and Karplus [9] and was further developed later [10–12]. As the name suggests, in this method, enhanced sampling is realized in selected local regions of the system without a significant increase in the overall computational demand. To achieve this goal, selected regions are represented in multiple copies. The copies do not feel the presence of each other during simulation and individually interact with the rest of system. On the other hand, the rest of the system feels the average effect of the multiple copies. It should be made clear that the average effect means the average energy, not the average coordinates. This idea is clearly shown in Fig. 1. Obviously, when the target regions and number

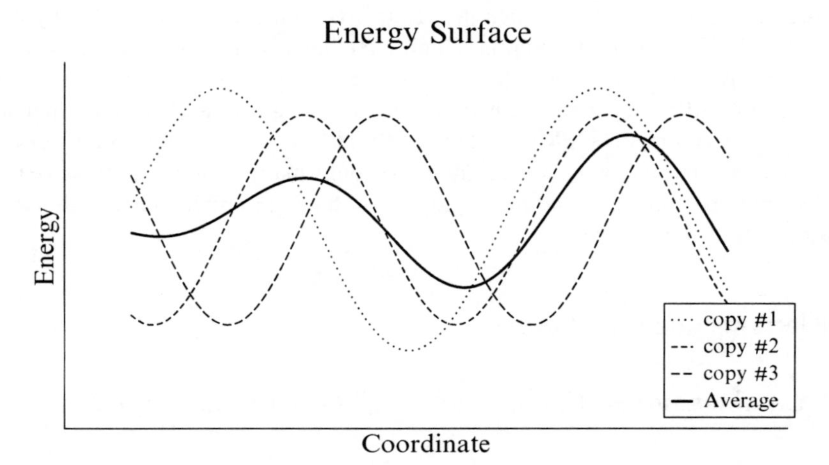

Fig. 1 Schematic illustration of the energy surface. The *thick solid line* is the average energy of three copies. The energy surface is flattened, which facilitates the energy barrier crossing

of copies are properly selected, more conformations can be sampled in those regions without a significant increase in computational time. There are two advantageous effects taking place in this technique: 1) multiple trajectories of the interested regions can be generated simultaneously, which directly increases the sampling space; and 2) the average effect of multiple copies smoothes the energy landscape, which reduces the energy barriers and allows the conformation to escape from local energy minima traps more easily (as shown in Fig. 1). With this technique, more conformations can be sampled in comparison with conventional MD.

2.2 *Grow-to-Fit Molecular Dynamics*

G2FMD is a recently developed *ab initio* method [8] for optimizing the side chains and local structures. In this method, a side chain grows from a reduced size to its regular size, allowing spontaneous selection of its most favorable conformation governed by the all-atom based physical interactions in a simulation. Rotamer-based methods have been developed for side chain assignment [13–20] based on experimental protein structures. In contrast, the grow-to-fit (G2F) method is complementary to these rotamer-based methods because it has the ability to remove atomic collisions from the structure predicted by the rotamer-based methods. Such atomic collisions often occur in structures predicted by rotamer-based methods and are a major limitation for such methods. In addition, the scaling of the potential energy term in G2F methods lowers the energy barriers and allows the structures to escape from their local energy traps.

The code of G2FMD is available on request and will be distributed in the next release of AMBER. The method is described in detail in the published paper [8] and is briefly outlined here. In this method, the bond length and van der Waals radii are simultaneously scaled by a scaling parameter $\lambda(0.6 \leq \lambda \leq 1.0)$ while the charges are scaled as $\sqrt{\lambda}$ such that the electrostatic potential is scheduled by λ. A typical G2FMD simulation comprises multiple (short) simulation time windows. In each window, the selected side chains are reduced to $\lambda = 0.6$ within 1 ps and then followed by a 10 ps segment in which the side chains grow back to the normal size ($\lambda = 1.0$). A complete G2FMD cycle (i.e., shrink and grow) is followed by energy minimization. The energies of the minimized conformations of two successive G2F cycles are to be compared and the conformation having a smaller energy should be retained for the next cycle.

3 Methods (also see Notes)

3.1 *Conformational Changes of Small Proteins and Peptides*

The conformational changes in small proteins and peptides often involve the entire protein that can be viewed in terms of unfolding or folding processes. Thus, those methods used in studying protein folding problems, including conventional MD simulations and REMD, are directly applicable to the studies of conformational changes of small proteins and peptides. The methodologies for setting up and running the simulation are discussed in detail elsewhere (see Chaps. 1 and 15). Here, we discuss the data analysis, which should be carried out to understand the conformational changes.

3.1.1 Analysis

Normal Mode and Principle Component and Essential Dynamics Analyses

Normal mode analysis (NMA) is frequently used to study the collective motions of proteins. In NMA, the energy profile (near the local minimum) is approximated as a quadratic energy surface. The frequency and displacements can be calculated from Hessian matrix of second derivatives of potential energy. Because it is very computationally intensive to calculate the second-order derivatives of potential energies for full-atom models, coarse-grained models are usually applied. We refer readers to **Note 1** for more information. Here, we focus on the analyses of MD simulation trajectories by the principle component analysis (PCA) and essential dynamics (ED) methods.

Principle components are a set of variables that define a projection that encapsulates the maximum amount of dynamic variation along a set of orthogonal vectors. Analogous to NMA, in PCA, the dynamics of the system is represented in the quadratic terms. Principle components can be obtained from diagonalizing the covariant matrix. The basic procedure is:

1. Calculate the coordinate variance–covariant matrix from the simulation trajectories.
2. Diagonalize the matrix to obtain eigenvalues and eigenvectors.
3. Project the conformations (trajectories) to the eigenvectors.

This can be done with the ptraj module in the AMBER package. A detailed explanation can be found in the ptraj section in the AMBER manual. One sample script is shown below:

```
trajin

rms first *

matrix covar name mcovar

analyze matrix mcovar out evec.pev vecs 25

go
```

and an example to generate the corresponding set of projections is shown below:

```
trajin

rms first *

projection modes evec.pev out proj.ppj beg 1 end 25

go
```

A key feature of the PCA is that a small reduced set of principle components can account for a significant portion of the dynamics and can be used to study the correlated movement of the molecular systems. This reduced set of principle components is also referred to as "ED" and can be used to study the conformational changes. On the other hand, because the reduced set only partially accounts for the total dynamics, it is important to examine the coverage, the fraction of the movements represented by the reduced set (see **Notes 2** and **3**). PCA analysis can also be applied to the systems discussed in next section, in which global motion of the protein is also of interest.

Clustering Analysis

Clustering analysis is a versatile tool that can be applied to analyze the conformations sampled during the simulations. A more detailed discussion is presented in Chap. 15. In the case of local conformational changes, a clustering analysis on the selected regions can sometimes be more informative than clusters of the entire protein. Depending on the size of the selected regions, one can use the clustering analysis to identify the possible conformations and to obtain semiquantitative information regarding the conformational preferences.

Secondary Structure Propensity and Dihedral Angle Distribution (Ramachandran Plot)

Similar to the methods described in Chap. 15, one can also analyze the secondary structure distributions observed in the simulations. For a detailed description, readers should follow Chap. 15. Conversely, one can also use dihedral angle distributions, which can provide (semiquantitative) information regarding the conformation preference of the individual residues. One example is shown in Fig. 2.

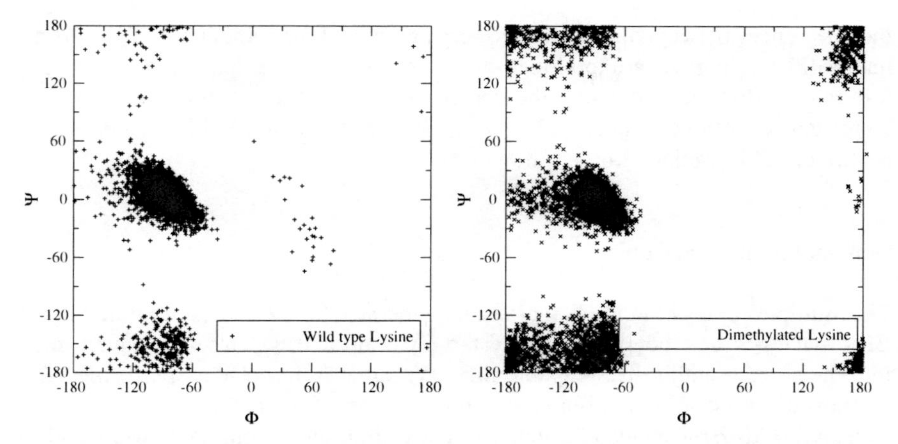

Fig. 2 Ramachandran plot. The methylation caused the conformation shift from α-helix to β-sheet region in the N-terminal tail peptide of histone H3

3.2 Conformational Changes of Large Protein with Flexible Regions

For large proteins, it is often not feasible to model the conformational changes of the entire protein. Thus, one should try to focus on the key regions whose dynamics can be closely linked to the functions of the protein. For example, in the case of HIV-1 integrase, the catalytic loop has been identified to play crucial roles in the enzymatic activity, and the surface loops of HPPK have also been linked to its function. Several methods have been developed to study the dynamics of focused regions. LES is one of such powerful techniques. In the following section, we describe the practical procedures for applying the LES module in the AMBER package.

3.2.1 LES Simulations

Define LES Regions

LES regions must be determined first based on prior knowledge regarding the structural and functional roles of the proteins. The most common case is that the target regions are the functionally important loops that are prone to conformational changes. In comparison, hydrophobic core and secondary structures are tightly packed. After identifying the LES regions, one needs to decide how many copies are needed and how long each segment should be. Too few copies cannot take full advantage of the LES technique; too many copies will flatten the energy landscape to the extent that important structural information may be lost because of the altered energy landscape. A typical choice is five copies for each region. The length of each segment is also crucial. If the segments are too short, the degrees of freedom are limited and the differences between the copies are very small, and the full benefit of LES is not realized. If the regions are too long, the copies may diverge, leading to completely different conformations among the copies, thereby, making it difficult to achieve convergence. A typical choice is between two and four amino acids. The basic procedure of LES simulation is as follows.

Generating Multiple Copies

The multiple copies of the LES regions are generated by the addles program in AMBER that takes the topology file and restart file of normal systems and converts them into the corresponding LES-capable files. We suggest that readers perform energy minimization and short equilibration before starting LES simulations. Readers may consult the AMBER manual for detailed instructions and examples on using addles. The output file of "addles" should be kept because it contains information regarding how the LES segments are partitioned that may be helpful in later analyses for decomposing the LES simulation data to conventional MD trajectories.

Here is a brief workflow for setting up systems with LES:

1. Build normal system without LES with tleap module.
2. Do energy minimization and system equilibration.
3. Generate LES regions with addles:

   addles < addles.in > addles.out
   Sample input file: **addles.in**

```
file rprm name=(test_wat.top) read (Input topology
file)

file rcvb name=(md5.rst) read (Input restart file;
after minimization and equilibration)

file wprm name=(les.parm7) wovr  (LES topology file)

file wcrd name=(les.rst7) wovr  (LES restart file)

action

(Note: the italic characters specify the file format;
read the manual carefully)

omas (Use original mass, otherwise, use 1/N of
original mass, N is the number of copies)

spac numc=5 pick #mon 1 5 done   (Five copies,
```

```
residue numbers 1 to 5 for this region)

spac numc=4 pick #mon 6 10 done   (Four copies,
residue numbers 6 to 10 for this region)
```

The brief explanations are noted in parentheses.

Simulation

LES simulations are performed by `Sander.LES` in AMBER instead of Sander. Please note that different initial velocities for different copies of LES regions are **required**. Otherwise, all copies will remain identical to one another. This means that, in a newly started simulation, the initial temperature should NOT be zero. If the initial temperature is greater than zero, then the difference in the LES copies is automatically accomplished by `Sander.LES` because it requires reassignment of the random velocities. If the velocity information is in the original coordinate file used for generating the LES files, one can set parameters in addles input to require different velocities for each copy.

Analysis

Because each LES simulation effectively represents multiple copies of the system, we need to decompose the multiple trajectories of LES regions into a single conventional MD simulation trajectories before further analysis. Carlos Simmerling has written a nice program to decompose LES trajectories, which runs on SGI machines. The command-line version works on Linux workstations [21]. Readers can also write a simple program based on the output file of addles to decompose the trajectory. Knowledge of the trajectory file format is important for readers to decompose the trajectories. In the case of AMBER, the format of LES restart file is shown in the following example:

```
2266 0.1600000E+02 !(Number of atoms, time in
picoseconds)

121.3869588 102.9172601  17.3051739 121.3869588
   102.9172601 17.3051739

121.3869588 102.9172601  17.3051739 121.3869588
   102.9172601 17.3051739

121.3869588 102.9172601  17.3051739 121.8295428
   102.2037738 16.7437825
```

```
121.8295428 102.2037738  16.7437825 121.8295428
   102.2037738 16.7437825

121.8295428 102.2037738  16.7437825 121.8295428
   102.2037738 16.7437825

121.7201291 103.7754014  16.8896120 121.7201291
   103.7754014 16.8896120

121.7201291 103.7754014  16.8896120 121.7201291
   103.7754014 16.8896120

121.7201291 103.7754014  16.8896120 120.3851944
   102.9844954 17.1954125

.......
```

Note that in the LES restart file example above, the coordinates of the first five atoms are the same because they represent the five copies of the same atom. The file format is:

```
(X1C1 Y1C1 Z1C1), (X1C2 Y1C2 Z1C2), (X1C3 Y1C3 Z1C3),
(X1C4 Y1C4 Z1C4), (X1C5 Y1C5 Z1C5), (X2C1 Y2C1 Z2C1),
(X2C2 Y2C2 Z2C2), (X2C3 Y2C3 Z2C3), (X2C4 Y2C4 Z2C4),
(X2C5 Y2C5 Z2C5), (X3C1 Y3C1 Z3C1), (X3C2 Y3C2 Z3C2),
(X3C3 Y3C3 Z3C3), (X3C4 Y3C4 Z3C4), (X3C5 Y3C5 Z3C5)

......
```

We can see that the coordinates are arranged according to atom index first, followed by the copy index. With this in mind, it is easy to write a program to decompose the trajectories. One can also use the ptraj to do the same work (ptraj can handle trajectories with single LES region in the AMBER9). After the LES trajectories are decomposed, analysis of the decomposed trajectories is the same as that in conventional MD simulations. Readers are referred to Sect. 3.1 of this chapter and other related chapters.

The amino acids at the boundaries of the LES regions play the role of hinges that link the neighboring LES regions. For instance, the loops will switch between open and closed modes. In this respect, Ramachandran plots of hinge residues will be very useful for analyzing the modes. Because the loop consisting of the multiple copies tends to move as a whole, simple structural analysis focused on the LES regions alone will not tell the whole story. Interaction and relative movement between the LES regions and the rest of the system also need to be studied. The way to analyze the relative motions will depend on the specifics of each case, but, in general, one can specify several quantities (e.g., distances and angles) to describe the relative positions of the LES regions to the rest of the system.

Generally speaking, this type of conformational change can also be studied with conventional MD simulations. However, as explained in Sect. 2 (Theory), LES is much more efficient in sampling larger conformational space. Proper selection of LES regions and assignment to length and number of copies are important for successful application of this technique.

3.3 Conformational Refinement of the Side Chains

The G2FMD method is a method for *ab initio* side chain assignment. It can also be applied to refine protein side chains. Setting up a G2FMD run is exactly the same as other conventional MD simulations with additional input parameters designed for G2FMD run. To prepare the input file, one first identifies the target residues whose side chain conformations are to be refined. A sample input file is provided below (also see **Note 4**):

```
&cntrl

imin=2, maxcyc=2000, nstlim=10000,

ntx=1, irest1=0

ntt=1, tempi=100.0, temp0=300, ig=123731,tautp=0.5,

ntc=3, ntf=1, nscm=500,

dt=0.001, ntb=0,

igb=5, saltcon=0.2, gbsa=1, rgbmax = 10.0,

cut=12.0, cut_inner=8.0,

intdiel=1.0, extdiel=78.5,

ntr=1, restraint_wt=5.0,restraintmask='@CA,HA,N,H,O,C',

ntpr=500, ntwx=500, ntwe=500, ntwr=500,

&end

&grown

grow0=0.6, grow1=1.0, igrbond=1, igrvdw=1, igrelec=1,
```

```
ngrowcyc=101, minntc=1,

nrgres=6,

rgres= 37 38 39 40 55 56

lgrow=

531, 532, 533, 534, 535, 536, 537, 538, 539, 540,

541, 542, 543, 544, 551, 552, 553, 554, 555, 556,

557, 558, 559, 560, 561, 568, 569, 570, 571, 572,

573, 574, 575, 576, 577, 578, 579, 580, 581, 588,

589, 590, 591, 592, 795, 796, 797, 798, 799, 800,

801, 802, 803, 804, 805, 812, 813, 814, 815, 816,

&end

# imin=2 Run G2F algorithm

# maxcyc=200 minimization after shrink and grow

# nstlim=10000 MD steps for grow

# &grown = namelist for G2FMD run

# grow0=0.6, grow1=1.0 Shrink to 60% of original size
then grow to 100%

# igrbond,igrvdw,igrelec: each equal to 1 when bond,van

der Waals and

# electrostatic terms are scaled; The value '0' will

switch off the

# scaling of particular term

# minntc=1 no SHAKE during minimization; eq 3 for SHAKE
```

```
# ngrowcyc=101 Number of G2FMD cycles.

# nrgres=6 Number of residues which will shrink and grow

# rgress = list for shrinking and growing

# lgrow = side chain atom list which will shrink and grow.
```

A recommended range for the scaling parameters is $0.6 \leq \lambda \leq 1.0$ (given by `grow0=0.6`, `grow1=1.0` in the input file). One may individually select the terms to be scaled, including electrostatic, van der Waals, or bond energy terms. However, it is usually a reasonable approach to scale them all together. The choice of number of G2FMD cycles (`ngrowcyc`) is at the user's discretion. If there is no change in the conformation or energy after 5 to 10 cycles, one may stop the run.

3.3.1 Analysis

When G2FMD is applied to side chain refinement, the analysis is focused on the side chain conformations, which are usually measured by the dihedral angles ($\chi 1, \chi 2$). Other measurements include the native contacts at the interface and RMSD (with respect to the native structure) of the side chain heavy atoms. For "blind" predictions, where the information regarding the native structure is unavailable, one may examine the increase or decrease of the number of contacts, hydrogen bonds, etc., at the interface. It should be checked whether the atomic collisions have been

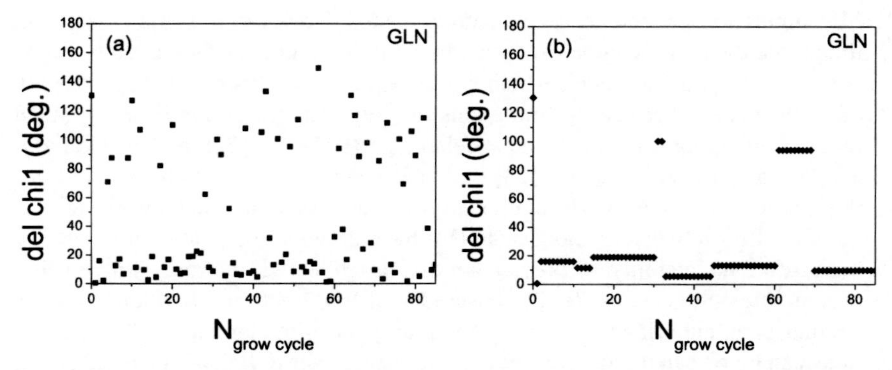

Fig. 3 Example of the refinement of $\chi 1$ dihedral angle for a GLN residue along a G2F trajectory. Shown is the angle difference del chi1 in degrees with respect to the initial structure. (**a**) Each point corresponds to an energy-minimized conformation obtained after a complete G2FMD cycle (shrink + grow). (**b**) Each point correspond to the lower energy conformation that is retained (for the next cycle of run) after comparison between two conformations obtained from two subsequent G2FMD cycles

removed or not. According to the scheme, the "Accepted" conformations after each cycle should have decreasing potential energy. Thus, the structure at the end of the simulation usually has the highest stability and is the optimal structure dictated by the underlying force field.

4 Conclusions

There has been a vast research effort related to the study of the conformational changes of proteins. It is, therefore, impossible to cover all aspects in one chapter. In this chapter, we have been able to discuss only a small portion of the effort, with the emphasis on recent development and practical applications. Many other important features of protein conformational changes need to be studied in detail. For example, it is not possible to model bond formation or breaking with classical MD, and hybrid MM/QM methods have to be used. These are beyond the scope of this chapter. Interested readers should refer to Chap. 3 and other literature.

MD simulations are still limited to study either the long time dynamics of small systems or short time dynamics of large systems. Yet, the application of special techniques, in conjunction with MD simulations, can significantly enhance our ability to tackle problems of biological relevance of both larger spatial and longer temporal scales. With the exponential growth in computer power and the development of new algorithms, MD simulations will be applied to increasingly broader research areas.

5 Notes

1. MD simulation methods can give atomic details. However, in some cases, long time scale dynamics can be examined by simplified models. One good example is the coarse-grained model, which has a long and rich history. A large portion of early models, because of limitations of computer power, are coarse-grained models, including some of the most celebrated HP lattice [22] models, Gō models [20], and Takada's models [23] that have been applied to study protein folding. These models can also be applied to study large systems and longer time simulations. Elastic network models (ENM) have drawn a lot of attention recently because of their ability to predict large-scale collective motions and protein–protein interactions [24–27]. The basic idea of ENM is to use harmonic springs to connect neighboring residues. By computing the normal modes, collective motions can be revealed. For a recent review, please refer to Bahar's paper [28].
2. The coverage of principle components is important to evaluate the PCA results. The top eigenvalues correspond to the most significant motions of the protein. One may inspect the eigenvalues to examine the contributions of each eigenvector, which usually decay exponentially. Then, the largest eigenvalues and the corresponding eigenvectors can be used for ED study.

3. John Mongan developed a plug-in for VMD to visualize the PCA results [29]. The usage is very straightforward. However, there is a compatibility problem with the new version of VMD, readers should try different versions (from our experience, it works well with VMD 1.8 on Linux workstations). Dynamite, developed by Noble and coworkers, is also very simple to use for nonspecialists [30, 31].
4. The implicit solvent model should be used during G2FMD run. A nonphysical cavity may transiently exist when the side chains are reduced, which may trap water molecules if explicit solvent is used. Restraining the backbone atoms of the protein during the run is optional, but it is good to keep them (relatively) fixed so that the scaling does not cause a collapse of the structure. In our code, we have made a provision to switch on or off the scaling of different terms in the energy expression. For example, it is possible to scale the van der Waals terms only and mildly (e.g., 90–95%), if that is sufficient for a particular case. Mild scaling of only the van der Waals term may allow working with a freely moving or mildly restrained backbone, in which the side chains refinement may couple with backbone conformational refinement. Another important concern is the accuracy of the force field. Because the optimization proceeds by comparing the energy of two conformation obtained from two successive G2FMD cycles, the final structure that will have the lowest energy may differ. Nevertheless, G2FMD has the potential to give the optimized structure with respect to a specific force field.

Acknowledgements This work was supported by research grants from the National Institutes of Health (NIH; GM64458 and GM67168 to YD).

References

1. Pawson, T., *Specificity in signal transduction: from phosphotyrosine-SH2 domain interactions to complex cellular systems.* Cell, 2004. **116**(2): p. 191–203.
2. Karplus, M. and J.A. McCammon, *Molecular dynamics simulations of biomolecules.* Nat. Struct. Biol., 2002. **9**(9): p. 646–652.
3. Karplus, M., *Molecular dynamics of biological macromolecules: a brief history and perspective.* Biopolymers, 2003. **68**(3): p. 350–358.
4. De Mori, G.M., M. Meli, L. Monticelli, and G. Colombo, *Folding and mis-folding of peptides and proteins: insights from molecular simulations.* Mini Rev. Med. Chem., 2005. **5**(4): p. 353–359.
5. Yang, R., M.C. Lee, H. Yan, and Y. Duan, *Loop conformation and dynamics of the Escherichia coli HPPK apo-enzyme and its binary complex with MgATP.* Biophys. J., 2005. **89**(1): p. 95–106.
6. Gumbart, J., Y. Wang, A. Aksimentiev, E. Tajkhorshid, and K. Schulten, *Molecular dynamics simulations of proteins in lipid bilayers.* Curr. Opin. Struct. Biol., 2005. **15**(4): p. 423–431.
7. Lee, M.C., J. Deng, J.M. Briggs, and Y. Duan, *Large scale conformational dynamics of the HIV-1 integrase core domain and its catalytic loop mutants.* Biophys. J., 2005. **88**: p. 3133–3146.
8. Zhang, W. and Y. Duan, *Grow to Fit Molecular Dynamics (G2FMD): an ab initio method for protein side-chain assignment and refinement.* Prot. Eng. Des. Sel., 2006. **19**(2): p. 55–65.

9. Elber, R. and M. Karplus, *Enhanced sampling in molecular-dynamics—use of the time-dependent Hartree approximation for a simulation of carbon-monoxide diffusion through myoglobin.* J. Am. Chem. Soc., 1990. **112**(25): p. 9161–9175.
10. Simmerling, C.L. and R. Elber, *Computer determination of peptide conformations in water—different roads to structure.* Proc. Natl. Acad. Sci. USA, 1995. **92**(8): p. 3190–3193.
11. Simmerling, C., T. Fox, and P.A. Kollman, *Use of locally enhanced sampling in free energy calculations: Testing and application to the alpha ->beta anomerization of glucose.* J. Am. Chem. Soc., 1998. **120**(23): p. 5771–5782.
12. Simmerling, C., J.L. Miller, and P.A. Kollman, *Combined locally enhanced sampling and Particle Mesh Ewald as a strategy to locate the experimental structure of a nonhelical nucleic acid.* J. Am. Chem. Soc., 1998. **120**(29): p. 7149–7155.
13. Ponder, J.W. and F.M. Richards, *Tertiary templates for proteins. Use of packing criteria in the enumeration of allowed sequences for different structural classes.* J. Mol. Biol., 1987. **193**(4): p. 775–791.
14. De Maeyer, M., J. Desmet, and I. Lasters, *All in one: a highly detailed rotamer library improves both accuracy and speed in the modelling of sidechains by dead-end elimination.* Fold. Des., 1997. **2**(1): p. 53–66.
15. Mendes, J., A.M. Baptista, M.A. Carrondo, and C.M. Soares, *Improved modeling of side-chains in proteins with rotamer-based methods: a flexible rotamer model.* Proteins: Structure, Function, and Bioinformatics, 1999. **37**(4): p. 530–543.
16. Looger, L.L. and H.W. Hellinga, *Generalized dead-end elimination algorithms make large-scale protein side-chain structure prediction tractable: implications for protein design and structural genomics.* J. Mol. Biol., 2001. **307**(1): p. 429–445.
17. Liang, S. and N.V. Grishin, *Side-chain modeling with an optimized scoring function.* Prot. Sci., 2002. **11**(2): p. 322–331.
18. Tsai, J., R. Bonneau, A.V. Morozov, B. Kuhlman, C.A. Rohl, and D. Baker, *An improved protein decoy set for testing energy functions for protein structure prediction.* Proteins: Structure, Function, and Bioinformatics, 2003. **53**(1): p. 76–87.
19. Peterson, R.W., P.L. Dutton, and A.J. Wand, *Improved side-chain prediction accuracy using an ab initio potential energy function and a very large rotamer library.* Prot. Sci., 2004. **13**(3): p. 735–751.
20. Ueda, Y., H. Taketomi, and N. Go, *Studies on Protein Folding, Unfolding, and Fluctuations by Computer-Simulation .2. 3-Dimensional Lattice Model of Lysozyme.* Biopolymers, 1978. **17**(6): p. 1531–1548.
21. Simmerling, C. and P. Kollman, *MOIL-View: A program for visualization of structure and dynamics of biomolecules.* Abst. of Papers of the American Chemical Society, 1996. **211**: p. 92–Comp.
22. Lau, K.F. and K.A. Dill, *A lattice statistical-mechanics model of the conformational and sequence-spaces of proteins.* Macromolecules, 1989. **22**(10): p. 3986–3997.
23. Yoshimi Fujitsuka, S.T.Z.A.L.-S.P.G.W., *Optimizing physical energy functions for protein folding.* Proteins: Structure, Function, and Bioinformatics, 2004. **54**(1): p. 88–103.
24. Chennubhotla, C., A.J. Rader, L.W. Yang, and I. Bahar, *Elastic network models for understanding biomolecular machinery: From enzymes to supramolecular assemblies.* Phys. Biol., 2005. **2**(4): p. S173–S180.
25. Kantarci, N., P. Doruker, and T. Haliloglu, *Cooperative fluctuations point to the dimerization interface of p53 core domain.* Biophys. J., 2006. **91**(2): p. 421–432.
26. Kozel, B.A., B.J. Rongish, A. Czirok, J. Zach, C.D. Little, E.C. Davis, R.H. Knutsen, J.E. Wagenseil, M.A. Levy, and R.P. Mecham, *Elastic fiber formation: A dynamic view of extracellular matrix assembly using timer reporters.* J. Cell. Phys., 2006. **207**(1): p. 87–96.
27. Liu, Y.M., M. Scolari, W. Im, and H.J. Woo, *Protein-protein interactions in actin-myosin binding and structural effects of R405Q mutation: A molecular dynamics study.* Proteins: Structure, Function, and Bioinformatics, 2006. **64**(1): p. 156–166.
28. Bahar, I. and A.J. Rader, *Coarse-grained normal mode analysis in structural biology.* Curr. Op. Struct. Biol., 2005. **15**(5): p. 586–592.

29. Mongan, J., *Interactive essential dynamics.* J. Comp. Aided Mol. Des., 2004. **18**(6): p. 433–436.
30. Barrett, C.P., B.A. Hall, and M.E. Noble, *Dynamite: a simple way to gain insight into protein motions.* Acta Cryst. D Biol. Cryst., 2004. **60**(Pt 12 Pt 1): p. 2280–2287.
31. Barrett, C.P. and M.E. Noble, *Dynamite extended: two new services to simplify protein dynamic analysis.* Bioinformatics, 2005. **21**(14): p. 3174–3175.

Chapter 15
Protein Folding and Unfolding by All-Atom Molecular Dynamics Simulations

Hongxing Lei and Yong Duan

Summary Computational protein folding can be classified into pathway and sampling approaches. Here, we use the AMBER simulation package as an example to illustrate the protocols for all-atom molecular simulations of protein folding, including system setup, simulation, and analysis. We introduced two traditional pathway approaches: *ab inito* folding and high-temperature unfolding. The popular replica exchange method was chosen to represent sampling approaches. Our emphasis is placed on the analysis of the simulation trajectories, and some in-depth discussions are provided for commonly encountered problems.

Keywords: Continuum solvation model · Explicit solvent · Molecular dynamics · Protein folding · Replica exchange method · Unfolding

1 Introduction

Understanding the molecular mechanisms governing the protein-folding process is fundamental to molecular and structural biology. In a nutshell, protein folding is a conformational search problem. Despite the astronomically large conformational space (3^N; N is the number of residues) [1], proteins need to reach the native state in a very short amount of time (microseconds to seconds). Therefore, exhaustive sampling is not permissive and specific pathway(s) must exist [2]. An enormous amount of effort has been devoted to search these intriguing pathways. Experimentally, various techniques have been developed for protein-folding studies, among which, the laser-induced rapid temperature jump experiments [3] and the rapid-mixing methods [4], including stopped flow, continuous flow, and quenched flow methods, are dominant. Effort in methodology development has been focused on reaching higher temporal resolution, from submillisecond to nanosecond, to enable studies of the very early events of protein folding. In folding experiments, the progression of protein folding is usually monitored by circular dichroism (CD) [5], fluorescence signals [6], nuclear magnetic resonance (NMR) [7], and x-ray scattering [8].

From: Methods in Molecular Biology, vol. 443, Molecular Modeling of Proteins
Edited by Andreas Kukol © Humana Press, Totowa, NJ

Because of the enormous complexity of the conformational space, computational protein folding demands extreme computing power. Thus, to reduce computing demand, protein folding has been traditionally studied by residue-level models [9, 10] with the potential to sacrifice the accuracy due the reduced detail of the representation. Go-type models, that consider only the native contacts at either the residue or atomic levels, have also been applied [11]. We limit our discussion in this chapter to the all-atom physics-based models that have gained increasing popularity recently. Direct simulations of protein folding from the fully denatured state using this approach is called *ab initio* folding, referring to the fact that the model does not contain any *a priori* information regarding the native protein structure because the parameters have been developed on short peptide fragments. Currently, the application of conventional molecular dynamics (MD) simulations on protein folding is still limited to a few model peptides and a dozen super-fast folding proteins [12]. Yet, with the exponential growth in computer power and the development of power sampling techniques, simulations of medium-size proteins become increasingly feasible. To overcome the conformational sampling problem, high-temperature unfolding [13] and mechanical unfolding [14] have been used to study the reverse process of protein folding—the unfolding process. Other strategies, such as targeted MD [15] and self-guided MD [16], have been developed to connect the folded and unfolded states, but the applications to protein folding have been limited because of technical challenges.

Complementary to the pathway-oriented approaches, enhanced-sampling techniques have also been developed that have been applied to study protein folding. The most popular one of these is the replica exchange method (REM) or parallel-tempering [17] method. A number of successful folding studies have been reported based on REM [18–20], which uses temperature hopping to overcome local energy barrier. One of the major advantages of this technique is that it naturally provides ways to analyze the temperature-dependent folding properties, such as melting curve [21]. Another technique is called umbrella sampling, which applies a biased potential to overcome energy barriers on the free energy landscape [22]. However, it has not been widely adopted because of the sophisticated technical details. Alternatively, one may perform large number of short simulations using the distributed computing platforms such as the folding@home projects, which can generate thousands to millions of short trajectories [23], where the challenge lies in the management of thousands of heteroplatform computers and the avalanche of data.

Based on the knowledge acquired from the experimental and computational studies, several models have been proposed for protein folding [24,25]. The hydrophobic collapse model states that folding is initiated by nonpolar residues moving away from the polar aqueous environment. Therefore, formation of the hydrophobic core occurs before the formation of secondary structures [26]. In contrast, the framework model states that formation of the tertiary contacts is preceded by the completion of the secondary structures [27]. In the diffusion–collision model, protein folding is interpreted as a stochastic process in which proteins form local structural elements (both native and nonnative) that coalesce, leading to the global folding. A key feature of the diffusion–collision model is that it does not require the completion of the secondary structures or stabilization of the native contacts during the diffusion–collision process [28]. The nucleation condensation model states that protein folding is initiated by the formation of a strong folding nucleus that leads to the propagation of both secondary structures and tertiary contacts [29, 30], and that formation of such a nucleus is the rate-limiting step. In the funnel landscape model, protein folding is considered as a heterogeneous process moving on the (hyper) free energy surface with numerous pathways and local traps [31, 32]. An important objective of folding and unfolding simulations is to validate and extend these existing models [33, 34] and to provide data for the development of new ones.

2 Theory

For more details regarding MD, please refer to Chap. 1. Here, we briefly outline the underlying theories for REM. In conventional MD simulations at 300 K, conformational transition is hindered by the slow barrier-crossing events. Higher temperatures can facilitate local energy barrier crossing and enhance sampling because of higher kinetic energy. REM implements a number of simulations running simultaneously at a range of temperatures. After certain steps (for example, 2,000 steps), simulations are halted and the energies of each replica are evaluated. The exchange of neighboring replicas is attempted based on the metropolis criterion:

$$\omega = \begin{cases} 1, & for\ \Delta \leq 0 \\ \exp(-\Delta), & for\ \Delta > 0 \end{cases}$$

$$\Delta = \left(\frac{1}{k_B T_i} - \frac{1}{k_B T_j}\right)(E_j - E_i),$$

where ω is the probability of exchange, T_i and T_j are the reference temperatures, and E_i and E_j are the instant energies at the time of exchange attempt. After each exchange, temperatures are adjusted to the new target temperatures, and the simulations continue. For more details, please refer to the work of Okamoto and coworkers [17, 35, 36].

3 Methods

In this section, we focus our descriptions on the three most popular techniques, including two pathway approaches (conventional folding and high temperature unfolding), and one sampling technique (REM). The availability of popular software packages, including AMBER [37], CHARMm [38], GROMACS [39], GROMOS [40], and OPLS [41], has greatly facilitated the studies of protein folding using all-atom physics-based models. Although file formats and technical details are usually different, the general procedures for system setup, simulation, and analysis are very similar. Please refer to the user manual of the specific simulation package for detailed instructions (see **Note 1**). For clarity, we use AMBER as an example to outline the procedures. Within AMBER package, the "tleap" program is for generating topology file and initial coordinate file, "sander" is for energy minimization and MD simulations, and program "ptraj" is for trajectory analysis. We put more emphasis on the analysis, which is typically less well covered in manuals, and, in many cases, may require some coding efforts from the readers.

3.1 Ab Initio *Folding by Conventional MD*

3.1.1 System Setup

Building Topology and Starting Structure

In a typical simulation, a set of files is needed. Some of the files are designed to keep track of the coordinates, whereas others are for the information of the system, such as force constants and connectivity of the molecules; the latter is often referred to as the "topology file." For simulations with continuum solvent models, such as generalized Born (GB) models (see **Note 2**) [42, 43], an extended chain is usually used as the starting structure. Utility programs can be found in simulation packages to generate extended polypeptide chain. A topology file should also be generated for the protein. The default protonation state for the system is for neutral pH condition. In case of nonneutral solvent pH, different protonation states shall be chosen for acidic (Asp and Glu) and basic (Lys and Arg) residues. Special attention should be paid to the protonation state of histidine because of the nearly neutral pK_a of the side chain (pK_a 6.04). Occasionally, blocking units are used for terminal residues to match experimental conditions. Although standard blocking units can be found in simulation packages, the nonstandard blocking units or amino acids need to be parameterized. Readers are advised to resort to the developers or user community of the simulation package for help. The following is an example for setting up a system for GB simulation using tleap:

>`source leaprc.ff03` (*specify force field FF03*)
>`mol = sequence {NMET GLY ASP PRO CPHE}` (*specify protein sequence*)
>`set default PBradii mbondi2` (*specify bond radii set for IGB* = 5)
>`saveamberparm mol protein.top protein.crd` (*write out topology and coordinates*)

For simulations with explicit solvent, water molecules are added to the system before building the topology and coordinate file, and the distance from protein atoms to the edge of water box should be at least 8 to 10 Å. Note that a truncated octahedron box is usually more efficient than a rectangular box that reduces the number of particles by approximately 40%. For proteins with net charges, counterions should be added to neutralize the system. Additional ions may be needed to bring the solution to the targeted salt concentration. Because of the large size of the water box, an extended chain is not recommended for the starting structure in simulations with explicit solvent (see **Note 3**). Rather, the starting structure typically comes from the collapsed chain generated by simulation with continuum solvation model, or from unfolding of the native structure.

Minimization

After the construction of the system, energy minimization is performed to remove potential steric problems that arose during the initial system building. This can be done in the Sander program of the AMBER package. The minimization method can be steepest descent itself or combined with conjugate gradient [44]. A simple input file needs to be prepared as follows for the minimization run. Notice that the optimization cycle has been set to a maximum of 1,000 steps here, because the purpose is to remove the potential steric problems rather than complete energy minimization.

```
&cntrl

 imin=1, maxcyc=1000, ncyc=500,

 ntpr=50, ntc=1,

 ntb=0, cut=10.0,

&end
```

When the input, topology, and coordinates files are ready, sander can be invoked as:

```
sander -O  -i min.in -o min.out -p protein.top -c

protein.crd -r protein.min.crd
```

Equilibration

After the minimization and before the production phase, an equilibration phase is needed to further “relax” the system. Because we do not intend to retain the initial structure in folding simulation, equilibration should be fairly straightforward. We can simply ramp up the temperature of the system to a target value (300 K, for example) with a 20 ps MD run. In case of multiple folding trajectories, multiple equilibration runs can be performed using different random number seeds to generate different trajectories. Alternatively, one can have only one equilibration run and use different random number seeds at the beginning of the production runs to achieve the same. This is also done in sander, which can be invoked the same way as in Sect. 3.1.1. A sample input file is given below:

```
&cntrl
 nstlim=10000, cut=12.0,
 ntpr=1000, ntwr=10000, ntt=1, ntwx=0,
 temp0=300.0, tempi =100.0,
 ntx=1, irest=0,
 ntc=2, ntf=2, tol=0.00005,
 dt=0.002, ntb=0, tautp=2.0, nscm=500,
 igb=5, gbsa=1, saltcon=0.2, rgbmax=12.0,
 intdiel=1.0, extdiel=78.5,
 nrespa=4, nrespai=2,
&end
```

3.1.2 Production Runs

Depending on the size of the system, the length of production phase usually ranges from 100 ns to 1 μs. Multiple trajectories are preferred to avoid the potential biased conclusion from a single trajectory. System coordinates can be saved every 10 ps. Simulations should be stopped and restarted every 1 ns (for continuum solvent) or 100 ps (for explicit solvent) to allow recovery in case of hardware or software failure (see **Note 4**). Whenever possible, parallel execution is always preferred over single CPU execution for folding simulations that are implemented in AMBER using message-passing interface (MPI). For simulation with explicit solvent on parallel platforms, the program pmemd is more efficient than sander. When using a computer cluster, simulation trajectories and output files should be saved on individual nodes to reduce the load to the network and to the shared storage area. Here is a sample input file:

```
&cntrl
 nstlim=500000, cut=12.0,

 ntpr=500, ntwr=500000, ntwx=5000,

 ntt=1, temp0=300, tempi =300,

 ntx=5, irest=1, ntb=0,

 ntc=2, ntf=2, tol=0.00005,

 dt=0.002, tautp=2.0, nscm=500,

 igb=5, gbsa=1, saltcon=0.2, rgbmax=12.0,

 intdiel=1.0, extdiel=78.5,

 nrespa=4,nrespai=2,

&end
```

The following is an example to run parallel sander with MPI (the nodes are specified in the file `$MF`):

```
mpirun -np 4 -machinefile $MF sander -O  -i input -p
protein.top\

-c run0045.rst -o run0046.out -x run0046.crd -r run0046.rst
```

3.1.3 Trajectory Analysis

Root Mean Square Deviation

The closeness to (or deviation from) the native structure is usually assessed by the root mean square deviation (RMSD) of C_α atoms, main chain atoms, or heavy atoms from the native structure. Although the RMSD criterion for folding depends on the system being studied, a distance of 3.0 to 4.0 Å RMSD with respect to the initial structure has been widely used as a typical indication for a reaching a folded conformation. In the case of partial folding, RMSD of certain fragments can be evaluated. When the protein reaches the folded state, one can obtain information regarding how long it stays in the folded state, whether reversible folding occurs, and, in the case of multiple trajectories, how many trajectories reach the folded state.

Clustering of Sampling Space

The conformational space sampled during the folding simulation can be evaluated by clustering. Hierarchical clustering [45] is a common approach, in which snapshots are separated into different clusters based on the pairwise RMSD. Again, the cutoff value depends on the system (the first try can be 3.0 Å), and a few terminal residues can be excluded during the clustering because of their general tendency to be disordered in the native structures. Clusters are ranked by population, and representative structures from highly populated clusters should be carefully examined. In the case of multiple long simulation trajectories, heuristic clustering methods [46] need to be applied because of the large memory consumption by hierarchical clustering.

Formation of Secondary Structures

α-helices and β-sheets are the most prominent secondary structures in proteins (see **Note 5**). The formation of secondary structures can be monitored by main chain hydrogen bonds, main chain dihedral angles, and local RMSD. The criterion for a hydrogen bond is usually a combination of distance cutoff (3.5 Å between donor and acceptor) and angle cutoff (120° for the donor hydrogen-acceptor angle). For helices, helicity can be measured based on $\phi/\psi(-57°, -47°) \pm 40°$. The observations include the folding time, initiation site, and dynamics of the secondary structure.

Development of Native Contacts

Native contacts refer to the close contacts between residues in the native state. There are a variety of ways to evaluate native contacts. Native contacts can be measured by the distances between C_α atoms with a 6- to 7-Å cutoff for nonadjacent residues, on which residue–residue contact maps can be constructed to reveal the structural features. For example, the continuous grids parallel and next to the diagonals are signs of helices and those perpendicular to the diagonals correspond to the anti-parallel β-sheets. One can also measure the native contacts by the closest distances between side chain atoms of two nonadjacent residues. Cutoff value can be based on van der Waals distance or a simple 5.0-Å distance between heavy atoms. Observations include the initiation and dynamics of each contact.

Change of Solvent-Accessible Surface Area

Folding process is usually accompanied by the significant decrease in solvent-accessible surface area (SASA) [47]. Therefore, the change of SASA (or the alternative "radius of gyration"), although not deterministic, is an indication of the folding process. Additionally, atoms can be separated into polar and nonpolar ones so that the burial of polar and nonpolar surfaces can be uncoupled. Furthermore, the SASA of individual residues can be monitored during the folding process. For simulations with explicit solvent, analyses that are more sophisticated can be performed on the solvent distribution, which could include the first and second solvation shell, residence time of solvent, etc.

Energetic Analysis

The most difficult analysis for protein folding lies in the dissection of the energetic contributions [48] because the close coupling of different energetic terms together determine the dynamics of the system, and protein folding is determined by free energy, which consists of potential energy and entropy. It is rather challenging to calculate the entropy change during folding (see **Note 6**) [49], which is further hindered by the insufficient sampling in most simulations. For simulations with explicit solvent, the potential energy can be separated into protein internal energy, solvent internal energy, and protein–solvent energy. Because of the existence of a large number of water molecules, when the total sums over the entire systems are under consideration, the energy components involving solvent tend to have large fluctuation and render more difficulty for interpretation.

Folding Landscape

Because protein folding is a multidimensional problem, analyses based on one-dimensional reaction coordinates often give an incomplete (and overly simplified) picture of the folding process. Therefore, various two-dimensional free energy landscapes have been contemplated to improve the pictorial presentation of the complex process. The choice of the two reaction coordinates, however, is rather arbitrary. The commonly used coordinates include RMSD, radius of gyration, and percentage of native contacts, as well as RMSD of structural segments. Other coordinates can be used as long as they can improve the understanding of the folding mechanism. It should be noted that any two-dimensional maps are still incapable of representing the hyperdimensional nature of the protein folding process [46]. Alternatively, one may apply principle component analysis [50]. However, the results are less intuitive.

Figure 1 shows a sample analysis of a folding trajectory of a three-helix protein villin headpiece subdomain. The analyses shown are the overall RMSD compared with the native structure, the formation of secondary structure (main chain hydrogen bonds) and tertiary contacts (side chain contacts), SASA, and the potential energy. A stepwise folding can be clearly seen from the RMSD and hydrogen bond plots.

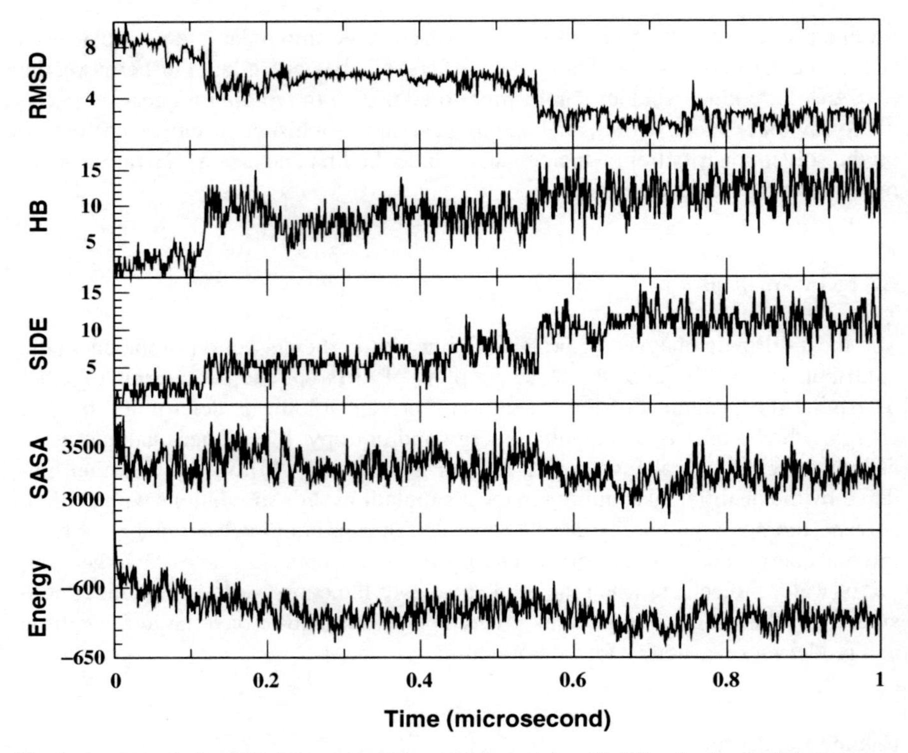

Fig. 1 An example for trajectory analyses from folding study with GB solvation model. From *top* to *bottom*, root mean square deviation (RMSD) using the native structure as reference (Å), number of native main chain hydrogen bonds (HB), number of native tertiary side chain contacts (SIDE), SASA in $Å^2$, and potential energy (kcal/mol)

3.2 High-Temperature Unfolding

Most of the procedures, including the analyses, should be similar to the ones described in Sect. 3.1. Here, we only emphasize the difference.

3.2.1 System Setup

For unfolding simulations, coordinates are usually taken from a PDB file. Before generating coordinates and topology files, the coordinates of the heavy atoms of the protein should be extracted from the PDB file. The target temperature is usually set to 400 to 500 K (see **Note 7**), and restraints should be applied during the minimization and equilibration process so that the protein can stay in the native conformation before the production run. This is especially important for simulations with continuum solvent. Longer and stepwise equilibration may be applied in case the initial system is not stable. It is also typical to run a comparative simulation at the room temperature.

3.2.2 Production Runs

The length for unfolding simulation is usually shorter than that of folding simulation, 10 to 50 ns should be sufficient. Multiple trajectories are still preferred to avoid biased observations of the unfolding process.

3.2.3 Trajectory Analysis

Average properties can be calculated for the evaluation of the overall trend. Special attention should be paid to the initiation of the unfolding: melting of secondary structure or loss of tertiary contacts. Because of better sampling in unfolding simulation than folding simulation, transition state may be easily identifiable from a two-dimensional landscape map (Fig. 2). However, because of the elevated temperature,

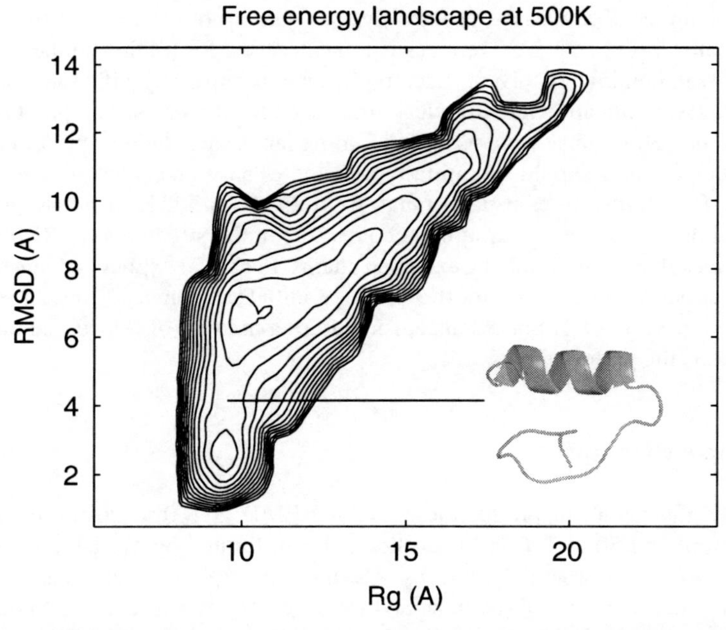

Fig. 2 A protein-unfolding landscape from simulation of a small designed ββα protein FSD with explicit solvent at 500 K shown as the free energy in dependence of RMSD with reference to the native structure (PDB code: 1FSD) and radius of gyration. The transition state can be identified and a representative structure is shown. According to the unfolding landscape, the native basin corresponds to the region with RMSD less than 3.0 Å, and the unfolded state corresponds to the region with RMSD greater than 5.0 Å. The transition state is the sparsely sampled region that links the native state and unfolded state

it is possible that the observed transition state is shifted in comparison with that at room temperature.

3.3 Folding by REM

In contrast to protein folding/unfolding simulations by conventional MD, replica exchange molecular dynamics (REMD) requires a large number of devoted nodes within the same computer cluster running simultaneously. In return, REMD gives much better sampling than conventional MD. Therefore, there is higher confidence regarding the derived folding landscape. However, convergency may still be a potential problem for REMD, which has not been extensively tested because of the limited computing resources [51]. Therefore, longer simulation is still preferred.

3.3.1 System Setup

The number of replicas should be chosen according to the size of the system. Ideally, the number of replicas is decided by the square root of atoms of the system. Realistically, a set of 20 to 30 replicas is typically used for folding studies of small proteins with continuum solvent. Each replica can occupy a specified node (two or four CPUs). Significantly more replicas are needed for REMD with explicit solvent. For example, 80 replicas were used in a folding landscape study of protein A [52]. Target temperatures should be carefully selected to have an exchange rate near 20 to 30%. The temperatures usually range from 250 K to 500 K or 600 K, with exponential distribution as the common choice. The initial structures for REMD with continuum solvent can simply be extended chains. In case of explicit solvent simulations, caution should be taken for the choice of initial structures for the same reason illustrated in Sect. 3.1.1. For instance, sometimes a mixture of folded and unfolded structures is the preferred choice.

3.3.2 Production Runs

Currently, the length of production runs for REMD is 100 to 200 ns for continuum solvent and 50 to 100 ns for explicit solvent, limited by available computing resource. *Ab initio* protein folding by REMD with explicit solvent has yet to be reported, other than a miniprotein named trp-cage [21]. Because of the frequent information exchange across the nodes, the stability of the network of the computer cluster is critical for the success of REMD. Thus, simulation trajectories should be saved on individual nodes to reduce the load of the network.

3.3.3 Trajectory Analysis

In REMD, the exchange is typically performed by swapping the temperatures to minimize the communication overhead. Therefore, each trajectory may record simulations at a range of temperatures. After the simulations, trajectories are collected from the individual nodes and sorted according to the recorded temperatures at each step so that each trajectory retains information of a specific target temperature. All of the analyses described in Sect. 3.1.3 are applicable to REMD. Temperature-dependent properties can also be obtained in a straightforward fashion where the time-dependent properties are masked by the altered kinetics. Analyses are first performed on each of the sorted trajectories individually. When those analyses are combined, the temperature-dependent folding should become apparent. The most important of which is the melting curve based on the population of folded and denatured states at each temperature where the folded and the denatured states are identified typically by RMSD in reference to the experimental native structure. Both ΔH and the melting temperature, T_m, can be derived by fitting the melting curve. In addition, the heat capacity profile can be generated from the fluctuation of the potential energy at each temperature in which the peak indicates the melting temperature (see Fig. 3).

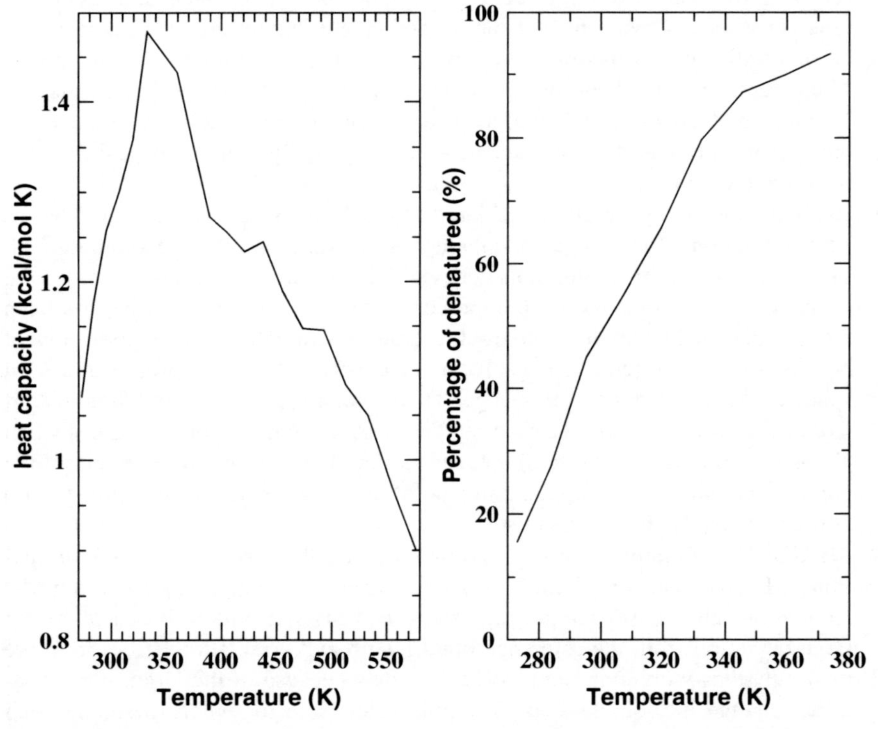

Fig. 3 Sample analyses from folding simulations with REMD, including a heat capacity profile (*left*) and a melting curve (*right*)

4 Notes

1. In folding simulations, as well as any other MD simulations, the choice of force field is critical. Unbalanced force field could lead to biased simulation results, such as excessive formation of α- or β-conformations. This is a particularly important issue in folding simulations. The most widely tested force fields include AMBER [53], CHARMm [54, 55], GROMOS [56, 57], and OPLS [58]. Even within AMBER, there are choices of various force fields, including FF94, FF96, FF99, and the more recent FF03. Extensive tests have shown that FF03 is more balanced than previous AMBER force fields. For a detailed comparison of force fields, please refer to Chap. 4 and other references [59–61].
2. For conventional folding simulations, another critical choice is the solvation model. Continuum solvent models provide a few advantages over explicit solvent, among which, the most important is the much longer accessible simulation time. The fastest folding known today is on the time scale of microseconds, which is still difficult to reach by simulations with explicit solvent. The easier access of microsecond time scale by simulations with continuum solvent models allows extensive testing and tuning of current force fields. Please note that it has been shown that the free energy landscapes of β-hairpins obtained from continuum solvent could be different than those from explicit solvent simulations [62], implying that the solvation effect in continuum models is somewhat less accurate. Nevertheless, with the constant improvement, the difference is expected to be reduced. The GB model is the most widely used continuum solvent model. The polar effect of solvent in GB model is parameterized based on the Poisson-Boltzmann electrostatic calculation. The nonpolar effect of solvent, on the other hand, is usually based on surface area and is typically more empirical than the polar counterpart.
3. Explicit solvent is generally considered as the "state of art" solvation model for MD simulation. However, despite the heroic microsecond folding simulation on villin headpiece [63], simulations with explicit solvation are still too slow for the observation of a complete folding process. *Ab initio* folding usually starts from an extended chain, which is also troublesome for simulation with explicit solvent because of the requirement of a solvent box covering the whole protein and extra space between protein and box edge. Therefore, a practical approach is to start from a somewhat collapsed structure(s). However, caution should be taken when choosing the initial structure(s), because it could take a long time to unfold the compact nonnative structure, and this unfolding process may have nothing to do with the actual folding mechanism.
4. SHAKE [64] failure is the most common and lethal problem in MD simulations. SHAKE is an algorithm for maintaining constant bond lengths during MD simulations allowing for larger time steps, and SHAKE failure is caused by the large deviation from the reference bond length and orientation. It occurs more in simulations with continuum solvent models because of the larger conformational change in a single step. A simple solution is to switch to smaller step size (1 fs) for a short period of time and then switch back to the original step size (2 fs). Careful scripting may be needed to maintain a non-interrupting simulation.

Alternatively, one may use the smaller step size for the entire simulation, which means much slower simulation. In REMD with continuum solvent, however, a typical approach is to use a small step size for the entire simulation because of the existence of high-temperature replicas and frequent exchange among replicas. In simulations with explicit solvent, SHAKE failure is usually caused by bad initial structure where atom overlaps exists. In this case, manual intervention is needed to remove atom overlap before simulation. This usually does not happen to folding simulation with explicit solvent because the initial structures are usually not compact.

5. VMD software [65] is widely used for the visualization of simulation trajectories. A movie of the folding process observed in simulation can be generated for presentation. Some of the analyses can also be performed by VMD, including the development of secondary structures in the simulation trajectory. PyMol is another visualization software [66] that can facilitate generation of publication quality graphics. In addition, RasMol [67] is also handy when a quick assessment on a specific structure is needed.
6. Quantitative information regarding entropy is usually extracted indirectly by subtracting enthalpy landscape from free energy landscape. When the extensive sampling of the conformational space is not available, estimation of entropy change can be made based on the loss of conformational freedom of residues in regular secondary structures, which can be separated into main chain and side chain entropy loss. Main chain entropy loss is roughly 4.8 cal/(mol K residue) [68] from fully denatured to fully folding state. There are various statistical indexes for the loss of side chain entropy on folding that is usually derived from protein structures in PDB [69, 70].
7. Folding mechanism could be obtained from unfolding simulations if folding follows the reverse pathway of the unfolding process. This hypothesis, however, has not been proven and is unlikely to be true for the folding of most proteins. Higher temperature is more efficient for unfolding simulations because it leads to faster unfolding. However, it has been demonstrated that, as the applied unfolding temperature changes in simulation, so does the unfolding landscape [71]. Therefore, readers are advised to stay away from excessive high temperatures if the goal is to understand the folding landscape at room temperature. On the other hand, the rapid growth in computer technology allows simulations to be conducted for much longer time. Thus, unfolding simulations at lower temperatures also become possible.

Acknowledgements This work was supported by research grants from the National Institutes of Health (NIH; GM64458 and GM67168 to YD).

References

1. Karplus, M., *The Levinthal paradox: yesterday and today.* Fold. Des., 1997. **2**(4): p. S69–75.
2. Dill, K.A. and H.S. Chan, *From Levinthal to pathways to funnels.* Nat. Struct. Biol., 1997. **4**(1): p. 10–19.

3. Ballew, R.M., J. Sabelko, and M. Gruebele, *Direct observation of fast protein folding: The initial collapse of apomyoglobin.* Proc. Natl. Acad. Sci. USA, 1996. **93**: p. 5759–5764.
4. Roder, H., K. Maki, H. Cheng, and M.C. Shastry, *Rapid mixing methods for exploring the kinetics of protein folding.* Methods, 2004. **34**(1): p. 15–27.
5. Roder, H., K. Maki, and H. Cheng, *Early events in protein folding explored by rapid mixing methods.* Chem. Rev., 2006. **106**(5): p. 1836–1861.
6. Eftink, M.R. and M.C. Shastry, *Fluorescence methods for studying kinetics of protein-folding reactions.* Methods Enzymol., 1997. **278**: p. 258–286.
7. Zeeb, M. and J. Balbach, *NMR spectroscopic characterization of millisecond protein folding by transverse relaxation dispersion measurements.* J. Am. Chem. Soc., 2005. **127**(38): p. 13207–13212.
8. Pollack, L., M.W. Tate, A.C. Finnefrock, C. Kalidas, S. Trotter, N.C. Darnton, L. Lurio, R.H. Austin, C.A. Batt, S.M. Gruner, and S.G. Mochrie, *Time resolved collapse of a folding protein observed with small angle x-ray scattering.* Phys. Rev. Lett., 2001. **86**(21): p. 4962–4965.
9. Dinner, A.R. and M. Karplus, *Is protein unfolding the reverse of protein folding? A lattice simulation analysis.* J. Mol. Biol., 1999. **292**(2): p. 403–419.
10. Socci, N.D., J.N. Onuchic, and P.G. Wolynes, *Stretching lattice models of protein folding.* Proc. Natl. Acad. Sci. USA, 1999. **96**(5): p. 2031–2035.
11. Paci, E., M. Vendruscolo, and M. Karplus, *Validity of Go models: comparison with a solvent-shielded empirical energy decomposition.* Biophys. J., 2002. **83**(6): p. 3032–3038.
12. Kubelka, J., J. Hofrichter, and W.A. Eaton, *The protein folding 'speed limit'.* Curr. Opin. Struct. Biol., 2004. **14**(1): p. 76–88.
13. Fersht, A.R. and V. Daggett, *Protein folding and unfolding at atomic resolution.* Cell, 2002. **108**(4): p. 573–582.
14. Rathore, N., Q. Yan, and J.J. De Pablo, *Molecular simulation of the reversible mechanical unfolding of proteins.* J. Chem. Phys., 2004. **120**(12): p. 5781–5788.
15. Ferrara, P., J. Apostolakis, and A. Caflisch, *Computer simulations of protein folding by targeted molecular dynamics.* Proteins, 2000. **39**(3): p. 252–260.
16. Wu, X., S. Wang, and B.R. Brooks, *Direct observation of the folding and unfolding of a beta-hairpin in explicit water through computer simulation.* J. Am. Chem. Soc., 2002. **124**(19): p. 5282–5283.
17. Mitsutake, A., Y. Sugita, and Y. Okamoto, *Generalized-ensemble algorithms for molecular simulations of biopolymers.* Biopolymers, 2001. **60**(2): p. 96–123.
18. Jang, S., E. Kim, and Y. Pak, *Free energy surfaces of miniproteins with a betabetaalpha motif: replica exchange molecular dynamics simulation with an implicit solvation model.* Proteins, 2006. **62**(3): p. 6663–6671.
19. Nguyen, P.H., G. Stock, E. Mittag, C.K. Hu, and M.S. Li, *Free energy landscape and folding mechanism of a beta-hairpin in explicit water: a replica exchange molecular dynamics study.* Proteins, 2005. **61**(4): p. 795–808.
20. Roe, D.R., V. Hornak, and C. Simmerling, *Folding cooperativity in a three-stranded beta-sheet model.* J. Mol. Biol., 2005. **352**(2): p. 370–381.
21. Zhou, R., *Trp-cage: folding free energy landscape in explicit water.* Proc. Natl. Acad. Sci. USA, 2003. **100**(23): p. 13280–13285.
22. Sheinerman, F.B. and C.L. Brooks, 3rd, *Calculations on folding of segment B1 of streptococcal protein G.* J. Mol. Biol., 1998. **278**(2): p. 439–456.
23. Pande, V.S., I. Baker, J. Chapman, S.P. Elmer, S. Khaliq, S.M. Larson, Y.M. Rhee, M.R. Shirts, C.D. Snow, E.J. Sorin, and B. Zagrovic, *Atomistic protein folding simulations on the submillisecond time scale using worldwide distributed computing.* Biopolymers, 2003. **68**(1): p. 91–109.
24. Dobson, C.M. and M. Karplus, *The fundamentals of protein folding: bringing together theory and experiment.* Curr. Opin. Struct. Biol., 1999. **9**(1): p. 92–101.

25. Onuchic, J.N. and P.G. Wolynes, *Theory of protein folding.* Curr. Opin. Struct. Biol., 2004. **14**(1): p. 70–75.
26. Chan, H.S., S. Bromberg, and K.A. Dill, *Models of cooperativity in protein folding.* Philos. Trans. R. Soc. Lond. B Biol. Sci., 1995. **348**(1323): p. 61–70.
27. Baldwin, R.L., *Intermediates in protein folding reactions and the mechanism of protein folding.* Annu. Rev. Biochem., 1975. **44**: p. 453–475.
28. Karplus, M. and D.L. Weaver, *Protein folding dynamics: the diffusion-collision model and experimental data.* Protein Sci., 1994. **3**(4): p. 650–668.
29. Fersht, A.R., *Optimization of rates of protein folding: the nucleation-condensation mechanism and its implications.* Proc. Natl. Acad. Sci. USA, 1995. **92**(24): p. 10869–10873.
30. Daggett, V. and A.R. Fersht, *Is there a unifying mechanism for protein folding?* Trends Biochem. Sci., 2003. **28**(1): p. 18–25.
31. Onuchic, J.N., Z. Luthey-Schulten, and P.G. Wolynes, *Theory of protein folding: the energy landscape perspective.* Annu. Rev. Phys. Chem., 1997. **48**: p. 545–600.
32. Wolynes, P.G., *Energy landscapes and solved protein-folding problems.* Philos. Transact. A Math. Phys. Eng. Sci., 2005. **363**(1827): p. 453–464; discussion 464–467.
33. Shea, J.E. and C.L. Brooks, 3rd, *From folding theories to folding proteins: a review and assessment of simulation studies of protein folding and unfolding.* Annu. Rev. Phys. Chem., 2001. **52**: p. 499–535.
34. Daggett, V. and A. Fersht, *The present view of the mechanism of protein folding.* Nat. Rev. Mol. Cell Biol., 2003. **4**(6): p. 497–502.
35. Mitsutake, A. and Y. Okamoto, *Replica-exchange extensions of simulated tempering method.* J. Chem. Phys., 2004. **121**(6): p. 2491–2504.
36. Okamoto, Y., *Generalized-ensemble algorithms: enhanced sampling techniques for Monte Carlo and molecular dynamics simulations.* J. Mol. Graph. Model., 2004. **22**(5): p. 425–439.
37. Case, D.A., et al., *AMBER 8.* 2004: University of California, San Francisco.
38. Brooks, B.R., R.E. Bruccoleri, B.D. Olafson, D.J. States, S. Swaminathan, and M. Karplus, *Charmm—a Program for Macromolecular Energy, Minimization, and Dynamics Calculations.* J. Comput. Chem., 1983. **4**(2): p. 187–217.
39. Van Der Spoel, D., E. Lindahl, B. Hess, G. Groenhof, A.E. Mark, and H.J. Berendsen, *GROMACS: fast, flexible, and free.* J. Comput. Chem., 2005. **26**(16): p. 1701–1718.
40. Christen, M., P.H. Hunenberger, D. Bakowies, R. Baron, R. Burgi, D.P. Geerke, T.N. Heinz, M.A. Kastenholz, V. Krautler, C. Oostenbrink, C. Peter, D. Trzesniak, and W.F. van Gunsteren, *The GROMOS software for biomolecular simulation: GROMOS05.* J. Comput. Chem., 2005. **26**(16): p. 1719–1751.
41. Tiradorives, J. and W.L. Jorgensen, *The Opls Force-Field for Organic and Biomolecular Systems.* Abstr. Pap. Am. Chem. S., 1992. **204**: p. 43–Comp.
42. Chen, J.H., W.P. Im, and C.L. Brooks, *Balancing solvation and intramolecular interactions: Toward a consistent generalized Born force field.* J. Amer. Chem. Soc., 2006. **128**(11): p. 3728–3736.
43. Zhu, J., E. Alexov, and B. Honig, *Comparative study of generalized Born models: Born radii and peptide folding.* J. Phys. Chem. B, 2005. **109**(7): p. 3008–3022.
44. Allwright, J.C., *Conjugate gradient versus steepest descent.* J. Optimiz. Theory App., 1976. **20**(1): p. 129–134.
45. Olson, C.F., *Parallel algorithms for hierarchical-clustering.* Parallel Comput., 1995. **21**(8): p. 1313–1325.
46. Krivov, S.V. and M. Karplus, *Hidden complexity of free energy surfaces for peptide (protein) folding.* Proc. Natl. Acad. Sci. USA, 2004. **101**(41): p. 14766–14770.
47. Richmond, T.J., *Solvent accessible surface-area and excluded volume in proteins—analytical equations for overlapping spheres and implications for the hydrophobic effect.* J. Mol. Biol., 1984. **178**(1): p. 63–89.

48. Griffiths-Jones, S.R., A.J. Maynard, and M.S. Searle, *Dissecting the stability of a beta-hairpin peptide that folds in water: NMR and molecular dynamics analysis of the beta-turn and beta-strand contributions to folding*. J. Mol. Biol., 1999. **292**(5): p. 1051–1069.
49. Creamer, T.P., *Conformational entropy in protein folding. A guide to estimating conformational entropy via modeling and computation*. Methods Mol. Biol., 2001. **168**: p. 117–132.
50. Mu, Y., P.H. Nguyen, and G. Stock, *Energy landscape of a small peptide revealed by dihedral angle principal component analysis*. Proteins, 2005. **58**(1): p. 45–52.
51. Zhang, W., C. Wu, and Y. Duan, *Convergence of replica exchange molecular dynamics*. J. Chem. Phys., 2005. **123**(15): p. 154105.
52. Garcia, A.E. and J.N. Onuchic, *Folding a protein in a computer: an atomic description of the folding/unfolding of protein A*. Proc. Natl. Acad. Sci. USA, 2003. **100**(24): p. 13898–13903.
53. Duan, Y., C. Wu, S. Chowdhury, M.C. Lee, G. Xiong, W. Zhang, R. Yang, P. Cieplak, R. Luo, T. Lee, J. Caldwell, J. Wang, and P. Kollman, *A point-charge force field for molecular mechanics simulations of proteins based on condensed-phase quantum mechanical calculations*. J. Comput. Chem., 2003. **24**(16): p. 1999–2012.
54. Patel, S. and C.L. Brooks, *CHARMM fluctuating charge force field for proteins: I parameterization and application to bulk organic liquid simulations*. J. Comput. Chem., 2004. **25**(1): p. 1–15.
55. Patel, S., A.D. Mackerell, and C.L. Brooks, *CHARMM fluctuating charge force field for proteins: II—Protein/solvent properties from molecular dynamics simulations using a nonadditive electrostatic model*. J. Comput. Chem., 2004. **25**(12): p. 1504–1514.
56. Oostenbrink, C., A. Villa, A.E. Mark, and W.F. van Gunsteren, *A biomolecular force field based on the free enthalpy of hydration and solvation: the GROMOS force-field parameter sets 53A5 and 53A6*. J. Comput. Chem., 2004. **25**(13): p. 1656–1676.
57. Oostenbrink, C., T.A. Soares, N.F. van der Vegt, and W.F. van Gunsteren, *Validation of the 53A6 GROMOS force field*. Eur. Biophys. J., 2005. **34**(4): p. 273–284.
58. Damm, W., T.A. Halgren, R.B. Murphy, A.M. Smondyrev, R.A. Friesner, and W.L. Jorgensen, *OPLS_2002: A new version of the OPLS-AA force field*. Abstr. Pap. Am. Chem. S., 2002. **224**: p. U471–U471.
59. Mu, Y.G., D.S. Kosov, and G. Stock, *Conformational dynamics of trialanine in water. 2. Comparison of AMBER, CHARMM, GROMOS, and OPLS force fields to NMR and infrared experiments*. J. Phys. Chem. B, 2003. **107**(21): p. 5064–5073.
60. Roterman, I.K., K.D. Gibson, and H.A. Scheraga, *A comparison of the Charmm, Amber and Ecepp potentials for peptides. 1. Conformational predictions for the tandemly repeated peptide (asn-ala-asn-pro)9*. J. Biomol. Struct. & Dyn., 1989. **7**(3): p. 391–419.
61. Roterman, I.K., M.H. Lambert, K.D. Gibson, and H.A. Scheraga, *A comparison of the Charmm, Amber and Ecepp potentials for peptides. 2. Phi-psi maps for N-acetyl alanine N′-methyl amide—comparisons, contrasts and simple experimental tests*. J. Biomol. Struct. & Dyn., 1989. **7**(3): p. 421–453.
62. Zhou, R., *Free energy landscape of protein folding in water: explicit vs. implicit solvent*. Proteins, 2003. **53**(2): p. 148–161.
63. Duan, Y. and P.A. Kollman, *Pathways to a protein folding intermediate observed in a 1-microsecond simulation in aqueous solution*. Science, 1998. **282**(5389): p. 740–744.
64. Ryckaert, J.P., G. Ciccotti, and H.J.C. Berendsen, *Numerical-integration of cartesian equations of motion of a system with constraints—molecular-dynamics of N-alkanes*. J. Comput. Phys., 1977. **23**(3): p. 327–341.
65. Humphrey, W., A. Dalke, and K. Schulten, *VMD: Visual molecular dynamics*. J. Mol. Graphics, 1996. **14**(1): p. 33–38.
66. DeLano, W.L. and J.W. Lam, *PyMOL: A communications tool for computational models*. Abstr. Pap. Am. Chem. S., 2005. **230**: p. U1371–U1372.
67. Sayle, R.A. and E.J. Milnerwhite, *Rasmol—biomolecular graphics for all*. Trends Biochem. Sci., 1995. **20**(9): p. 374–376.
68. Thompson, J.B., H.G. Hansma, P.K. Hansma, and K.W. Plaxco, *The backbone conformational entropy of protein folding: experimental measures from atomic force microscopy*. J. Mol. Biol., 2002. **322**(3): p. 645–652.

69. Doig, A.J. and M.J. Sternberg, *Side-chain conformational entropy in protein folding.* Protein Sci., 1995. **4**(11): p. 2247–2251.
70. Pal, D. and P. Chakrabarti, *Estimates of the loss of main-chain conformational entropy of different residues on protein folding.* Proteins, 1999. **36**(3): p. 332–339.
71. Lei, H.X. and Y. Duan, *The role of plastic beta-hairpin and weak hydrophobic core in the stability and unfolding of a full sequence design protein.* J. Chem. Phys., 2004. **121**(23): p. 12104–12111.

Chapter 16
Modeling of Protein Misfolding in Disease

Edyta B. Małolepsza

Summary A short review of the results of molecular modeling of prion disease is presented in this chapter. According to the "one-protein theory" proposed by Prusiner, prion proteins are misfolded naturally occurring proteins, which, on interaction with correctly folded proteins may induce misfolding and propagate the disease, resulting in insoluble amyloid aggregates in cells of affected specimens. Because of experimental difficulties in measurements of origin and growth of insoluble amyloid aggregations in cells, theoretical modeling is often the only one source of information regarding the molecular mechanism of the disease. Replica exchange Monte Carlo simulations presented in this chapter indicate that proteins in the native state, N, on interaction with an energetically higher structure, R, can change their conformation into R and form a dimer, R_2. The addition of another protein in the N state to R_2 may lead to spontaneous formation of a trimer, R_3. These results reveal the molecular basis for a model of prion disease propagation or conformational diseases in general.

Keywords: Amyloid · Molecular dynamics · Molecular modeling · Monte Carlo dynamics · Peptide · Prion disease · Replica Exchange Monte Carlo

1 Introduction

The prion diseases became widely known a few years ago, when a human version of mad cow disease, Creutzfeldt-Jakob disease, appeared. A prion disease can be caused, for example, by eating infected bovine meat and is, therefore, pertinent to a wide group of people. Presently, there are no vaccines or medicines that prevent the disease and no effective treatments to halt or slow its progress. This disease is one of the neurodegenerative diseases, like Alzheimer's or Parkinson disease. A common feature of such illnesses is the accumulation of misfolded proteins in cells that leads to the appearance of insoluble amyloid plaques and, as a result, to

From: Methods in Molecular Biology, vol. 443, Molecular Modeling of Proteins
Edited by Andreas Kukol © Humana Press, Totowa, NJ

various clinical manifestations. The mechanism of amyloid aggregation is still unknown, however, experimental and theoretical investigations point to the refolding of cellular proteins from their native conformations into other isoforms with very different three-dimensional (3D) structures as a main cause of neurodegenerative diseases. The reason behind such behavior may be the interaction of native proteins with disease-causing agents leading to wrongly folded proteins. In this chapter, a model of amyloid propagation induced by interactions with such misfolded protein is presented.

The model is based on the assumption that the presence of other species, in this case misfolded proteins, can induce large conformational changes in the studied species. Thus, the energetic features of an isolated molecule (monomer) and its oligomers (dimer and trimer) will be presented in the context of prion disease propagation.

1.1 Protein Structure

Proteins are built from amino acids linked through peptide bonds. The polypeptide chain forms a backbone structure in proteins (Fig. 1).

Four levels of a protein's structure are defined:

1. Primary structure: the amino acid sequence.
2. Secondary structure: locally defined substructures (motifs): α-helix and β-sheet.
3. Tertiary structure: the overall structure of a single protein molecule (the spatial relationship of the secondary motifs to each other).
4. Quaternary structure: the structure that is a result of cooperation of more than one protein molecule.

1.2 Prions and Prion Diseases

The native structure of a protein, which is necessary for its biological activity, is not a unique conformation, rather, there are a number of possible structures very close in energies and very similar in shapes. They collectively form the so-called

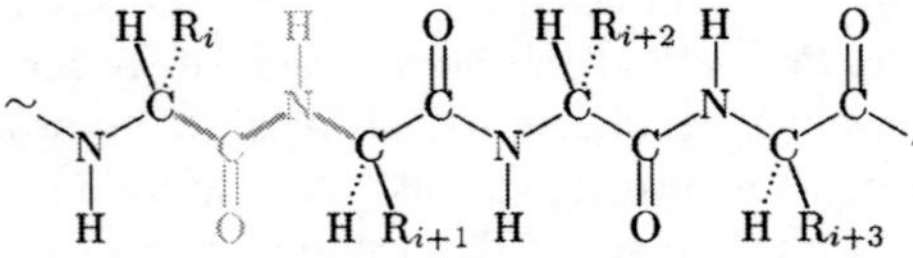

Fig. 1 The polypeptide chain forms a backbone structure in proteins. The *gray-colored* part of the protein scheme is an example of a peptide bond. R_i describes a side chain of amino acid (there are 20 different amino acids in natural proteins)

energy basin. Within the potential energy hypersurface of the protein, many other minima are possible. Geometries of the protein corresponding to various minima can differ widely in their 3D structures. When the protein adopts a structure beyond a basin of native conformation, it usually looses its activity or becomes active in another direction. Such a behavior may be destructive for a host organism. The theory of protein folding states that globular proteins possess smooth, funnel-like conformational energy landscapes, and folding leads to stable native states [1], and, fortunately, the amino acid sequence specifies one biological active conformation.

In mammals, prions are normal cellular prion proteins called PrP^C [2–4] located on the extracellular face of the plasma membrane. In contrast with other proteins they can adopt two structures, coming from different energy basins, which may act in cells. Such an event is in disagreement with the theory of protein states (funnel effect) and their relations to biological activity. The native (healthy) forms of prion proteins, PrP^C, can convert into conformational disease-causing isoforms called scrapie prion proteins, PrP^{Sc} [5–15]. Both forms have exactly the same amino acid sequence, but completely different 3D structures [16–19]. The PrP^{Sc} isoform is formed from the cellular protein PrP^C by a posttranslational process [20–23]. Prions in PrP^{Sc} form can be characterized by a resistance to protease. PrP^{Sc} is partially hydrolyzed by proteases to form fragments, which are called PrP 27–30 (their molecular weights are between 27 and 30 kDa), whereas PrP^C is completely degraded under the same conditions [2, 16]. The protease-resistant fragments of PrP^{Sc} aggregate [6, 20] and produce insoluble amyloid plaques [6, 24–28] which are typical for neurodegenerative diseases.

The main structural differences between native and scrapie forms of prion proteins are explain in a paper by Prusiner [29]. The structure of the native protein was solved experimentally (using NMR and x-ray techniques), but, because of problems obtaining a PrP^{Sc} crystal, the scrapie form is presented as a model of its conformation (PrP^{Sc} appears as insoluble amyloids). A fragment of Syrian hamster prion protein PrP^C contains three α-helices and a small piece of β-sheet. The model of its scrapie isoform, PrP^{Sc}, is less rich in α-helices and has many more β-sheets. The model of PrP^{Sc} isoform is based on experiments by Pan et al. [7], Gasset et al. [30,31], Huang et al. [32,33], and others [34–42]. Their results indicate that PrP^C is highly helical (42%) with some β-sheet structure (3%). In contrast, PrP^{Sc} contains a large amount of β-structure (42%) and less helical structure (30%). Changes in protein conformation on the conversion from α-helices to β-sheets are typical for the prion diseases [3, 6, 43–46].

Such abnormalities in protein structures, especially when referring to neural cells or, in general, to cells forming the central nervous system, lead, in particular, to neurodegenerative diseases to which the prion disease belongs [28, 47–49]. The human prion diseases are: kuru, Creutzfeldt-Jakob disease (CJD) and its new variant (vCJD), Gerstmann-Sträussler-Scheinker syndrome (GSS) and the fatal familial insomnia, whereas other neurodegenerative diseases are: Alzheimer's, Parkinson, Pick, and Huntington diseases, amyotrophic lateral sclerosis, and frontotemporal dementia. Neuropathology, etiology, and pathogenesis of prion diseases have been presented in numerous papers [9, 50–64].

The feature that distinguishes prion diseases from other neurodegenerative illnesses is their infectiousness: prions are infectious proteins [65]. The events of transmissions of such proteins between the same or different animal species (sheep, goats, chimps, mice, hamsters, and others) that have succumbed to prion diseases have been presented in many papers [66–78]. The other mentioned illnesses are not infectious and have not been transmitted to the bodies of laboratory animals [29].

Prions are interesting from the medical point of view for other reasons. They do not use nucleic acid to propagate in host organisms, unlike every other infectious pathogen with their duplication control based on DNA or RNA [79, 80]. The clinical manifestations of prion diseases are the most numerous among diseases caused by single factor: dementia (a condition of deteriorated mentality, often with emotional apathy), ataxia (an inability to coordinate voluntary muscular movements), insomnia, paraplegia (paralysis of the lower half of the body including both legs), and paresthesia (a sensation of pricking, tingling, or creeping on the skin that has no objective cause). Each of these are related to damage to the central nervous system that corresponds to the accumulation of wrongly folded proteins. The scrapie forms of prion proteins may occur in various wrongly folded conformations [81–84] and each of these seem to cause a specific disease version.

In healthy cells, the misfolded proteins are broken down and removed from cells, whereas, in neurodegenerative diseases, the wrongly folded proteins or their protease-resistant fragments aggregate into insoluble plaques [6, 16, 17, 20, 24]. Therapy for prion diseases is very difficult because presymptomatic diagnosis is impossible and the earliest possible diagnosis is based on clinical manifestations and symptoms. However, the diagnosis of prion diseases have improved recently [85–88], with the search for therapy agents very well expressed in the following statement taken from a review paper by Wiessmann and Aguzzi [89]: "Based on the assumption that PrP^{Sc} is the infectious agent, or at least the pathogenic entity, compounds have been sought that in a cell-free system would stabilize PrP^{C}, destabilize PrP^{Sc}, or prevent conversion and thereby decrease the level of PrP^{Sc}." This paper also contains a table with representative compounds used in therapy. Details of therapies may be found in [90–93].

The prion diseases problem has been widely discussed in many review articles written by Prusiner's group [84, 94–98] and others [4, 48, 99–101].

1.3 Molecular Background of Prion Diseases

Prion diseases and other neurodegenerative diseases are induced by propagation of wrong conformations of cellular proteins and by the formation of hard, insoluble plaques. Conformational conversion of the two most popular motifs of the protein secondary structure, α-helix and β-sheet, and the formation of amyloids, should be considered as the main cause of prion disease [6–9, 11, 13, 14, 45, 47, 102]. Therefore, prion disease can be called a conformational disease, although the propagation mechanism is still unknown [7, 48, 103, 104]. Prion stability experiments take into account temperature, pressure, pH, and detergent dependence [8, 105–113], as well as sequence substitution effects [114, 115]. General considerations of protein folding, misfolding, and aggregation may be found in a review article by Markossian and Kurganov [116].

Theories regarding the causes of prion disease were first constructed from medical reports. The first one was proposed by Bjørn Sigurdsson in 1954 who indicated so-called "slow viruses" as a cause of scrapie and visna in sheep [117] because of a long incubation time. Many scientists also pointed out some unknown viruses. However, detailed studies made by Prusiner allowed him to introduce "proteinaceous and infectious" particles that lack nucleic acid [2] and this discovery of the prion protein accelerated his original "protein only hypothesis." It stated that the posttranslational misfolding of prion proteins in cells caused the prion diseases. The work of Prusiner and its importance were honored by the Nobel Committee in 1997 [118]. The other theory assumes the presence of an unknown protein X, "a factor defined by molecular genetic studies that binds to PrP^{C} and facilitates PrP^{Sc} formation" [47, 119, 120]. It is possible that protein X is a chaperone, a protein that assists other proteins in achieving proper folding. The next hypothesis, contradictory to the Prusiner "protein only hypothesis" and less popular, supposes that PrP^{C} is a receptor for a hitherto unknown virus, whose ablation induces antiviral resistance. In a paper by Soto [121], some critical arguments against Prusiner's hypothesis are presented including protease-resistant forms of prion proteins interacting with RNA, and a lack of infectious particles in other neurodegenerative diseases. Details can be found in [122–124].

Two models of conformational conversion of PrP^{C} into PrP^{Sc} are based on Prusiner's theory of prion disease [102]. The first model is called template-directed refolding [33, 125], and the second is called seeded nucleation [126–130].

The template-directed refolding model supposes that the PrP^{Sc} isoform is a template for the PrP^{C} isoform and the presence of scrapie conformation leads to structural changes of the native prion protein. The seeded nucleation assumes that both isoforms exist in equilibrium, but strongly shifted toward the PrP^{C} isoform. A single PrP^{Sc} is harmless, however, it can be a bud in aggregation of other PrP^{Sc} proteins and becomes an infectious agent.

1.4 Theoretical Investigations

Many papers present theoretical considerations at various levels regarding unclear aspects of prion disease.

1.4.1 Mathematical Models of Prion Diseases

The simplest model of prion disease, based on linear nucleated polymerization, was presented by Masel and coworkers [131]. PrP^{C} and PrP^{Sc} proteins were treated as mathematical units (without any physical features) and their abundances were calculated according to differential equations. Quantifying parameters of the model, they, e.g., estimated minimal nucleation size and the mean length of polymer. Their model was extended by Pöschel et al. [132] to include time dependence of the monomer concentration. This allowed them to observe several possible evolutions of prion disease, namely the unlimited exponential growth of fibril, their complete disappearance, the initial exponential growth of a number of PrP^{Sc} polymers, and, finally, numerical saturation. Slepoy and coworkers [133] applied a simple two-dimensional (2D) lattice-cell model of prion disease: PrP^{C} and PrP^{Sc} isoforms were represented by boxes with a very simple energy landscape. The authors constructed a hypothesis that a long prion protein incubation time is related to fluctuation-dominated growth, seeded by a few nanometer scale aggregates. This work was later extended [134, 135]. Another mathematical model of prion propagation, explored by Rieger and coworkers [136], was based on differential equations describing several processes. After including the activity of chaperones, the authors observed the aggregation of scrapie form as a bistable process.

1.4.2 Molecular Dynamics Simulations

The next group of papers relate to the molecular dynamics (MD) of prion proteins. The starting geometries were taken from the Protein Data Bank (PDB). The force fields came from the AMBER, CHARM, Encad, and Discover programs, being the most popular tools in the MD field, with protein representation varying from all-atom to united-atom models. Such studies allowed predictions of scrapie isoform conformations, unobtainable through experiment, by examining prion protein stability in various conditions (e.g., changes in pH) and allowed a protein flexibility [114, 137–147].

A paper published in 1995 by Kazmirski et al. [114] presented results of MD simulations of a human prion peptide (residues 109–122) in water with respect to mutation in position 117. The authors suggested that alanine to valine mutation in this position destabilizes the PrP^{C} isoform and would expose the hydrophobic core to solvent, which may facilitate conversion into the scrapie form. Similar mutations were examined by Okimoto and coworkers [141]. Zuegg and Gready [137, 140] indicated that oligosaccharides connected to N terminus of the human prion protein stabilize native isoform, PrP^{C}, whereas a glycosyl–phosphatidyl–inositol anchor linking the prion protein to membrane (GPI anchor) has a much smaller influence on the structure of PrP; however, its flexibility allows freedom for the orientation of the prion protein. Guilbert et al. [139] also stated that the "N-terminal region of prion protein could feature in the conversion to PrP^{Sc} by template-assisted formation of β-structure," whereas the "β-sheet region of PrP^{C} may be the nucleation site for the conformational transition to the infectious."

During the simulation of eight-amino acid fragments of Syrian hamster PrP, Ma and Nussinov [145] observed that amyloid aggregation begins with an initial slow lag phase followed by a subsequent rapid growth of aggregate with octamer as a stabile template for fibril propagation. Similar aggregations were presented by Kuwata et al. [143] for 10-amino acid fragments of mouse PrP. Barducci et al. [147] have performed simulations of mouse prion protein in hydrophobic solvent (CCl_4) to prove that monomer fold is hydrophobic. DeMarco and Daggett [144] presented results of MD simulations for an all-atom protein representation of Syrian hamster prion protein in low pH. They observed an increase of β-structures and a decrease of helical parts. Starting from these conformations, they modeled the formation of fibrils and then they reproduced the electron crystallography image of PrP 27–30 crystals.

Armen and coworkers [148] have simulated the formation of α-pleated sheets (carbonyl groups are located on one side of the sheet and amide groups on the other side) from prion protein in low pH, which were protected against denaturation conditions (urea). The authors suggested that such structures can be intermediates in amyloid formation. Stork and coworkers [149] studied the stability of small β-helices in an aqueous solution and, according to experimental results, fibrils of poly-glutamines were formed from such blocks. Langella et al. [146] presented the results of simulations of human PrP in acidic pH. Protonation of four histidines in such conditions lead to significant destruction of one of the helices. They suggested that conformational changes of this helix are crucial in the formation of the scrapie isoform. Simulations carried out by Tsai et al. [150] pointed out interactions between asparagines as a "glue" in the formation of fibrils from fragments of human calcitonin, a protein that is able to form ordered amyloids. The importance of such interactions was verified when mutation of asparagine to alanine significantly decreased the stability of β-sheets.

1.4.3 Monte Carlo Simulations: Mathematical Model of Amino Acids

Another treatment is presented in the works of Harrison et al. [151], Derreumaux [152], Peng and Hansmann [153], Dima and Thirumalai [154], Kammerer et al. [155], and Ding and coworkers [156]. Instead of MD, they used Monte Carlo dynamics. Harrison, Chan, Prusiner, and Cohen [151] used a very simple 2D lattice model of 16-amino acid proteins built from four types of monomers (hydrophobic, polar, and positively and negatively charged) with pair interaction energies as an energetic score. They took into account three processes of protein refolding:

$$2N \rightarrow R_2 \quad (1)$$

$$R_2 + N \rightarrow R_3 \quad (2)$$

$$3N \rightarrow R_3 \quad (3)$$

N indicates the native conformation (the structure with the lowest free energy), R indicates stable conformation with energy higher than N. The first and third reactions are spontaneous refolding of native states N into dimer R_2 and trimer R_3, respectively. Equation 2 describes template propagation of the wrong form (R); native conformation N refolds into R in the presence of dimer R_2. In the presented model, the process in Eq. 2 was five times slower than the process in Eq. 1, and five times faster than the process in Eq. 3. The authors showed that, without R_2, the template conversion of N state into R takes infinite time, thus, the process of propagation of R structures is autocatalytic. One of the suggestions in [151] was that "the R_2 dimer should be in the dimeric ground state to enable efficient propagation of the R conformation to a third chain." A very small ratio of decay to formation rates of R_3 allowed them to conclude that the R structure can propagate to R_n and form amyloid.

Dima and Thirumalai [157] used the model of Harrison et al. [151] to formulate general characteristics of a self-propagating model:

1. Native conformation is preferentially formed at temperatures below the folding temperature (definition is given below).
2. Besides the native state, there exists another state that can polymerize.
3. Polymerization is induced by interaction with other proteins.
4. Conformational changes of proteins should be permitted.

The simple model applied by Dima and Thirumalai allowed exact calculation of partitioning functions and thermodynamic properties for reactions:

$$R_{n-1} + U \rightarrow R_n;\ n = 3\text{–}7, \tag{4}$$

where U represents the unfolded state. In the template assembly model, the growth rate is independent of protein concentration C, whereas, in the nucleated-polymerization model, it increases as C^{α} (α unknown). The model considered by Dima and Thirumalai did not allow them to distinguish between the two models. They determined dimer R_2 as a minimal template for aggregation and calculated a phase diagram in the temperature–concentration plane for two-state folding. They also obtained several distinct dimers separated by energetic barriers and observed that the choice of structures occurs early in the dynamics and depends on the initial conditions.

1.4.4 Monte Carlo Simulations: Real Amino Acids

Peng and Hansmann [153] have observed the folding of seven-amino acid peptides in the presence of a β-sheet and resulting in aggregation. Because they used just one chain in the simulation, two interacting proteins were represented by one peptide composed of two seven-amino acid fragments connected by several glycines. In fact, such a model of two interacting proteins does not allow the free movement of both studied fragments.

Dynamics (united atom representation of protein, force field as a sum of short- and long-range interactions) performed by Derreumaux [152] have shown that the middle part of mouse prion protein (PrP 127–164) can form two distinct and isoenergetic conformations with β- or αβ-like structures. Kammerer with coworkers [155] and Ding et al. [156] have presented results of Monte Carlo dynamics of a de novo designed 17-amino acid peptide that can be called a molecular switch: in low temperatures, it adopts the α-shape, whereas, at higher temperatures, it forms amyloid fibrils.

2 Model of Conformational Disease

The main hypothesis of the chapter is that the presence of spatial restrictions may cause different properties of the molecule. The spatial restrictions may also be reinforced by the intermolecular electrostatic, induction, and dispersion interactions. In cases of conformational change, this leads to what is known as the template effect.

Based on experimental and theoretical investigations mentioned in the Introduction, in particular, the one-protein theory of Prusiner and work by Dima and Thirumalai [157], a hypothesis of the model of conformational diseases was built and presented in [158]. The model is based on the hypothesis that some misfolded proteins may induce misfolding of other protein molecules.

The main assumption of this hypothesis is that wrongly folded proteins that interact with native proteins may cause them to fold or refold into misfolded structures and, in some cases, the misfolded conformations can aggregate. Here are the most important facts concerning prion diseases:

- The most probable cause of prion disease is the change in spatial conformations of cellular prion proteins, PrP, such that the native prion protein, PrP^{C}, refolds into its scrapie form, PrP^{Sc}
- Native PrP^{C} isoforms contain more α-helices than scrapie PrP^{Sc}, whereas PrP^{Sc} have more β-sheets than PrP^{C}
- Scrapie isoforms of prion proteins are partly degraded by protease into fragments that aggregate
- Amyloid aggregates found in infected cells are insoluble and difficult to examine experimentally, however, it is known that they contain β-sheets

The approach presented in [158] postulates that a protein can have the following properties:

- It rapidly folds into its global energy minimum structure
- It can exist in a metastable conformation with higher energy and a different structure
- It exhibits a strong autocatalytic effect leading to the formation of a metastable structure when in the presence of another protein

The presence of proteins with the properties mentioned above may lead to conformational diseases like the prion diseases. To prove the hypothesis, a series of proteins with the required properties were designed, and then subjected to *in silico* simulations of their folding.

In contrast to many attempts in the past, the present approach aims to predict the 3D structures of the proteins, because 2D models have been excluded as too simple. The prediction follows from extensive Monte Carlo simulations that have proved to be very successful in blind prediction of the 3D structures of real proteins, and from powerful numerical techniques that have allowed the study of the total conformational space of the proteins composed of several dozen amino acids.

For the sake of simplicity, a two α-helix bundle was chosen for the native state and a higher-energy β-hairpin for the metastable structure. Thus, the proposed secondary structures correspond to what has been found in the prion proteins: α-helices prevail in the native form and β-like structures prevail in the scrapie form.

The third postulate is the most important and the basis of the hypothesis. Namely, a high-energy minimum in a potential energy hypersurface of the protein (misfolded protein, β-sheet) decreases its energy when it interacts with another molecule (template), so that the energy of the wrongly folded protein and template is lower than the energy of the native conformation and template protein.

The model of conformational disease described will be presented with details in the next section.

3 Methods

3.1 Protein Representation

Protein molecules are usually composed of hundreds of amino acid residues. The prion proteins have approximately 200 to 300 amino acid residues [159]. In the PDB, where protein structures are collected, 105- to 140-amino acid fragments of prion protein are available. These fragments seem to be responsible for conformational changes leading to amyloid aggregation and they are usually used in modeling.

The length of prion protein chains or their fragments indicate that the number of atoms is on the order of thousands. This number is prohibitively large if the exploration of the entire conformational space is to be considered. Therefore, it is necessary to simplify the protein molecule in such a way as to make such calculations feasible.

First, 32 amino acid protein chains were considered, and de novo designed for the calculations presented in this chapter. Second, the united-atom approach was used to reduce the number of atoms. A simple explanation of the reduced protein model is shown in Fig. 2, with the details in [160, 161].

Because the peptide bonds connecting neighboring amino acids are very rigid, the protein backbone can be represented as a chain of Cα atoms linked only by virtual bonds with constant lengths equal to 3.8 Å. Side chains of amino acids are represented by one or two united atoms, depending on their size.

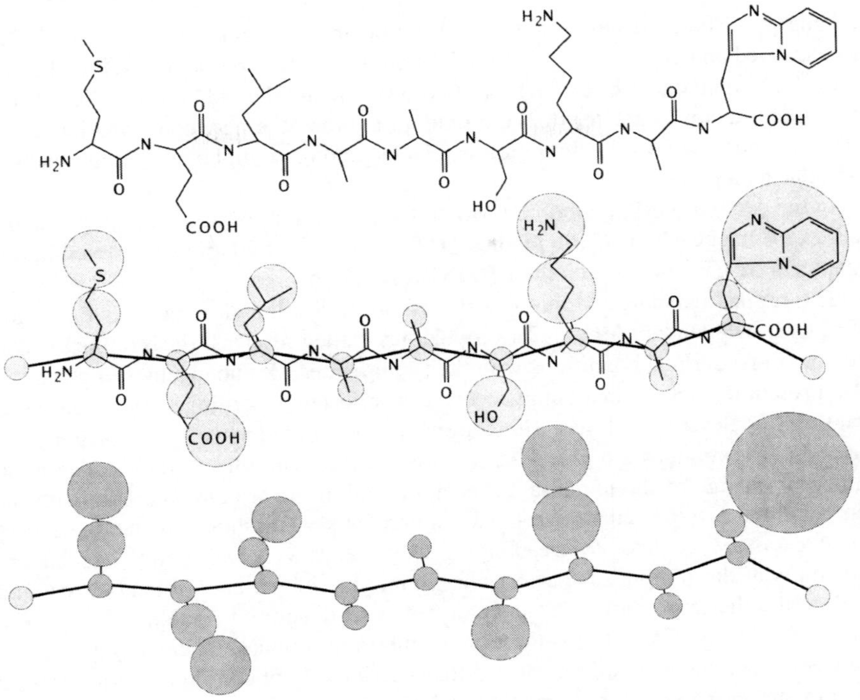

Fig. 2 A united-atoms representation of a protein. *Top*: the scheme of a 9-amino acid peptide. *Middle*: Cα atoms are connected by virtual bonds, side chain atoms are grouped into united atoms, whereas glycines on both chain ends (instead of NH_2 and COOH groups) are needed to describe the orientation of these parts. *Bottom*: united-atoms model of the protein used in the present work. Courtesy of Dr. Michal Boniecki

3.2 Replica Exchange Monte Carlo Dynamics

Monte Carlo dynamics (see Chap. 5) is a technique to explore the whole conformational space of molecules and was used to perform the first numerical simulation of a molecular system [162]. The method is based on sampling the conformational space by drawing conformations with random changes of atom positions using a random number generator. In Monte Carlo simulations, the so-called "importance sampling technique" described by Metropolis [163] is usually used. This technique prefers low-energy configurations, which make a large contribution to the partition function because of the high probability of such states—large values of Boltzmann factor, $\exp(-E/k_B T)$, where E is the energy of the configuration, k_B is the Boltzmann constant, and T is the temperature of the dynamics. Each conformation can be characterized by a corresponding energy computed using a force field. In each iteration, a new configuration is generated. If the new configuration is lower in energy than the old configuration, then the new configuration is accepted and becomes the starting configuration for next iteration. If the energy of the new conformation is higher than the energy of its predecessor, then the Boltzmann factor equal to $\exp(-\Delta E/k_B T)$ is calculated and compared with a random number between 0 and 1. ΔE is the difference between energies of the new configuration and the old conformation. If the Boltzmann factor is greater than the random number, the new conformation is accepted. If not, the new configuration is rejected and the initial configuration is used in the next step.

In the present work, a modified Monte Carlo technique was used, known as the replica exchange Monte Carlo method [160, 161]. The simulations presented in this chapter were carried out with the REFINER program.

The replica exchange Monte Carlo method allows the simultaneous execution of several independent Monte Carlo dynamics, called replicas. Each replica is carried out in isothermal conditions with their temperature T, and changes in geometry of a protein in each replica being accepted or rejected according to the Metropolis criterion, as described above. Simulations are performed for a series of temperatures and the replicas are ordered according to increasing temperature. The magnitude of change of the united-atom position within the proteins increase with the temperature. A wide range of replica temperatures were chosen to facilitate very small geometry changes (corresponding to proteins with well-defined secondary structures at the bottom of the low-energy minima of the energy landscape), and to describe the large movement characteristics of denatured proteins. Every fixed number of Monte Carlo steps, the two neighboring replicas are randomly chosen. Between them, the generalized Metropolis criterion is applied. This criterion is used to compare energies of conformations taken from replicas performed at different temperatures (in contrast to the standard Metropolis criterion comparing energies at the same temperature). If the energy E_2 of the protein in the higher-temperature T_2 replica is lower than the energy E_1 of the structure in the lower-temperature T_1 replica, their geometries are exchanged. In the other case, exchange is performed with a probability equal to:

$$\exp(E_2/k_B T_2 - E_1/k_B T_1). \tag{5}$$

If, during the folding, the protein is trapped at the local minimum, the geometry exchanges mentioned above allow it to escape, thus, increasing the chance of finding the global minimum.

To obtain reliable results, each run consisted of at least one million Monte Carlo steps. Each type of dynamic (differing by range of temperatures, number of proteins, or secondary structure starting geometries) was carried out by performing many runs with different random seeds.

3.3 Force Field

The force field used in calculations presented in this chapter was a statistical force field designed by the Kolinski group, Faculty of Chemistry, Warsaw University. The force field has a complex formula and is an effective potential taking into account the following physical effects:

- Electrostatics
- Induction
- Dispersion
- Hydrogen bonds
- Hydrophobic interactions
- Interactions with solvent molecules (water)
- Other effects

The total effect is expressed by the frequency of appearance of amino acid pairs in chosen sets of proteins and by the characteristics of the local 3D geometry of these molecules. This statistical analysis of structural regularities in crystallographic structures of globular proteins taken from the PDB was done for a set of approximately 1,500 proteins. The energy form is described in detail in [164, 165].

In general, the terms of the force field can be divided into two groups:

- Short-range potentials that depend, for example, on the identity of interacting amino acids and the distances between ith and $i+3$, ith and $i+4$, and ith and $i+5$ Cα atoms along the chain and between torsional angles
- Long-range potentials, for example, contact-type potentials of side chain atoms (*inter alia* hydrophobic interactions)

In the presented approach, a solvent (water) does not enter explicitly into the formula for energy, but is treated as an effective medium. Because structures of proteins taken from PDB to obtain the potentials were solved for crystals containing water molecules, the water influence is effectively included in the force field and, therefore, potentials mimic "averaged" physiological conditions.

The quality of the force field used was examined in the worldwide protein structure prediction experiment (CASP-6). In this contest, the Kolinski group is placed in the highest positions [166].

4 Results

Three main types of simulation were performed: Monte Carlo dynamics with one, two, and three chains. One-chain dynamics were carried out to design sequences with requested properties. Two- and three-chain simulations were essentially aimed at studying intermolecular interactions. All calculations were performed using the program and force field designed by Boniecki and Kolinski from Laboratory of Theory of Biopolomers at Warsaw University.

Replica exchange Monte Carlo dynamics with 10 replicas, generally, were used to perform the simulations. Each type of dynamics was repeated with different random seeds, with different initial protein conformations and with different mutual positions of protein starting geometries in cases of multichain dynamics. The temperatures of the replicas were distributed around the folding temperature corresponding to the maximum of the first derivative of mean energy with respect to the temperature (constant-volume heat capacity). Thus, a few initial dynamics for every type of simulation were performed to determine the folding temperature. Figure 3 presents representative plots for one-chain dynamics. Detailed information regarding the folding temperature and properties of replicas are presented in the next section.

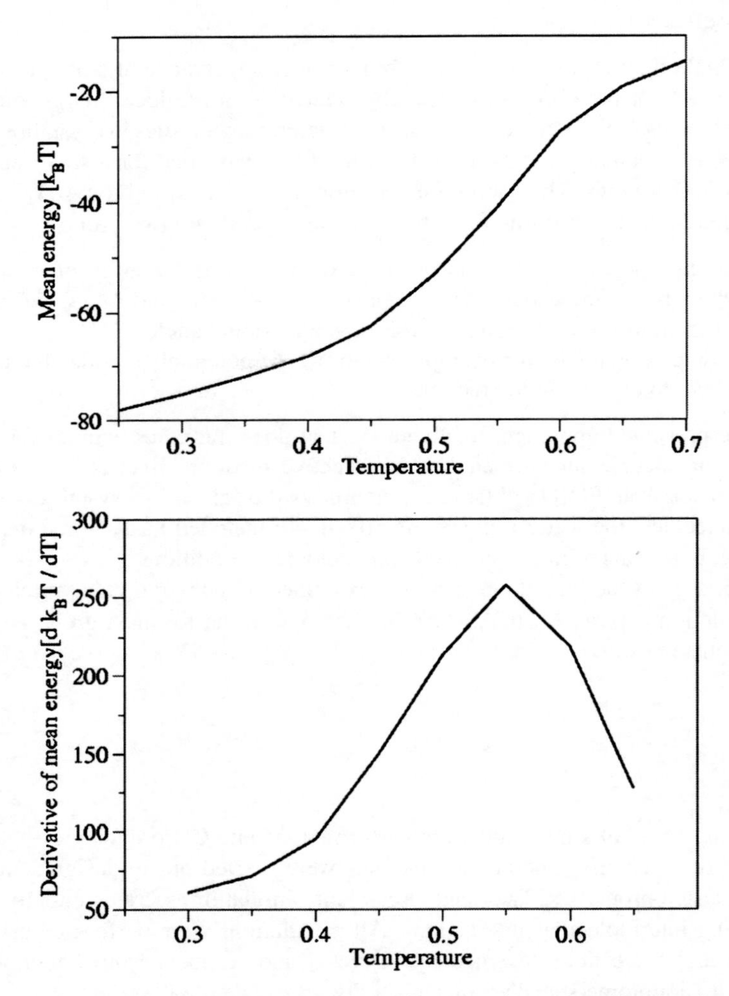

Fig. 3 Representative plot of mean energy (*left*) and heat capacity (*right*) of replicas in one-chain dynamics as a function of temperature

4.1 Designing the Sequence: One-Chain Dynamics

Because of computational costs, the length of protein was limited to 32 amino acids. The goal was to design a sequence of 32 amino acids that simulates a two-helix bundle in the global minimum conformation, and a β-sheet in the metastable conformation. When designing the amino acid sequence, some empirical facts were taken into account:

- Natural amino acids appear in α and β structures in proteins with different frequency:
  - Alanine, leucine, and glutamic acid are known as α-helix makers
  - Valine, isoleucine, and threonine are known as β-structure makers
- Short peptides fold easier to α-helices than to β-forms, thus, the required sequence should contain a larger number of β-type amino acids than α-helical-type amino acids
- The presence of glycines should hopefully facilitate the formation of loops, which are necessary for α-helix bundles and β-sheets
- α-helix and β-sheet structures are known to have different polar–hydrophobic patterns: $(HPPHHPP)_n$ and $(HP)_{nn}$, respectively, where P and H stand for polar and hydrophobic amino acids, respectively

Based on this information, some sequences were proposed and examined in Monte Carlo simulations regarding whether they satisfied the structural and energetic requirements according to the hypothesis. When the global minimum for the chosen sequence corresponded to the α-helix, but the β-sheet was not stable enough even at low temperatures, or when the β-sheet had lower energy than the α-helix, or when the energy difference between both conformations were too small or too large, a trial-and-error method was used to change the amino acids by point mutation to

Table 1 Thirty-two amino acid sequences designed to fulfill two conditions: the global minimum corresponds to an α-helix, and a metastable β-sheet local minimum exists

1	GVEIAVKGAEVAAKVGGVEIAVKAGEVAAKVG
2	GVEIAVKGGEVAAKVGGVEIAVKGGEVAAKVG
3	GVEIAIKGGEIAAKVGGVEIAVKGGEVAAKVG
4	GVEIAVKAGEVAAKVGGVEIAVKAGEVAAKVG
5	GVEIAIKGGEIAAKVGGVEIAVKGGEIAAKVG
6	GVEIAVKGGEVAAKVGGVEIAVKGGEIAAKVG
7	IKVAIEVGGVKAAVEGGKVAIEVGGVKAAVEI
8	IKVAIEVAGVKAAVEGGKVAIEVAGVKAAVEI
9	IKVAIEVAGVKAAVEGGKVAIEVGAVKAAVEI
10	IKVAIEIGGVKAAVEGGKVAIEVGGVKAAVEI
11	IKVAIEVGGIKAAVEGGKVAIEVGGVKAAVEI
12	IKVAIEVGGIKAAVEGGKVAIEIGGVKAAVEI
13	IKVAIEVAGVKGAVEGGKVAIEVGGVKAAVEI
14	IKVAIEVAGVKAAVEGGKVAIEVAGVKAAIEI

A one-character code describes the amino acids: G, glycine; A, alanine; V, valine; E, glutamic acid; I, isoleucine; K, lysine

improve their behavior. As a result, 14 sequences were preselected. They are collected in Table 1.

All of the sequences in Table 1 fulfilled the first and second hypothesis assumptions (see Section 2). Regardless of the starting geometry (α-helix, β-sheet, or extended chain) and a random seed, the lowest energy conformation and the most frequent structure for all sequences was an α-helix. The β-sheet conformations for all sequences were obtained in low-temperature dynamics using β-sheets as the starting geometry. β-sheets from the sequences in Table 1 were stable at low temperatures, however, when a wide range of dynamics were applied (higher temperatures were available) they unfolded and refolded into α-helices. This result proves that the β-form basin is separated from the global minimum basin by a barrier.

Both geometries are presented in Fig. 4 as an example of Sequence #1 in Table 1.

The energies of both structures for Sequence #1, equal to −85.3 and −75.8 in k_BT units for an α-helix and a β-sheet, respectively, were taken from the lowest temperature replica (temperature = 0.25 units).

Fig. 4 The global minimum, α-helix, and metastable minimum, β-sheet conformations, found for Sequence #1. *White* corresponds to glycine (G); *pink*, valine (V); *blue*, lysine (K); *green*, alanine (A); *dark green*, isoleucine (I); and *crimson*, glutamic acid (E). For clarity, only the backbone of the protein is shown

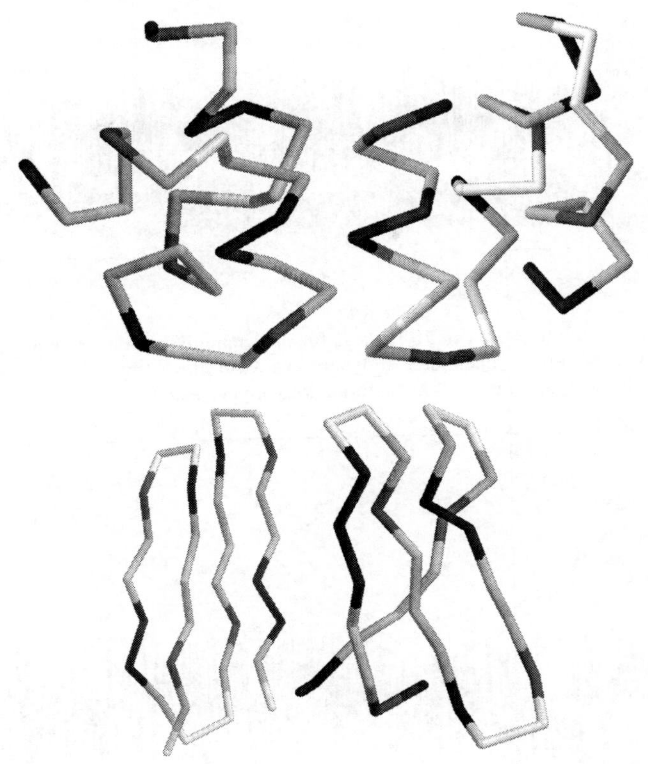

Fig. 5 Some of the millions of conformations (α-helices and β-sheets) obtained in one-chain dynamics

The global minima for the chosen sequences usually correspond to two-helix bundles, however, other α-helical structures were also found. Similarly, a few other β-type conformations were obtained. Four of them are presented in Fig. 5. In the first structure, two fragments of the α-helix are coiled, in the second, secondary protein structure contains three α-helices. The third conformation differs from β-sheets in Fig. 4, with different chain flexions and, in the last structure, the protein fragments form one flat sheet built to form a three-member β-barrel instead of a two-stranded sheet as in Fig. 4.

The determination of the folding temperature is important to the dynamics presented in this chapter, because, in temperatures close to the folding temperature, the most important processes are observed. Usually, the dynamics with a wide range of temperatures were performed in the first instance to find the best temperature interval. As an example, the energies of structures during some of the one-chain dynamics for Sequence #1 with an α-helix as the starting geometry are presented in Fig. 6 (for better clarity, 4 replicas out of 10 are presented). In Fig. 7, the energies of structures in the folding temperature replica in this dynamic are shown with some representative conformations.

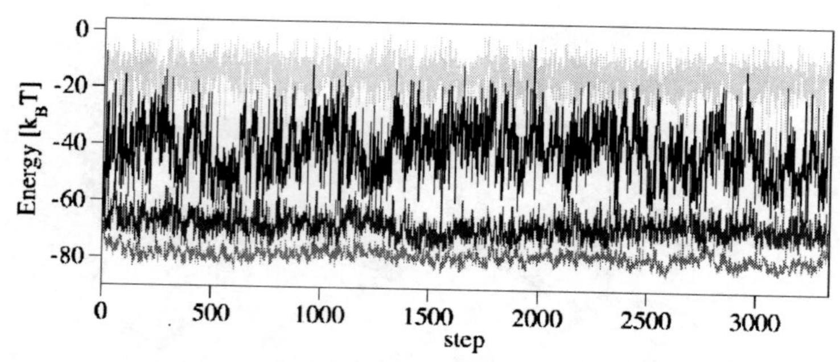

Fig. 6 Energies of structures in 4 of 10 replicas for one-chain dynamics for Sequence #1 over a wide range of temperature. Temperatures of dynamics are as follows (from the bottom): 0.25, 0.4, 0.55, and 0.7. The energies of every 300 conformations are presented

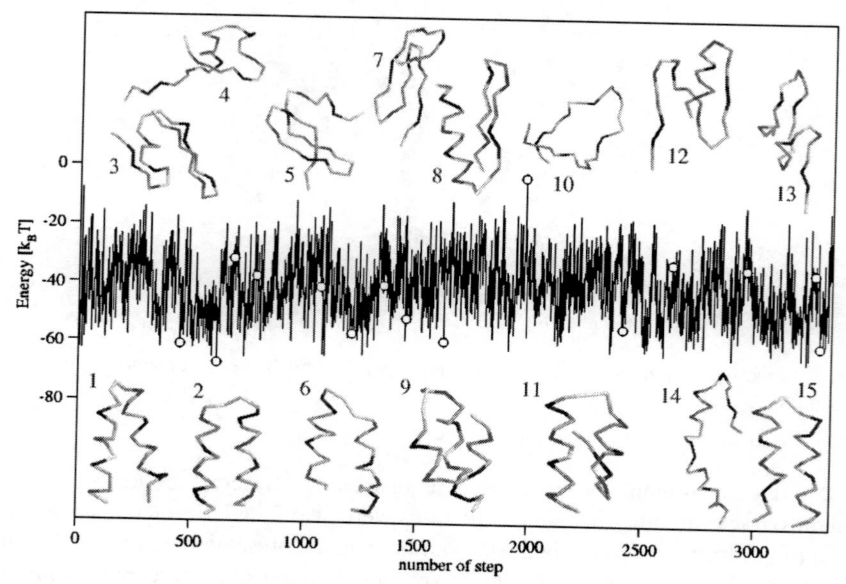

Fig. 7 Conformations and the corresponding energies in the folding temperature (0.55) replica in one-chain dynamics for Sequence #1. The presented conformations correspond to the *white circles*. Energies of every 300 conformations are shown

Figure 6 shows the energies of conformations for various temperature replicas. The mean energy increases with the temperature of the dynamics. For the lowest temperature replica (0.25 units), after a few steps from the beginning of the dynamics, the protein is trapped in a local minimum of the α-type. Because the changes of geometry depend on the temperature, the protein changes its conformations slowly (small changes between energies of neighboring structures). For the highest-temperature replica (0.7 units), the protein is fully denatured and no secondary structures (α-helix, β-sheet) are observed. The changes of geometry between the subsequent structures are large; too large to allow second-order structures to form. The standard deviation of energy is very small.

The most important changes happen in the replica that is at the folding temperature or close to it (second line from the top in Fig. 6). The folding temperature is high enough to allow proteins to refold (to change local minimum or, in the other words, to pass through the saddle of the energy hypersurface), but is low enough to preserve low-energy motifs (α-helix, β-sheet). Conformations in this replica vary from very well-defined helices or sheets to geometries corresponding to denatured molecules. Therefore, a large standard deviation of energy is observed. Figure 7 presents conformations corresponding to chosen points in the dynamics in the 0.55 temperature replica (second line from the top in Fig. 6).

The lowest-energy structures are α-helix bundles (Structures #1, 2, 6, 9, 11, and 15). Geometry #10 is an example of a completely denatured chain without even small helical or β-sheet fragments. Conformations #4 and 12 contain fragments of α-helix, whereas structures #3, 5, 7, and 13 have small amounts of β-sheet. Low-lying conformation #8 (however, still higher energetically than α-helix bundles) is an example of a mixed structure. It contains well-defined α-helix and β-sheet fragments and, therefore, has relatively low energy.

Similar pictures of dynamics appear for the remaining 13 sequences from Table 1. The results prove that the hypersurfaces of potential energy for the designed sequences have the required profiles, as shown in Fig. 8.

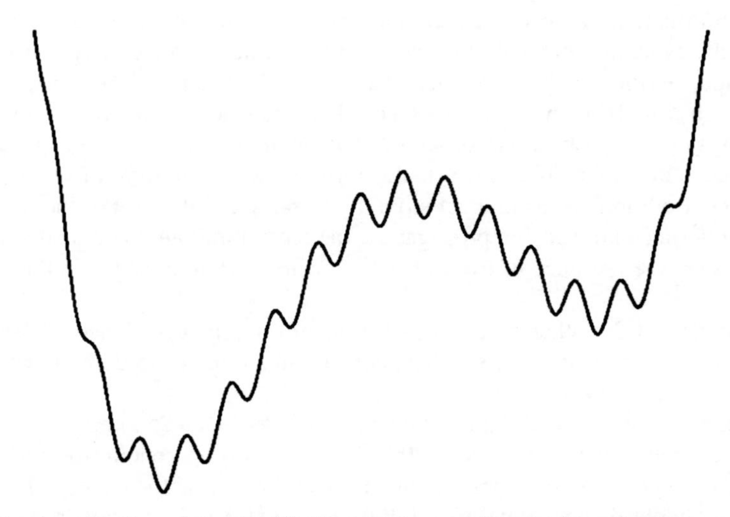

Fig. 8 A schematic view of the simplified potential energy hypersurface of each sequence. The lowest minimum corresponds to the native structure, α-helix bundle, whereas the higher corresponds to the β-sheet

4.2 Two-Chain Dynamics

The next step was the dynamics of two interacting proteins.

The results presented in this chapter were obtained using a program written by Boniecki and Kolinski allowing simulations of several chains, not just a simple one. Such an approach leads to results that are more reliable because the folding proteins can freely change their mutual positions and they are not restricted by the amino acids connecting them, as in work in reference [153].

One may assume that 1 of the 32 amino acid proteins is frozen in its metastable β-type conformation. This freezing may occur for a number of reasons, such as a chemical link [167]. The second protein molecule (of identical amino acid sequence) folds in the presence of the frozen metastable form.

Such dynamics of a single chain in the presence of the frozen protein were performed for all sequences in Table 1. For Sequences #2 to #14, a freely moving protein folded into an α-helix. The presence of a β form did not strongly influence the behavior of these proteins. For Sequence #1, the situation was completely different. When the second chain was a β-sheet, the folding of this peptide led to the formation of the next β-sheet (Fig. 9).

The folding into a β-sheet for Sequence #1 was independent of the starting geometry. In the case of a β form as an initial conformation, in high-temperature replicas (above the folding temperature), the chain remained unfolded. However, in low-temperature replicas, only β-sheet conformations appeared for a long time, and only these conformations were stable and independent of a random seed, the number of steps in the dynamics, or small differences in the replica's temperature distribution. In addition, starting from an extended chain led to the β-dimer. The crucial test referred to an α-helix as the initial geometry. For Sequence #1, the refolding from a helical form into a β-sheet was observed. This numerical experiment suggests that the presence of a misfolded (metastable) form causes misfolding of other protein molecules. Both molecules interact strongly, forming a β dimer. For the presented model of wrong conformation propagation, the most important factors affecting the dynamics are the presence of frozen β-sheets with an α-helix as the initial conformation.

In the case of the other sequences, they manifested independence of the initial conditions: independent of the starting geometry, the proteins fold into α-helices in the presence of their β forms.

Energies of well-formed β dimers for Sequence #1 in a 0.25 temperature replica are located between −195.5 and −200.4 in k_BT units, depending on the details of the side chains. These values are much lower than twice the energy of a single α-helix or single β-sheet equal to −170.6 and −151.6 in k_BT units, respectively. This comparison indicates that the interactions between both β-sheets forming a dimer are strong and may decrease the energy of the two-monomer system.

The next point of calculation was the examination of ββ and αα dimer stability. Dynamics runs were carried out for both dimers, and both dimer chains were allowed to move. For all sequences, except Sequence #1, α-helical dimers were stable, but β-sheet dimers occurred only at the beginning of the simulations and then

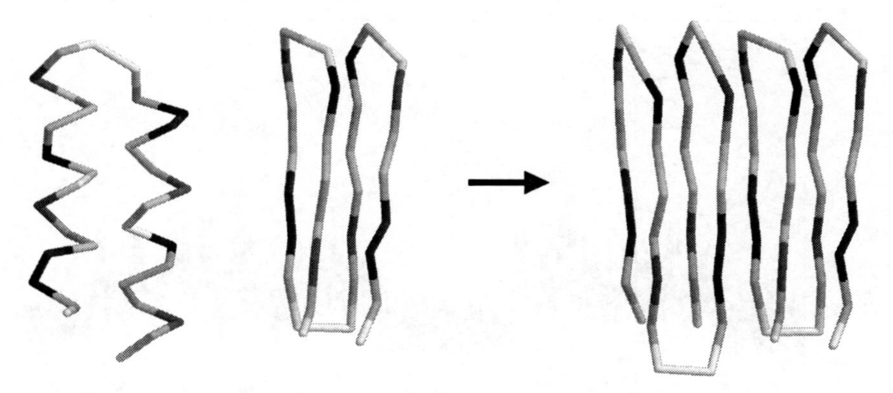

Fig. 9 β-type dimer as the lowest energy conformation for two-chain dynamics for Sequence #1. A freely moving α-helix in the presence of a frozen β-sheet (starting point of dynamics in *left panel*) folds into a β-sheet forming a β dimer. For clarity, only the protein backbones are shown

either two α-helices or mixed structures were formed. Mixed structures are conformations in which some protein fragments exhibit helical shape, whereas other protein fragments are similar to the β form. For Sequence #1, both dimers were stable over very long simulations and the conversion from one dimer to another in replicas performed at temperatures lower than the folding temperature was not observed in any direction. Such a situation can be explained by a high energy barrier between energy basins of both dimers.

Figure 10 presents energies of conformations and selected structures obtained during a 0.56 temperature replica run for two-chain dynamics for Sequence #1 with an α-helix dimer as the initial structure. Both molecules were allowed to move. In low-temperature replicas (T lower than the folding temperature which equals ~ 0.55), only two α-helices were observed. At the folding temperature, the α-helix dimer had the lowest energy and the longest lifetime (Structures #1, 4, 6, 8, 10, and 13). However, other conformations were observed at this temperature: some of them corresponding to denatured proteins (#5 and 11), with others having β-sheet fragments (#3 and 7) or α-helix origins (#2 and 7).

Dynamics for Sequence #1 with a β-sheet dimer as a starting conformation at the folding temperature gives a similar picture, but this time, the lowest energy structures are the β-sheet dimers.

4.3 Three-Chain Dynamics

The conclusion of two-chain dynamics for 14 designed sequences is the ability to form stable β-sheet dimers using Sequence #1. The other sequences fold to two α-helical structures. To answer the question of how stable longer β-sheet clusters are and how the folding proceeds for more than one chain in the presence of a template

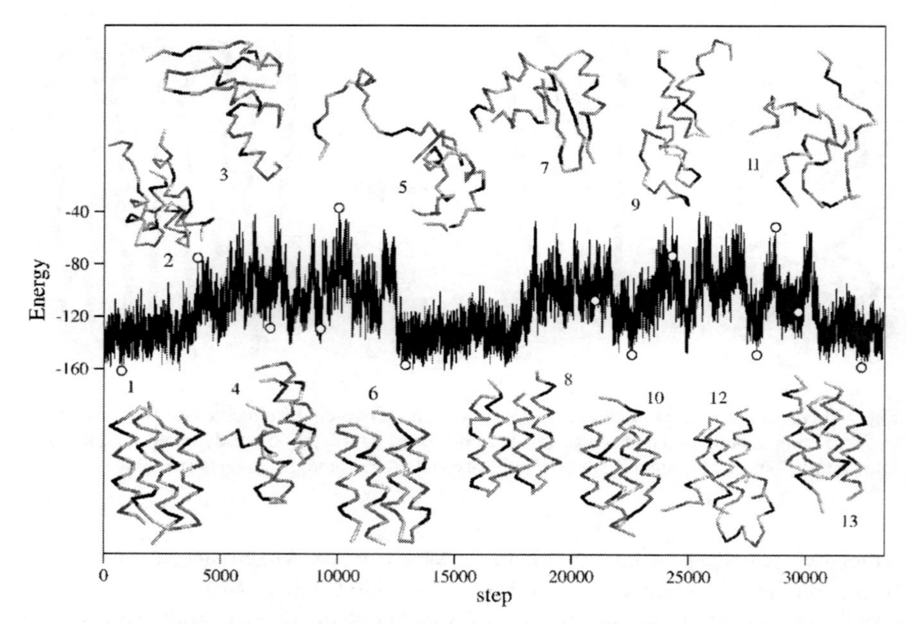

Fig. 10 α-helix dimer stability in two-chain dynamics in a 0.56 temperature replica for Sequence #1. Energies of every 100 conformations are presented. The numbers of the structures correspond to the *white circles* in the plot

(metastable conformation containing a β-sheet), three-chain dynamics were carried out.

Four types of conditions were taken into account:

1. As the first one, dynamics with one frozen molecule, a β-sheet, and two freely moving proteins were performed. Initial geometries were chosen as either one α-helix and one β-sheet or two α-helices, as is shown in Fig. 11.

   For Sequence #1, the lowest energy conformation and the most stable structure was a β trimer, whereas, for the other sequences, some mixed conformations formed. Progress of the 0.65 temperature replica (close to the folding temperature) for Sequence #1 is presented in Fig. 12, with representative conformations obtained during one of the dynamics.

   The starting geometry (Structure #1) was chosen as two α-helices and one β-sheet, the last molecule was frozen in the dynamics. The frozen protein was placed in the middle of the trimer or as an external molecule. The system quickly formed a β-sheet trimer (Structure #2). These conformations occurred in several mutual orientations (Structure #2, 4, 6, and 8). As previously noted, denatured conformations were observed (Structure #3) as were the formation of helical origins (Structure #5, 7, and 9).
2. One α-helix and one β-sheet dimer, with all molecules allowed to move, represents the second type of three-chain dynamics (initial conformation as per the right panel of Fig. 11). For Sequence #1, with replicas at temperatures lower than

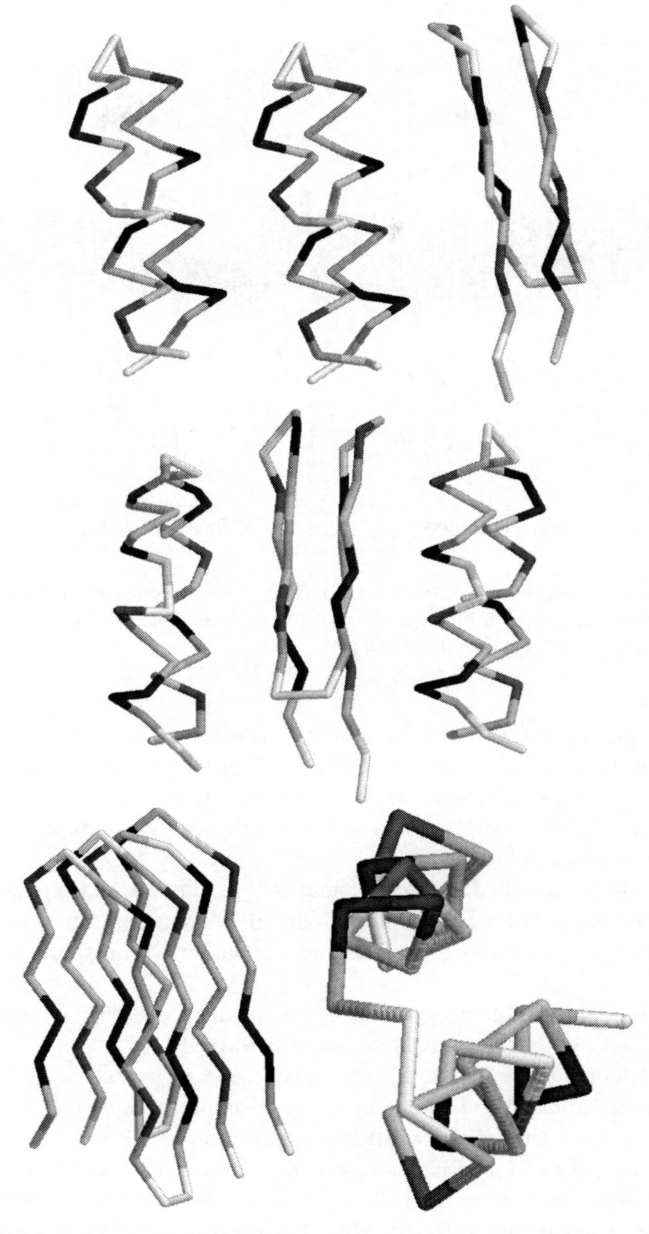

Fig. 11 Three initial geometries of molecules in three-chain dynamics with a frozen β-sheet

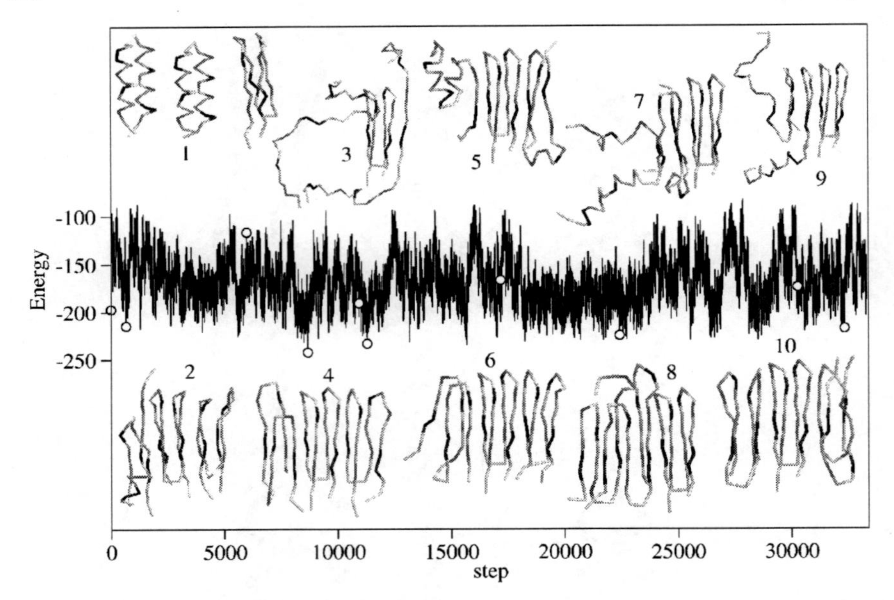

Fig. 12 Three-chain dynamics for Sequence #1 (0.65 temperature replica) when one molecule was frozen in β-sheet conformation. Energies of every 100 conformations are presented. Numbers of structures correspond to the *white circles* in a plot

the folding temperature, the formation of a β trimer was observed, except during the initial steps of dynamics. In the folding temperature replica, the other structures were also formed. The β-sheet dimer was stable and the third molecule either formed a β-sheet or had helical fragments. Two representative β-sheet trimers are shown in Fig. 13.

The trimer presented in the left panel of Fig. 13 has a very regular structure. The monomers aggregated, forming a cuboid. Monomers in the second trimer (right panel) exhibited, to a small degree, a disorder with respect to the cuboid structure.

The remaining sequences did not aggregate to a β-sheet trimer, rather, mixed conformations with fragments of both motifs formed.

3. The third dynamics had two α-helices and one β-sheet as starting geometry, with all molecules allowed to move. The energies and representative conformations obtained by a 0.6 temperature replica are shown in Fig. 14.

   The structures formed in such dynamics were more often mixed conformations compared with cases 1 and 2, however, the β-sheet trimers were stable for Sequence #1 when they occurred. The other sequences formed α-helices.
4. The goal of the last type of three-chain dynamics was to check the stability of β trimers. Such trimers were stable for Sequence #1 during very long dynamics. Very surprisingly, similar properties were demonstrated by the β-sheet trimer for Sequence #2. Trimers for the other sequences dissociated and refolded into separate α-helices or mixed structures. This means that sequence #1 can aggregate

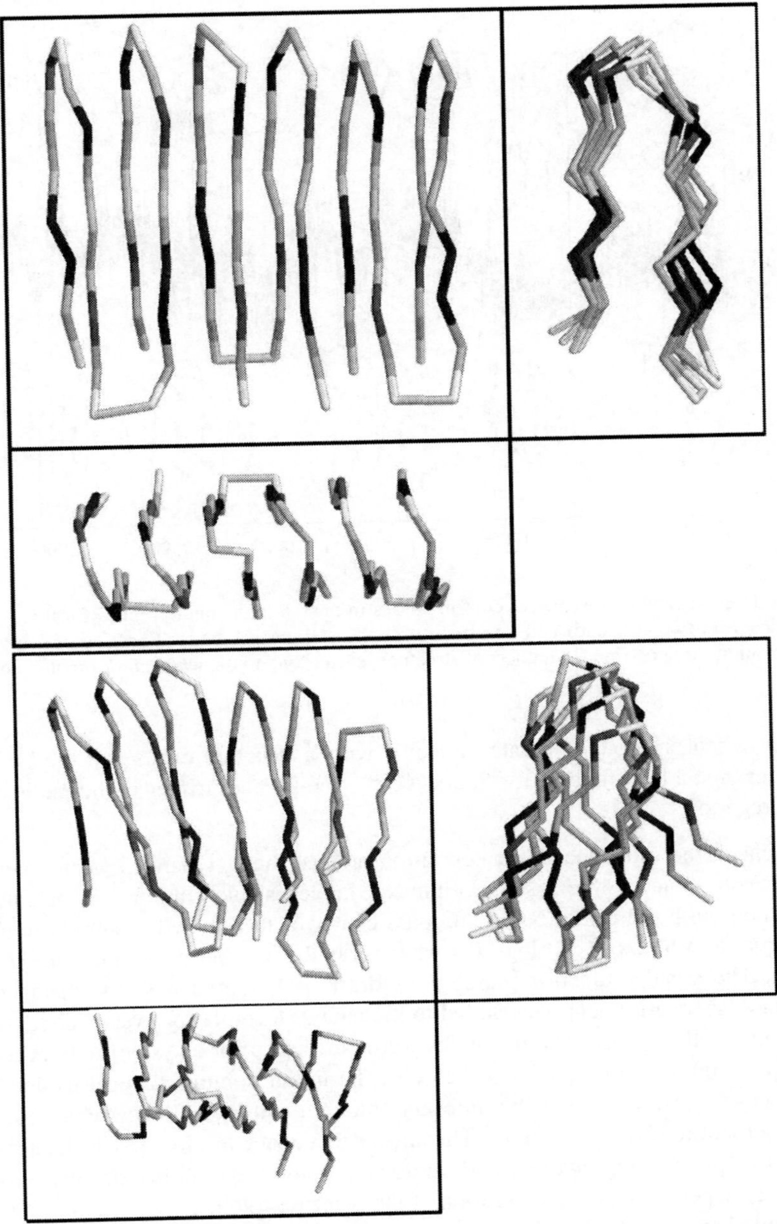

Fig. 13 Two representative β-sheet trimers obtained in three-chain dynamics with all molecules freely moving and with an α-helix and β-sheet dimer as the starting conformation. For both structures, two side and top views are shown

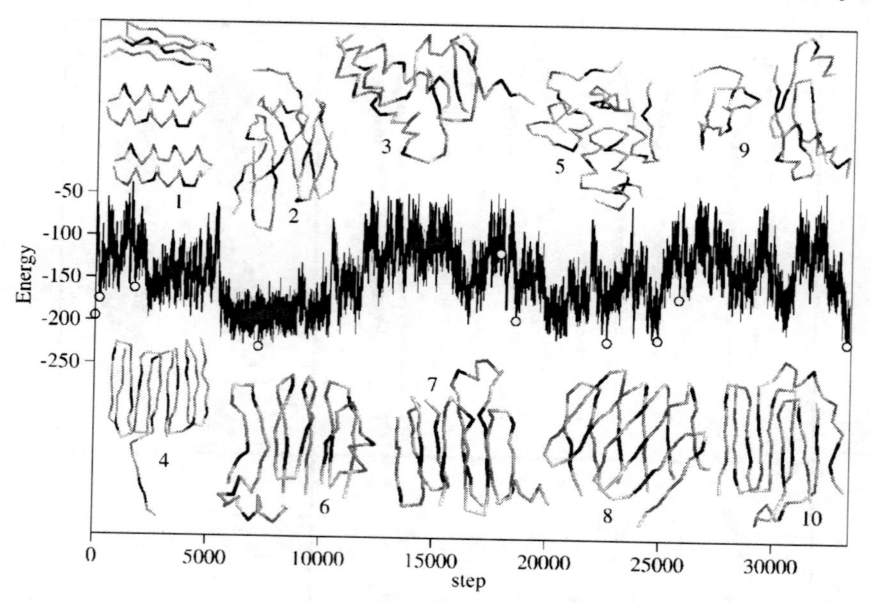

Fig. 14 Energies and representative conformations in the 0.6 temperature replica for three-chain dynamics with two α-helices and one β-sheet as the starting geometry. Energies of every 100 conformations are presented. Numbers of structures correspond to the *white circles* in the plot

even when only one template molecule with β-structure exists and the β-sheet dimer is stable, whereas, for Sequence #2, the β-sheet trimer is the nucleus of aggregation.

To check the influence of concentration effects, these dynamics were repeated in increasingly accessible system volumes. The accessible volume was calculated as a sphere with the center being the mass center of the studied system with a radius equal to 5.5* exp[0.38* log(N)], where N is the number of amino acids in the system. The penalty function, being a quadratic polynomial with the distance to the sphere as an argument, was added to the energy formula for every united atom lying outside the sphere. The dynamics were performed for a system with accessible volume increased by 50% and 125% of the initial volume. β-sheet trimers for Sequences #1 and #2 were stable in every situation, whilst the other sequences refolded into more helical structures. This means that concentration may influence the kinetics of protein aggregation, and, more importantly, the ability of amyloids to form may depend on the concentration of the β-form proteins.

Detailed studies of all designed sequences and their hydrophobic–polar patterns from the point of view of their tendency to aggregate, cannot indicate the features responsible for this ability. It seems that the final effect results from a subtle interplay among various interactions and sequence features.

5 Conclusions

The common feature of degenerative diseases is the presence of hard and insoluble amyloid plaques in neural cells. The main component of such plaques is protein fragments rich in β-sheet structures. The "one-protein theory" proposed by Prusiner states that they come from normal cellular prion proteins. After their synthesis and folding into native forms in cells, they may succumb to refolding into their isoforms, which contain more β-sheets and are resistant to protease activity; an enzyme that breaks peptide bonds between protein amino acids to remove wrongly folded peptides.

The results presented in this chapter point out that some proteins considered as monomers adopt a structure, called the native state, N, whereas, in the presence of an energetically higher structure, R, they can change their native geometry into R and form a dimer R_2. The energies of both kinds of dimers, N_2 and R_2, are similar, however, between respective minima in potential energy hypersurface, a high energetic barrier is located, with both dimers being long-lived structures. The addition of another protein in the N state to the system containing R_2 dimers leads to spontaneous formation of trimer R_3. Such behavior can be a model of prion disease propagation, or, more generally, a model of a conformational disease.

References

1. Nymeyer, H., Garcia, A. E. and Onuchic, J. N. (1998) Folding funnels and frustration in off-lattice minimalist protein landscapes. *Proc. Natl. Acad. Sci. USA* **95,** 5921
2. Prusiner, S. B. (1982) Novel proteinaceous infectious particles cause scrapie. *Science* **216,** 136
3. Prusiner, S. B. (1991) Molecular biology of prion diseases. *Science* **252,** 1515
4. Aguzzi, A. and Polymenidou, M. (2004) Mammalian prion biology: One century of evolving concepts. *Cell* **116,** 313
5. Prusiner, S. B., Bolton, D. C., Groth, D. F., Bowman, K. A., Cochran, S. P. and McKinley, M. P. (1982) Further purification and characterization of scrapie prion. *Biochemistry* **21,** 6942
6. Prusiner, S. B., McKinley, M. P., Bowman, K. A., Bolton, D. C., Bendheim, P. E., Groth, D. F. and Glenner, G. G. (1983) Scrapie prions aggregate to form amyloid-like birefringent rods. *Cell* **35,** 349
7. Pan, K.-M., Baldwin, M., Nguyen, J., Gasset, M., Serban, A., Groth, D., Mehlhorn, I., Huang, Z., Fletterick, R. J., Cohen, F. E. and Prusiner, S. B. (1993) Conversion of alpha-helices beta-sheets features in the formation of the scrapie prion proteins. *Proc. Natl. Acad. Sci. USA* **90,** 10962
8. Safar, J., Roller, P. P., Gajdusek, D. C. and Gibbs, C. J. Jr. (1993) Thermal stability and conformational transitions of scrapie amyloid (prion) protein correlate with infectivity. *Protein Sci.* **2,** 2206
9. Prusiner, S. B. (1994) Biology and genetics of prion diseases. *Annu. Rev. Microbiol.* **48,** 655
10. Kocisko, D. A., Come, J. H., Priola, S. A., Chesebro, B., Raymond, G. J., Lansbury, P. T. and Caughey, B. (1994) Cell-free formation of protease-resistant prion protein. *Nature* **370,** 471
11. Nguyen, J., Baldwin, M. A., Cohen, F. E. and Prusiner, S. B. (1995) Prion protein peptides induce alpha-helix to beta-sheet conformational transitions. *Biochemistry* **34,** 4186

12. Pergami, P., Jaffe, H. and Safar, J. (1998) Semipreparative chromatographic method to purify the normal cellular isoform of the prion protein in nondenatured form. *Anal. Biochem.* **236,** 63
13. Harrison, P. M., Bamborough, P., Daggett, V., Prusiner, S. B. and Cohen, F. E. (1997) The prion folding problem. *Curr. Opin. Struct. Biol.* **7,** 53
14. Horwich, A. L. and Weissman, J. S. (1997) Deadly conformations misfolding in prion disease. *Cell* **89,** 499
15. Bossers, A., Rigter, A., de Vries, R. and Smits, M. A. (2003) In vitro conversion of normal prion protein into pathologic isoforms. *Clin. Lab. Med.* **23,** 227
16. Oesch, B., Westaway, D., Walchli, M., McKinley, M. P., Kent, S. B. H., Aebersold, R., Barry, R. A., Tempst, P., Teplow, D. B., Hood, L. E., Prusiner, S. B. and Weissmann, C. (1985) A cellular gene encodes scrapie PrP 27–30 protein. *Cell* **40,** 735
17. Meyer, R. K., McKinley, M. P., Bowman, K. A., Braunfeld, M. B., Barry, R. A. and Prusiner, S. B. (1986) Separation and properties of cellular and scrapie prion proteins. *Proc. Natl. Acad. Sci. USA* **83,** 2310
18. Caughey, B. W., Dong, A., Bhat, K. S., Ernst, D., Hayes, S. F. and Caughey, W. S. (1991) Secondary structure analysis of the scrapie-associated protein PrP 27–30 in water by infrared spectroscopy. *Biochemistry* **30,** 7672
19. Kocisko, D. A., Priola, S. A., Raymond, G. J., Chesebro, B., Lansbury, P. T. Jr. and Caughey, B. (1995) Species specificity in the cell-free conversion of prion protein to protease-resistant forms: A model for the scrapie species barrier. *Proc. Natl. Acad. Sci. USA* **92,** 3923
20. Borchelt, D. R., Scott, M., Taraboulos, A., Stahl, N. and Prusiner, S. B. (1990) Scrapie and cellular prion proteins differ in their kinetics of synthesis and topology in cultured cells. *J. Cell Biol.* **110,** 743
21. Borchelt, D. R., Taraboulos, A. and Prusiner, S. B. (1992) Evidence for synthesis of scrapie prion proteins in the endocytic pathway. *J. Biol. Chem.* **267,** 16188
22. Taraboulos, A., Raeber, A. J., Borchelt, D. R., Serban, D. and Prusiner, S. B. (1992) Synthesis and trafficking of prion proteins in cultured cells. *Mol. Biol. Cell* **3,** 851
23. Caughey, B. and Raymond, G. J. (1991) The scrapie-associated form of PrP is made from a cell surface precursor that is both protease- and phospholipase-sensitive. *J. Biol. Chem.* **266,** 18217
24. DeArmond, S. J., McKinley, M. P., Barry, R. A., Braunfeld, M. B., McColloch, J. R. and Prusiner, S. B. (1985) Identification of prion amyloid filaments in scrapie-infected brain. *Cell* **41,** 221
25. Kitamoto, T., Tateishi, J., Tashima, T., Takeshita, I., Barry, R. A., DeArmond, S. J. and Prusiner, S. B. (1986) Amyloid plaques in Creutzfeldt-Jakob disease stain with prion protein antibodies. *Ann. Neurol.* **20,** 204
26. Roberts, G. W., Lofthouse, R., Allsop, D., Landon, M., Kidd, M., Prusiner, S. B. and Crow, T. J. (1988) CNS amyloid proteins in neurodegenerative diseases. *Neurology* **38,** 1534
27. Tagliavini, F., Prelli, F., Ghiso, J., Bugiani, O., Serban, D., Prusiner, S. B., Farlow, M. R., Ghetti, B. and Frangione, B. (1991) Amyloid protein of Gerstmann-Straussler-Scheinker disease (Indiana kindred) is an 11 kD fragment of prion protein with an N-terminal glycine at codon 58. *EMBO J.* **10,** 513
28. Lynn, D. G. and Meredith, S. C. (2000) Review: Model peptides and the physicochemical approach to beta-amyloids. *J. Structur. Biol.* **130,** 153
29. Prusiner, S. B. (2001) Neurodegenerative diseases and prions. *N. Engl. J. Med.* **344,** 1516
30. Gasset, M., Baldwin, M. A., Lloyd, D. H., Gabriel, J.-M., Holtzman, D. M., Cohen, F., Fletterick, R. J. and Prusiner, S. B. (1992) Predicted alpha-helical regions of the prion protein when synthesized as peptides form amyloid. *Proc. Natl. Acad. Sci. USA* **89,** 10940
31. Gasset, M., Baldwin, M. A., Fletterick, R. J. and Prusiner, S. B. (1993) Perturbation of the secondary structure of the scrapie prion protein under conditions that alter infectivity. *Proc. Natl. Acad. Sci. USA* **90,** 1

32. Huang, Z., Gabriel, J.-M., Baldwin, M. A., Fletterick, R. J., Prusiner, S. B. and Cohen, F. E. (1994) Proposed three-dimensional structure for the cellular prion protein. *Proc. Natl. Acad. Sci. USA* **91,** 7139
33. Huang, Z., Prusiner, S. B. and Cohen, F. E. (1995) Scrapie prions: a three-dimensional model of an infectious fragment. *Fold. Des.* **1,** 13
34. Zhang, H., Kaneko, K., Nguyen, J. T., Livshits, T. L., Baldwin, M. A., Cohen, F. E., James, T. L. and Prusiner, S. B. (1995) Conformational transitions in peptides containing two putative alpha-helices of the prion protein. *J. Mol. Biol.* **250,** 514
35. James, T. L., Liu, H., Ulyanov, N. B., Farr-Jones, S., Zhang, H., Donnes, D. G., Kaneko, K., Groth, D., Mehlhorn, I., Prusiner, S. B. and Cohen, F. E. (1997) Solution structure of a 142-residue recombinant prion protein corresponding to the infectious fragment of the scrapie isoform. *Proc. Natl. Acad. Sci. USA* **94,** 10086
36. Liu, H., Farr-Jones, S., Ulyanov, N. B., Llinas, M., Marqusee, S., Groth, D., Cohen, F. E., Prusiner, S. B. and James, T. L. (1999) Solution structure of Syrian hamster prion protein rPrP(90-231). *Biochemistry* **38,** 5362
37. Garcia, F. L., Zahn, R., Riek, R. and Wuthrich, K. (2000) NMR structure of the bovine prion protein. *Proc. Natl. Acad. Sci. USA* **18,** 8334
38. Calzolai, L., Lysek, D. A., Gunter, P., von Schroetter, C., Riek, R., Zahn, R. and Wuthrich, K. (200) NMR structures of three single-residue variants of the human prion protein. *Proc. Natl. Acad. Sci. USA* **97,** 8340
39. Zahn, R., Liu, A., Luhrs, T., Riek, R., von Schroetter, C., Garcia, F. L., Billeter, M., Calzolai, L., Wider, G. and Wuthrich, K. (2000) NMR solution structure of the human prion protein. *Proc. Natl. Acad. Sci. USA* **97,** 145
40. Knaus, K. J., Morillas, M., Swietnicki, W., Malone, M., Surewicz, W. K. and Yee, V. C. (2001) Crystal structure of the human prion protein reveals a mechanism for oligomerization. *Nat. Struct. Biol.* **8,** 770
41. Riek, R. and Luhrs, T. (2003) Three-dimensional structures of the prion protein and its doppel. *Clin. Lab. Med.* **23,** 209
42. Gossert, A. D., Bonjour, S., Lysek, D. A., Fiorito, F. and Wuthrich, K. (2005) Prion protein NMR structures of elk and of mouse-elk hybrids. *Proc. Natl. Acad. Sci. USA* **102,** 646
43. Prusiner, S. B., Groth, D., Serban, A., Koehler, K., Foster, D., Torchia, M., Burton, D., Yang, S.-L. and DeArmond, S. J. (1993) Ablation of the prion (PrP) gene in mice prevents scrapie and facilitates production of anti-PrP antibodies. *Proc. Natl. Acad. Sci. USA* **90,** 10608
44. Prusiner, S. B., Groth, D., Serban, A., Stahl, N. and Gabizon, R. (1993) Attempts to restore scrapie prion infectivity after exposure to protein denaturants. *Proc. Natl. Acad. Sci. USA* **90,** 2793
45. Cohen, F. E., Pan, K.-M., Huang, Z., Baldwin, M. and Fletterick, R. J. (1994) Structural clues to prion replication. *Science* **264,** 530
46. Benzinger, T. L. S., Gregory, D. M., Burkoth, T. S., Miller-Auer, H., Lynn, D. G., Botto, R. E. and Meredith, S. C. (1998) Propagating structure of Alzheimer's beta-amyloid (10–35) is parallel beta-sheet with residues in exact register. *Proc. Natl. Acad. Sci. USA* **95,** 13407
47. Prusiner, S. B. (1998) Prions. *Proc. Natl. Acad. Sci. USA* **95,** 13363
48. DeArmond, S. J. (2004) Discovering the mechanisms of neurodegeneration in prion diseases. *Neurochem. Res.* **29,** 1979
49. Safar, J. G., Geschwind, M. D., Deering, C., Didorenko, S., Sattavat, M., Sanchez, H., Serban, A., Vey, M., Baron, H., Giles, K., Miller, B. L., DeArmond, S. J., Prusiner, S. B. (2005) Diagnosis of human prion disease. *Proc. Natl. Acad. Sci. USA* **102,** 3501
50. Prusiner, S. B. (1996) Molecular biology and pathogenesis of prion diseases. *Trends Biochem. Sci.* **21,** 482
51. Telling, G. C., Parchi, P., DeArmond, S. J., Cortelli, P., Montagna, P., Gabizon, R., Mastrianni, J., Lugaresi, E., Gambetti, P. and Prusiner, S. B. (1996) Evidence for the conformation of the pathologic isoform of the prion protein enciphering and propagating prion diversity. *Science* **274,** 2079
52. Prusiner, S. B. (1997) Prion diseases and BSE crisis. *Science* **278,** 245

53. Cashman, N. R. (1997) A prion primer. *Can. Med. Assoc. J.* **157,** 1381
54. Riek, R., Wider, G., Billeter, M., Hornemann, S., Glockshuber, R. and Wuthrich, K. (1998) Prion protein NMR structure and familial human spongiform encephalopathies. *Proc. Natl. Acad. Sci. USA* **95,** 11667
55. Scott, M. R., Will, R., Ironside, J., Nguyen, H.-O. B., Tremblay, P., DeArmond, S. J. and Prusiner, S. B. (1999) Compelling transgenetic evidence for transmission of bovine spongiform encephalopathy prions to humans. *Proc. Natl. Acad. Sci. USA* **96,** 15137
56. Mastrianni, J. A. and Roos, R. P. (2000) The prion diseases. *Semin. Neurol.* **20,** 337
57. Pedersen, N. S. and Smith, E. (2002) Prion diseases: Epidemiology in man. *Acta Pathol. Mic. Sc.* **110,** 14
58. Dalsgaard, N. J. (2002) Prion diseases. An overview. *Acta Pathol. Mic. Sc.* **110,** 3
59. Jacobsen, J. S. (2002) Alzheimer's disease: An overview of current and emerging therapeutic strategies. *Curr. Top. Med. Chem.* **2,** 343
60. DeArmond, S. J. and Prusiner, S. B. (2003) Perspectives on prion biology, prion disease pathogenesis, and pharmacologic approaches to treatment. *Clin. Lab. Med.* **23,** 1
61. Mastrianni, J. A. (2004) Prion diseases. *Clin. Neurosci. Res.* **3,** 469
62. Beghi, E., Gandolfo, C., Ferrarese, C., Rizzuto, N., Poli, G., Tonini, M. C., Vita, G., Leone, M., Logroscino, G., Granieri, E., Salemi, G., Savettieri, G., Frattola, L., Ru, G., Mancardi, G. L. and Messina, C. (2004) Bovine spongiform encephalopathy and Creutzfeldt-Jakob disease: facts and uncertainties underlying the causal link between animal and human diseases. *Neurol. Sci.* **25,** 122
63. Harder, A., Gregor, A., Wirth, T., Kreuz, F., Schulz-Schaeffer, W. J., Windl, O., Plotkin, M., Amthauer, H., Neukirch, K., Kretzschmar, H. A., Kuhlmann, T., Braas, R., Hahne, H. H. and Jendroska, K. (2004) Early age of onset in fatal familial insomnia. Two novel cases and review of the literature. *J. Neurol.* **251,** 715
64. Olsen, S. B., Sheikh, A., Peck, D. and Darzi, A. (2005) Variant Creutzfeldt-Jakob disease, a cause for concern. Review of the evidence for risk of transmission through abdominal lymphoreticular tissue surgery. *Surg. Endosc.* **19,** 747
65. Silveira, J. R., Raymond, G. J., Hughson, A. G., Race, R. E., Sim, V. L., Hayes, S. F. and Caughey, B. (2005) The most infectious prion protein particles. *Nature* **437,** 257
66. Bueler, H., Aguzzi, A., Sailer, A., Greiner, R.-A., Autenried, P., Aguet, M. and Weissmann, C. (1993) Mice devoid of PrP are resistant to scrapie. *Cell* **73,** 1339
67. Race, R. and Chesebro, B. (1998) Scrapie infectivity found in resistant species. *Nature* **392,** 770
68. Bons, N., Mestre-Frances, N., Belli, P., Cathala, F., Gajdusek, D. C. and Brown, P. (1999) Natural and experimental oral infection of nonhuman primates by bovine spongiform encephalopathy agents. *Proc. Natl. Acad. Sci. USA* **96,** 4046
69. Sweeney, T., Kuczius, T., McElroy, M., Parada, M. G. and Groschup, M. H. (2000) Molecular analysis of Irish sheep scrapie cases. *J. Gen. Virol.* **81,** 1621
70. Houston, F., Foster, J. D., Chong, A., Hunter, N. and Bostock, C. J. (2000) Transmission of BSE by blood transfusion in sheep. *Lancet* **356,** 999
71. Crozet, C., Flamant, F., Bencsik, A., Aubert, D., Samarut, J. and Baron, T. (2001) Efficient transmission of two different sheep scrapie isolates in transgenic mice expressing the ovine PrP gene. *J. Virol.* **75,** 5328
72. Blattler, T. (2002) Transmission of prion disease. *Acta Pathol. Mic. Sc.* **110,** 71
73. Hunter, N., Foster, J., Chong, A., McCutcheon, S., Parnham, D., Eaton, S., MacKenzie, C. and Houston, F. (2002) Transmission of prion diseases by blood transfusion. *J. Gen. Virol.* **83,** 2897
74. Galbraith, D. N. (2002) Transmissible spongiform encephalopathies and tissue cell culture. *Cytotechnology* **39,** 117
75. Williams, E. S. (2003) Scrapie and chronic wasting disease. *Clin. Lab. Med.* **23,** 139
76. Nonno, R., Esposito, E., Vaccari, G., Conte, M., Marcon, S., Bari, M. D., Ligios, C., Guardo, G. D. and Agrimi, U. (2003) Molecular analysis of cases of Italian sheep scrapie and comparison with cases of bovine spongiform encephalopathy (BSE) and experimental BSE in sheep. *J. Clin. Microbiol.* **41,** 4127

77. Baron, T., Crozet, C., Biacabe, A.-G., Philippe, S., Verchere, J., Bencsik, A., Madec, J.-Y., Calavas, D. and Samarut, J. (2004) Molecular analysis of the protease-resistant prion protein in scrapie and bovine spongiform encephalopathy transmitted to ovine transgenic and wild-type mice. *J. Virol.* **78,** 6243
78. Browning, S. R., Mason, G. L., Seward, T., Green, M., Eliason, G. A. J., Mathiason, C., Miller, M. W., Williams, E. S., Hoover, E. and Telling, G. C. (2004) Transmission of prions from mule deer and elk with chronic wasting disease to transgenic mice expressing cervid PrP. *J. Virol.* **78,** 13345
79. Alper, T., Haig, D. A. and Clarke, M. C. (1966) The exceptionally small size of the scrapie agent. *Biochem. Bioph. Res. Co.* **22,** 278
80. Latarjet, R., Muel, B., Haig, D. A., Clarke, M. C. and Alper, T. (1985) Inactivation of the scrapie agent by near monochromatic ultraviolet light. *Nature* **227,** 1341
81. DeArmond, S. J., Yang, S.-L., Lee, A., Bowler, R., Taraboulos, A., Groth, D. and Prusiner, S. B. (1993) Three scrapie prion isolates exhibit different accumulation patterns of the prion protein scrapie isoform. *Proc. Natl. Acad. Sci. USA* **90,** 6449
82. Safar, J., Wille, H., Itri, V., Groth, D., Serban, H., Torchia, M., Cohen, F. E. and Prusiner, S. B. (1998) Eight prion strains have PrP^{Sc} molecules with different conformations. *Nat. Med.* **4,** 1157
83. Scott, M. R., Groth, D., Tatzelt, J., Torchia, M., Tremblay, P., DeArmond, S. J. and Prusiner, S. B. (1997) Propagation of prion strains through specific conformers of the prion protein. *J. Virol.* **71,** 9032
84. Cohen, F. E. and Prusiner, S. B. (1998) Pathologic conformations of prion proteins. *Annu. Rev. Biochem.* **67,** 793
85. Bennion, B. J. and Daggett, V. (2000) Protein conformation and diagnostic tests: The prion protein. *Clin. Chem.* **48,** 2105
86. Kubler, E. and Bruno, O. and Raeber, A. J. (2003) Diagnosis of prion diseases. *Brit. Med. Bull.* **66,** 267
87. Kretzschmar, H. A. (2003) Diagnosis of prion diseases. *Clin. Lab. Med.* **23,** 109
88. Aguzzi, A., Heikenwalder, M. and Miele, G. (2004) Progress and problems in the biology, diagnostics, and therapeutics of prion diseases. *Clin. Invest.* **114,** 153
89. Wiessmann, C. and Aguzzi, A. (2005) Approaches to therapy of prion diseases. *Annu. Rev. Med.* **56,** 321
90. Soto, C. (1999) Alzheimer's and prion disease as disorders of protein conformation: implications for the design of novel therapeutic approaches. *J. Mol. Med.* **77,** 412
91. Soto, C., Kascsak, R. J., Saborio, G. P., Aucouturier, P., Wisniewski, T., Prelli, F., Kascsak, R., Mendez, E., Harris, D. A., Ironside, J., Tagliavini, F., Carp, R. I. and Frangione, B. (2000) Reversion of prion protein conformational changes by synthetic beta-sheet breaker peptides. *Lancet* **355,** 192
92. Rymer, D. L. and Good, T. A. (2000) The role of prion peptide structure and aggregation in toxicity and membrane binding. *J. Neurochem.* **75,** 2536
93. Korth, C., May, B. C. H., Cohen, F. E. and Prusiner, S. B. (2001) Acridine and phenothiazine derivatives as pharmacotherapeutics for prion disease. *Proc. Natl. Acad. Sci. USA* **98,** 9836
94. Prusiner, S. B. (1989) Scrapie prions. *Annu. Rev. Microbiol.* **43,** 345
95. Stahl, N. and Prusiner, S. B. (1991) Prions and prion proteins. *FASEB J.* **5,** 2799
96. Prusiner, S. B. and DeArmond, S. J. (1994) Prion diseases and neurodegeneration. *Annu. Rev. Neurosci.* **17,** 311
97. Prusiner, S. B. and Scott, M. R. (1997) Genetics of prions. *Annu. Rev. Genet.* **31,** 139
98. Prusiner, S. B., Scott, M. R., DeArmond, S. J. and Cohen, F. E. (1998) Prion protein biology. *Cell* **93,** 337
99. Ziesche, P., Smith, V. H. Jr., Ho, M., Rudin, S. P., Gersdorf, P. and Taut, M. (1999) Theoretical methods and algorithms—the He isoelectronic series and the Hooke's law model: correlation measures and modifications of Collins' conjecture. *J. Chem. Phys.* **110,** 6135
100. Harris, D. A. (1999) Cellular biology of prion diseases. *Clin. Microbiol. Rev* **12,** 429
101. DeArmond, S. J. and Bouzamondo, E. (2002) Fundamentals of prion biology and diseases. *Toxicology* **181–182,** 9

102. Tuite, M. F. and Koloteva-Levin, N. (2004) Propagating prions in fungi and mammals. *Mol. Cell* **14,** 541
103. Cordeiro, Y., Kraineva, J., Winter, R. and Silva, J. L. (2005) Volume and energy folding landscape of prion revealed by pressure. *Braz. J. Med. Biol. Res.* **38,** 1195
104. Weissmann, C. (2005) Birth of a prion: Spontaneous generation revisited. *Cell* **122,** 165
105. Swietnicki, W., Petersen, R., Gambetti, P. and Surewicz, W. K. (1997) pH-dependent stability and conformation of the recombinant human prion protein PrP(90–231). *J. Biol. Chem.* **272,** 27517
106. Brown, P., Rau, E. H., Bacote, A. E., Gibbs, C. J. and Gajdusek, D. C. (1999) New studies on the heat resistance of hamster-adapted scrapie agent: Threshold survival after ashing at 600 suggests an inorganic template of replication. *Proc. Natl. Acad. Sci. USA* **97,** 3418
107. Ferreo-Gonzales, A. D., Souto, S. O., Silva, J. L. and Foguel, D. (2000) The preaggregated state of an amyloidogenic protein: Hydrostatic pressure converts native transthyretin into the amyloidogenic state. *Proc. Natl. Acad. Sci. USA* **97,** 6445
108. Torrent, J., Alvarez-Martinez, M. T., Heitz, F., Liautard, J.-P., Balny, C. and Lange, R. (2003) Alternative prion structural changes revealed by high pressure. *Biochemistry* **42,** 1318
109. Brown, P., Meyer, R., Cardone, F. and Pocchiari, M. (2003) Ultra-high-pressure inactivation of prion infectivity in processed meat: A practical method to prevent human infection. *Proc. Natl. Acad. Sci. USA* **100,** 6093
110. Martins, S. M., Chapeaurouge, A. and Ferreira, S. T. (2003) Folding intermediates of the prion protein stabilized by hydrostatic pressure and low temperature. *J. Biol. Chem.* **278,** 50449
111. Cordeiro, Y., Kraineva, J., Ravindra, R., Lima, L. M. T. R., Gomes, M. P. B., Foguel, D., Winter, R. and Silva, J. L. (2004) Hydration and packing effects on prion folding and beta-sheet conversion: high pressure spectroscopy and pressure perturbation calorimetry studies. *J. Biol. Chem.* **279,** 32354
112. Silva, J. L., Foguel, D., Suarez, M., Gomes, A. M. O. and Oliveira, A. C. (2004) High-pressure applications in medicine and pharmacology. *J. Phys.-Condens. Mat.* **16,** 0
113. Garcia, A. F., Heindl, P., Voigt, H., Buttner, M., Butz, P., Tauber, N., Tauscher, B. and Pfaff, E. (2005) Dual nature of the infectious prion protein revealed by high pressure. *J. Biol. Chem.* **280,** 9842
114. Kazmirski, S. L., Alonso, D. O. V., Cohen, F. E., Prusiner, S. B. and Daggett, V. (1995) Theoretical studies of sequence effects on the conformational properties of a fragment of the prion protein: implications for scrapie formation. *Curr. Biol.* **2,** 305
115. Inouye, H., Bond, J., Baldwin, M. A., Ball, H. L., Prusiner, S. B. and Kirschner, D. A. (2000) Structural changes in a hydrophobic domain of the prion protein induced by hydratation and by Ala -> Val and Pro -> Leu substitutions. *J. Mol. Biol.* **300,** 1283
116. Markossian, K. A. and Kurganov, B. I. (2004) Protein folding, misfolding, and aggregation. Formation of inclusion bodies and aggresomes. *Biochemistry (Moscow)* **69,** 1196
117. Sigurdsson, B. R. (1954) A chronic encephalitis of sheep with general remarks on infections which develop slowly and some of their special characteristics. *Brit. Vet. J.* **110,** 341
118. The Nobel Assembly at the Karolinska Institute (1997), http://nobelprize.org/nobel_prizes/medicine/laureates/1997/press.html.
119. Kaneko, K., Zulianello, L., Scott, M., Cooper, C. M., Wallace, A. C., James, T. L., Cohen, F. E. and Prusiner, S. B. (1997) Evidence for protein X binding to a discontinuous epitope on the cellular prion protein during scrapie prion propagation. *Proc. Natl. Acad. Sci. USA* **94,** 10069
120. Fisher, E., Telling, G. and Collinge, J. (1998) Prions and the prion disorders. *Mamm. Genome* **9,** 497
121. Soto, C. and Castilla, J. (2004) The controversial protein-only hypothesis of prion propagation. *Nat. Med.* **10,** 0
122. Zahn, R. (1999) Prion propagation and molecular chaperones. *Q. Rev. Biophys.* **4,** 309
123. Jackson, G. S. and Collinge, J. (2000) Prion disease—the propagation of infectious protein topologies. *Microbes Infect.* **2,** 1445

124. Liebman, S. W. (2002) Progress toward an ultimate proof of the prion hypothesis. *Proc. Natl. Acad. Sci. USA* **99,** 9098
125. Cohen, F. E. (1999) Protein misfolding and prion diseases. *J. Mol. Biol.* **293,** 313
126. Jarrett, J. T. and Lansbury, P. T. (1993) Seeding "one-dimensional crystallization" of amyloid: A pathogenic mechanism in Alzheimer's disease and scrapie? *Cell* **73,** 1055
127. Harper, J. D. and Lansbury, P. T. Jr. (1997) Models of amyloid seeding in Alzheimer's disease and scrapie: Mechanistic truths and physiological consequences of the time-dependent solubility of amyloid proteins. *Annu. Rev. Biochem.* **66,** 385
128. Morrissey, M. P. and Shakhnovich, E. I. (1999) Evidence for the role of PrP^C helix 1 in the hydrophilic seeding of prion aggregates. *Proc. Natl. Acad. Sci. USA* **96,** 11293
129. Weissmann, C. (1999) Molecular genetics of transmissible spongiform encephalopathies. *J. Biol. Chem.* **274,** 3
130. Hortschansky, P., Schroeckh, V., Christopeit, T., Zandomeneghi, G. and Fandrich, M. (2005) The aggregation kinetics of Alzheimer's beta-amyloid peptide is controlled by stochastic nucleation. *Protein Sci.* **14,** 1753
131. Masel, J., Jansen, V. A. A. and Nowak, M. A. (1999) Quantifying the kinetic parameters of prion replication. *Biophys. Chem.* **77,** 139
132. Poschel, T., Brilliantov, N. V. and Frommel, C. (2003) Kinetics of prion growth. *Biophys. J.* **85,** 3460
133. Slepoy, A., Singh, R. R. P., Pazmandi, F., Kulkarni, R. V. and Cox, D. L. (2001) Statistical mechanics of prion diseases. *Phys. Rev. Lett.* **87,** 58101
134. Kulkarni, R. V., Slepoy, A., Singh, R. R. P., Cox, D. L. and Pazmandi, F. (2003) Theoretical modeling of prion disease incubation. *Biophys. J.* **85,** 707
135. Mobley, D. L., Cox, D. L., Singh, R. R. P., Kulkarni, R. V. and Slepoy, A. (2003) Simulations of oligomeric intermediates in prion diseases. *Biophys. J.* **85,** 2213
136. Rieger, T. R., Morimoto, R. I. and Hatzimanikatis, V. (2005) Bistability explains threshold phenomena in protein aggregation both in vitro and in vivo. *Biophys. J.* **in press**
137. Zuegg, J. and Gready, J. E. (1999) Molecular dynamics simulations of human prion protein: importance of correct treatment of electrostatic interactions. *Biochemistry* **38,** 13862
138. Parchment, O. G. and Essex, J. W. (2000) Molecular dynamics of mouse and Syrian hamster PrP: Implications for activity. *Proteins* **38,** 327
139. Guilbert, C., Ricard, F. and Smith, J. C. (2000) Dynamic simulation of the mouse prion protein. *Biopolymers* **54,** 406
140. Zuegg, J. and Gready, J. E. (2000) Molecular dynamics simulation of human prion protein included both N-linked oligosaccharides and the GPI anchor. *Glycobiology* **10,** 959
141. Okimoto, N., Yamanaka, K., Suenaga, A., Hata, M. and Hoshino, T. (2002) Computational studies on prion proteins: Effect of Ala117->Val mutation. *Biophys. J.* **82,** 2746
142. Gsponer, J., Haberthur, U. and Caflisch, A. (2003) The role of side-chain interactions in the early steps of aggregation: Molecular dynamics simulations of an amyloid-forming peptide from the yeast prion Sup35. *Proc. Natl. Acad. Sci. USA* **100,** 5154
143. Kuwata, K., Matumoto, T., Cheng, H., Nagayama, K., James, T. L. and Roder, H. (2003) NMR-detected hydrogen exchange and molecular dynamics simulations provide structural insight into fibril formation of prion protein fragment 106–126. *Proc. Natl. Acad. Sci. USA* **100,** 14790
144. DeMarco, M. L. and Daggett, V. (2004) From conversion to aggregation: Protofibril formation of the prion protein. *Proc. Natl. Acad. Sci. USA* **101,** 2293
145. Ma, B. and Nussinov, R. (2004) Molecular dynamics simulations of alanine rich beta-sheet oligomers: Insight into amyloid formation. *Protein Sci.* **11,** 2335
146. Langella, E., Improta, R. and Barone, V. (2004) Checking the pH-induced conformational transition of prion protein by molecular dynamics simulations: Effect of protonation of histidine residues. *Biophys. J.* **87,** 3623
147. Barducci, A., Chelli, R., Procacci, P. and Schettino, V. (2005) Misfolding pathways of the prion protein probed by molecular dynamics simulations. *Biophys. J.* **88,** 1334

148. Armen, R. S., DeMarco, M. L., Alonso, D. O. V. and Daggett, V. (2004) Pauling and Corey's alpha-pleated sheet structure may define the prefibrillar amyloidogenic intermediate in amyloid disease. *Proc. Natl. Acad. Sci. USA* **101,** 11622
149. Stork, M., Giese, A., Kretzschmar, H. A. and Tavan, P. (2005) Molecular dynamics simulations indicate a possible role of parallel beta-helices in seeded aggregation of poly-Gln. *Biophys. J.* **88,** 2442
150. Tsai, H.-H., Reches, M., Tsai, C.-J., Gunasekaran, K., Gazit, E. and Nussinov, R. (2005) Energy landscape of amyloidogenic peptide oligomerization by parallel-tempering molecular dynamics simulation: Significant role of Asn ladder. *Proc. Natl. Acad. Sci. USA* **102,** 8174
151. Harrison, P. M., Chan, H. S., Prusiner, S. B. and Cohen, F. E. (2001) Conformational propagation with prion-like characteristics in a simple model of protein folding. *Protein Sci.* **10,** 819
152. Derreumaux, P. (2001) Evidence that the 127–164 region of prion proteins has two equienergetic conformations with beta or alpha features. *Biophys. J.* **81,** 1657
153. Peng, Y. and Hansmann, U. H. E. (2003) Helix versus sheet formation in a small peptide. *Phys. Rev. E* **68,** 41911
154. Dimeo, R. M., Neumann, D. A., Glanville, Y. and Minor, D. B. (2002) Pore-size dependence of rotational tunneling in confined methyl iodide. *Phys. Rev. B* **66,** 104201
155. Kammerer, R. A., Kostrewa, D., Zurdo, J., Detken, A., Garcia-Echeverría, C., Green, J. D., Muller, S. A., Meier, B. H., Winkler, F. K., Dobson, C. M. and Steinmetz, M. O. (2004) Exploring amyloid formation by a de novo design. *Proc. Natl. Acad. Sci. USA* **101,** 4435
156. Ding, F., LaRocque, J. J. and Dokholyan, N. V. (2005) Direct observation of protein folding, aggregation, and a prion-like conformational conversion. *J. Biol. Chem.***280,** 40235
157. Dima, R. I. and Thirumalai, D. (2002) Exploring protein aggregation and self-propagation using lattice models: Phase diagram and kinetics. *Protein Sci.* **11,** 1036
158. Malolepsza, E., Boniecki, M., Kolinski, A. and Piela, L. (2005) Theoretical model of prion propagation: A misfolded protein induces misfolding. *Proc. Natl. Acad. Sci. USA* **102,** 7835
159. Lysek, D. A., Nivon, L. G. and Wuthrich, K. (2004) Amino acid sequence of the *Felis catus* prion protein. *Gene* **341,** 249
160. Skolnick, J., Zhang, Y., Arakaki, A. K., Kolinski, A., Boniecki, M., Szilagyi, A. and Kihara, D. (2003) A unified approach to protein structure prediction. *Proteins* **53 (Suppl. 6),** 469
161. Boniecki, M., Rotkiewicz, P., Skolnick, J. and Kolinski, A. (2003) Protein fragment reconstruction using various modeling techniques. *Curr. Comp.-Aided Drug Design* **17,** 725
162. Leach, A. R. (1999) *Molecular modelling. Principles and applications.* Pearson Education Limited, Edinburgh Gate
163. Metropolis, N., Resenbluth, A. W., Rosenbluth, M. L., Teller, A. H. and Teller, E. (1953) Equation of state calculations by fast computing machines. *J. Chem. Phys.* **21,** 1087
164. Kolinski, A. (2004) Protein modeling and structure prediction with a reduced representation. *Acta Biochim. Pol.* **51,** 349
165. Kolinski, A. and Skolnick, J. (2004) Reduced models of proteins and their applications. *Polymer* **45,** 511
166. Critical assessment of techniques for protein structure prediction (2004) http:// predictioncenter.org/casp6/Casp6.html.
167. Tompa, P., Tusnady, G. E., Friedrich, P. and Simon, I. (2002) The role of dimerization in prion replication. *Biophys. J.* **82,** 1711

Part VI
Applications to Drug Design

Chapter 17
Identifying Putative Drug Targets and Potential Drug Leads
Starting Points for Virtual Screening and Docking

David S. Wishart

Summary The availability of three-dimensional (3D) models of both drug leads (small molecule ligands) and drug targets (proteins) is essential to molecular docking and computational drug discovery. This chapter describes an emerging methodology that can be used to identify both drug leads and drug targets using three newly developed web-accessible databases: 1) DrugBank; 2) The Human Metabolome Database; and 3) PubChem. Specifically, it illustrates how putative drug targets and drug leads for exogenous diseases (i.e., infectious diseases) can be readily identified and their 3D structures selected using only the genomic sequences from pathogenic bacteria or viruses as input. It also illustrates how putative drug targets and drug leads for endogenous diseases (i.e., non-infectious diseases or chronic conditions) can be identified using similar databases and similar sequence input. This chapter is intended to illustrate how bioinformatics and cheminformatics can work synergistically to help provide the necessary inputs for computer-aided drug design.

Keywords: Bioinformatics · Chemical similarity · Disease · Drug · Drug target · Endogenous disease · Exogenous disease · Metabolite · Sequence comparison

1 Introduction

As most readers have already seen, protein modeling is a rapidly developing field that allows many interesting biological questions to be addressed using only a computer. Insights gained through computational modeling have helped us better understand proteins and their many important structure/function relationships. Although macromolecular modeling has helped enormously to advance basic biology, one of the central justifications for the enormous resources that have gone into this field during the past 25 years is the hope that molecular modeling could, one day, accelerate drug discovery [1–3]. Computational drug discovery is a subfield of macromolecular modeling that involves the docking or virtual screening of one or more

From: Methods in Molecular Biology, vol. 443, Molecular Modeling of Proteins
Edited by Andreas Kukol © Humana Press, Totowa, NJ

small molecule compounds against a chosen protein target. The small molecule ligands are called *drug leads* and the protein of interest is called the *drug target*. Both computer-aided docking and virtual screening use a variety of algorithms that allow the small molecule(s) to be rapidly rotated and translated around the protein surface or active site and scored on the basis of their steric fit and/or predicted free energy [4–7]. In more advanced packages, the ligand (and even the protein) is allowed to exhibit some conformational flexibility. When an optimal orientation is found or a particularly high-scoring molecule is identified, a drug lead or a drug mechanism is said to have been "discovered." The results of these computational experiments are used in an iterative fashion by synthetic organic chemists to help design or select improved lead compounds.

What distinguishes virtual screening from docking is the number of molecules used (screening uses 1000s, docking uses 1), the objective of the search process (screening identifies drug leads, docking identifies active sites or mechanisms), and the robustness or complexity of the docking energy function (docking uses a complex force field, screening does not). There are now many excellent docking and/or virtual screening software packages, such as Dock [8], AutoDock [9], Gold [10], and Glide [11]. Almost all are freely available. These will be discussed in more detail in the next chapters.

However, it is important to remember that before either virtual screening or macromolecular docking can begin, a protein target needs to be identified (and modeled) and a set of potential drug leads needs to be assembled. This chapter describes how both drug targets and drug leads can be identified through several easily accessible web resources. Specifically, this section shows how putative drug targets for pathogenic viruses or bacteria can be identified directly from their genomic sequences and how the three-dimensional (3D) structures of putative drug leads and drug targets can be subsequently extracted from two newly developed databases: DrugBank [12] and PubChem [13]. This chapter also illustrates how human drug targets and potential drug leads for prostate cancer can be similarly identified and extracted for docking/screening programs using the Human Metabolome Database (HMDB) [14], DrugBank, and PubChem. The intent of this chapter is to give readers the necessary input files and knowledge to proceed to the next steps (docking and screening) in computational drug discovery.

2 Theory

When medicinal chemists or pharmaceutical scientists think about drugs and drug targets they generally classify them into two separate groups: 1) those that are associated with "endogenous" human diseases and 2) those that are associated with infectious or "exogenous" diseases. Endogenous diseases are typically chronic human disorders or conditions that arise because of germ-line mutations (genetic diseases), somatic mutations (cancer), age (atherosclerosis, immune disorders), or other internal factors. On the other hand, exogenous diseases are typically temporary diseases or conditions that arise from external, nonhuman agents, such as viruses, bacteria, fungi, protozoans, or poisonous animals (snakes, insects). The vast majority of drug targets (97%) and drugs (89%) are associated with endogenous diseases, whereas only a tiny minority of drug targets (3%) and drugs (11%) are actually associated with exogenous or infectious diseases [12, 15].

2.1 Identifying Drug Targets and Drug Leads for Exogenous Diseases

The identification of putative drug targets and drug leads for exogenous diseases can take one of two paths, both of which depend substantially on bioinformatics and sequence database comparisons. One can either attempt to identify a completely novel drug target/drug lead or one can attempt to identify a drug target/drug lead that is similar (or even identical) to an existing class of drug targets or drug leads. In both cases, one needs either the complete protein or DNA sequence of the pathogen of interest. Fortunately, these days the entire DNA sequence of many infectious agents of interest is already known or can be determined in as little as a week.

If one chooses to identify a completely novel drug target or drug lead, the task is then to identify those genes or proteins in the genome that are: 1) essential to viability; 2) disease causing; or 3) presented on the surface of the organism. Surface-bound proteins may be identified by sequence analysis by looking for trans-membrane segments using such tools as TMHMM [16] or PSortB [17]. Essential genes, especially for bacteria, may be identified by comparing sequences with existing databases of essential genes, such as in the Database of Essential Genes [18]. Likewise, disease-causing genes can be identified by comparing sequences between nonpathogenic forms of the microbe with pathogenic forms (say *Escherichia coli* O157 versus *E. coli* MG1655). Alternately, essential genes or disease-causing genes may be experimentally identified through knock-out mutations or deletions. Generally, all viral genes in a viral genome are essential whereas only 200 to 300 bacterial genes in a given bacterial genome are essential. Furthermore, among most pathogens, only a small fraction of proteins or genes (<20) are typically disease causing. Once these "druggable" genes or protein targets are identified, one must select for those that are sufficiently different (<35% identity) from any human homologs. This prevents any cross-reactivity or potentially adverse drug interactions. After these nonhomologous protein targets are found, one can either search or screen for an inhibitory molecule or develop a vaccine (using parts of the surface proteins). When working with completely novel drug targets, it is often difficult to know which lead compounds might work, therefore, widespread chemical library screening is often used.

If one wishes to find matches to an existing class of drug targets or drug leads, the task involves identifying those genes or proteins in the genome of the organism that are similar to known drug targets. The underlying assumption is that if a novel virus or a newly identified pathogenic bacterium shares some significant sequence similarity to a protein that is a known drug target from another organism, then the same (or similar) drugs may be used to combat or kill this pathogen. Alternately, these previously known drugs may serve as potential drug leads for further synthetic modification to develop therapies that are more effective for the organism of interest. What is needed for this process to work is a database of known drug target sequences, each of which is linked to a set of associated drugs. Ideally, this database should also include the 3D structures (known or predicted) of the drug targets and the drugs themselves. Fortunately, such a database exists. It is called DrugBank [12]. DrugBank is an online database that combines detailed drug (i.e., chemical) data with comprehensive drug target (i.e., protein) information. The database contains more than 4,300 drug entries including more than 1,100 FDA-approved small molecule and biotech drugs as well as more than 3,200 experimental drugs. Each compound entry contains detailed structure files in SDF, MOL, and PDB formats. Additionally, more than 16,000 protein or drug target sequences are linked to these drug entries, many of which have 3D structures or 3D homology models associated with them.

DrugBank supports standard BLAST sequence queries, including appropriately formatted multiple sequence inputs (i.e., complete proteomes). The output from these queries includes the name(s) and hyperlinks to the associated drugs and the 3D structures of the drug targets. Once the drugs are identified, it is possible to use DrugBank again to search for similar drugs (based on structure similarity) or to use PubChem to search for similar chemicals (also based on structure similarity). The structures of all of these chemical "hits" may be downloaded, either as PDB files (DrugBank) or as SDF files (PubChem). PubChem's SDF files can then be converted to PDB files using the freely available tools MolConverter (ChemAxon), CACTVS [19], or the Cactus online converter (http://cactus.nci.nih.gov/services/translate/). Thus, by using both DrugBank and PubChem, it is possible to rapidly obtain 3D structures of putative drug targets and the 3D structures of 100s or even 1,000s of drug leads. These data sets would obviously serve as the basis for docking or virtual screening studies.

2.2 *Identifying Drug Targets and Drug Leads for Endogenous Diseases*

Identifying drug targets for endogenous diseases is often far more challenging than identifying drug targets for infectious or exogenous diseases. This is because most endogenous human diseases have a complex etiology. With the exception of approximately 400 [20, 21] relatively rare, monogenic (single gene) disorders, the vast majority of endogenous diseases are multifactorial or polygenic (multigene) in origin. Nevertheless, with the advent of such techniques as microarray analysis or high-throughput proteomics, it is now possible to identify large numbers of disease-associated genes relatively rapidly [22]. Likewise, high-throughput sequencing, comparative genomic hybridization (CGH) arrays, and single nucleotide polymorphism (SNP)-typing are also allowing the rapid identification of the genes associated with many inherited monogenic (and even polygenic) diseases. This information is being cataloged in many online databases such as OMIM [20], the HMDB [14], and GeneCards [23]. It is also possible to find disease–gene associations directly through PubMed-Entrez [13], or other web servers, such as MedGene [24] and PolySearch [25]. Current estimates put the number of "druggable" gene targets in humans at approximately 3,050 [26].

Once a list of genes or proteins associated with a given disease is available (along with their sequences), then it is a matter of performing a series of similar kinds of sequence searches against DrugBank, as described for Sect. 2.1. However, it is also possible to find additional or even novel drug leads by looking through the HMDB. The HMDB, like DrugBank, is a multipurpose bioinformatics–cheminformatics informatics database containing detailed information regarding metabolites, their associated enzymes or transporters, and their disease-related properties. The usefulness of the HMDB in drug discovery lies in the fact that most drugs are actually analogs of existing metabolites, cofactors, or signaling molecules. Therefore, if one identifies a protein or proteins in a disease-specific pathway that requires a certain metabolite or cofactor, then these proteins may prove to be good drug targets and their cofactors or metabolites could prove to be good drug leads. Indeed, many inborn errors of metabolism (phenylketonuria, alkaptonuria, and galactosemia) are treated through the addition or removal of metabolites in the diet.

Both DrugBank and the HMDB support single and multiple protein sequence queries, and both produce results that include the name(s) and hyperlinks to the associated drugs or metabolites and the 3D structures of the corresponding proteins. Once the small molecule leads are identified, it is possible to use DrugBank or HMDB again to search for structurally similar drugs or metabolites. Alternately, one may use PubChem to search for similar chemicals (also based on structure similarity). The structures of all of these chemical "hits" may be downloaded, either as PDB files (DrugBank and HMDB) or as SDF files (PubChem). PubChem's SDF files can then be converted to PDB files using the freely available tools MolConverter (ChemAxon) or the Cactus web server. Thus, by judiciously using DrugBank, HMDB, and PubChem, it is possible to rapidly obtain 3D structures of putative drug targets and the 3D structures of numerous drug leads for endogenous diseases.

2.3 Sequence Matching and Chemical Compound Matching

This chapter focuses on using two different matching protocols, one for sequence matching and another for chemical structure matching. Sequence matching, or sequence alignment, is a central feature to much of bioinformatics, whereas chemical structure matching is a central feature to much of cheminformatics.

Sequence alignment is often based on a technique called dynamic programming. Strictly speaking, dynamic programming is an efficient mathematical technique that can be used to find optimal "paths" or routes to multiple destinations or to locate paths that could be combined to score some maximum value. The application of dynamic programming to sequence alignment was first demonstrated longer than 35 years ago by Needleman and Wunsch [27]. As these two researchers demonstrated, dynamic programming allows two or more sequences to be efficiently and automatically lined up, permitting gaps to be inserted, extended, or deleted to make an optimal pairwise alignment. In dynamic programming, the two sequences being compared (say, sequence A and sequence B) are put on either axis of a table. Sequence A might be on the x-axis, whereas sequence B might be on the y-axis. Each letter in the query sequence is compared to each letter the reference sequence and a number (based on a scoring matrix and a special recursive function) is placed in every box or cell that intersects each pair of letters. Once the table of numbers is filled out, a second stage (called the traceback stage) is undertaken, wherein the table is scanned in a diagonal fashion from the lower right to upper left to look for the highest scores. The path that is chosen is actually a series of "maximum" numbers. When all of the scores in this optimal path are added together, it gives a quantitative measure of the pairwise sequence similarity, while, at the same time, defining which letters in sequence A should be matched with the letters in sequence B.

Dynamic programming is a relatively slow, memory intensive process. However, it can be sped up considerably. For instance, if look-up tables are used, if advanced statistics are used, if more than one letter at a time (a "tuple") is scored, and if the traceback search is limited to sections close to the diagonal line, then you have the essence of the BLAST algorithm [28]. This is the very fast algorithm used to perform most alignments against large sequence databases. It is also the algorithm used in the sequence searches for DrugBank and HMDB.

Chemical compound matching shares some similarities to sequence matching. Thanks to the development of standardized text representations of chemical compounds through IUPAC International Chemical Identifer (InChI) strings and SMILES strings [29], it is possible to give every chemical a unique character string. In other words, InChI and SMILES strings uniquely define chemical compounds much like a protein can be uniquely defined by its sequence. Therefore, if one converts a standard chemical database (which may have been assembled using chemical names or structural MOL files) into a collection of SMILES strings or InChI identifiers, then it is possible to use character string comparison to perform compound matching. Several web-based conversion sites, including the Molecular Structure File Converter (http://iris12.colby.edu/~www/sconv.cgi), the Cactus Structure File Converter (http://cactus.nci.nih.gov/services/translate/), and the file converter at InChI.info (http://inchi.info/converter_en.html) are now available to facilitate conversion between MOL, SDF, PDB, SMILES, and InChI formats. The actual search algorithm requires that both the query compound and the database of searchable compounds be expressed in SMILES or InChI strings. By using simple string parsing and string matching utilities, it is often possible to identify structurally similar compounds by looking for shared character strings or substrings between the query and the database compounds. Unfortunately, this approach is not always fool proof. The scoring schemes for chemical substring matching are not yet as sophisticated as they are with sequence matching algorithms. Likewise, there are several different "flavors" and dialects of SMILES strings, which makes it difficult to exchange databases or search algorithms.

More sophisticated chemical structure matching algorithms also exist. These are based on the idea of matching substructures. This is somewhat similar to the idea of structure superposition, which is done with protein structure comparison. However, because the structures of chemical compounds are far more diverse than what is seen for proteins, the structure-matching utilities in chemistry have to be slightly more sophisticated. In particular, chemists must use the concept of subgraph isomorphisms [30] and adjacency matrices to identify chemical similarity. For substructure searching, the two-dimensional (2D) chemical structures of both the query and database compounds must be recast as tables that indicate the bond connectivity between each pair of atoms. These tables, which have 1s for connected atoms and 0s for unconnected atoms are called adjacency matrices. The name comes from the fact that they indicate which atoms are adjacent (connected) to each other. Once prepared, the adjacency matrix from the query structure is compared with every adjacency matrix in the database. If substantial sections of the query matrix match to an adjacency matrix (or portion thereof) in the database, then it is likely that the two structures are similar. Different scoring schemes and adjustable threshold cutoffs may be used to distinguish strong matches from weak matches or to identify compounds with particularly important substructures.

As will be seen in the examples to follow, both sequence searching and chemical structure searching play an important role in drug target and lead compound identification.

3 Methods

In this section, we describe two protocols. One describes the identification of drug targets and drug leads for a novel retrovirus that exhibits strong similarity to the acquired immune deficiency syndrome (AIDS) virus (human immunodeficiency virus [HIV]). The other describes the identification of drug leads (from a preexisting list of putative drug targets) for prostate cancer.

3.1 *Identifying Drug Targets and Drug Leads for a Novel Virus*

In this example, we use sequence data derived from a recently sequenced, but unnamed virus that exhibits strong sequence similarity to HIV. This particular virus has a total of 16 identifiable open reading frames, which have been fully translated. We will use this sequence information, in combination with DrugBank, to identify several drug targets, several drug leads, and the necessary coordinate files to conduct rational drug design efforts via docking and virtual screening (see **Note 1**).

1. Start your local web browser and go to the DrugBank website at http://www.drugbank.ca/. The DrugBank homepage should be visible, as should the blue menu bar located near the top of the page with the eight clickable titles: **Home**, **Browse**, **PharmaBrowse**, **ChemQuery**, **TextQuery**, **SeqSearch**, **Data Extractor**, and **Download**.
2. Click on the **SeqSearch** link. A window with the title "DrugBank BLAST Search" should appear (Fig. 1). As seen in Fig. 1, the window contains a standard online BLAST search form with a text box window, **Submit** and **Reset**

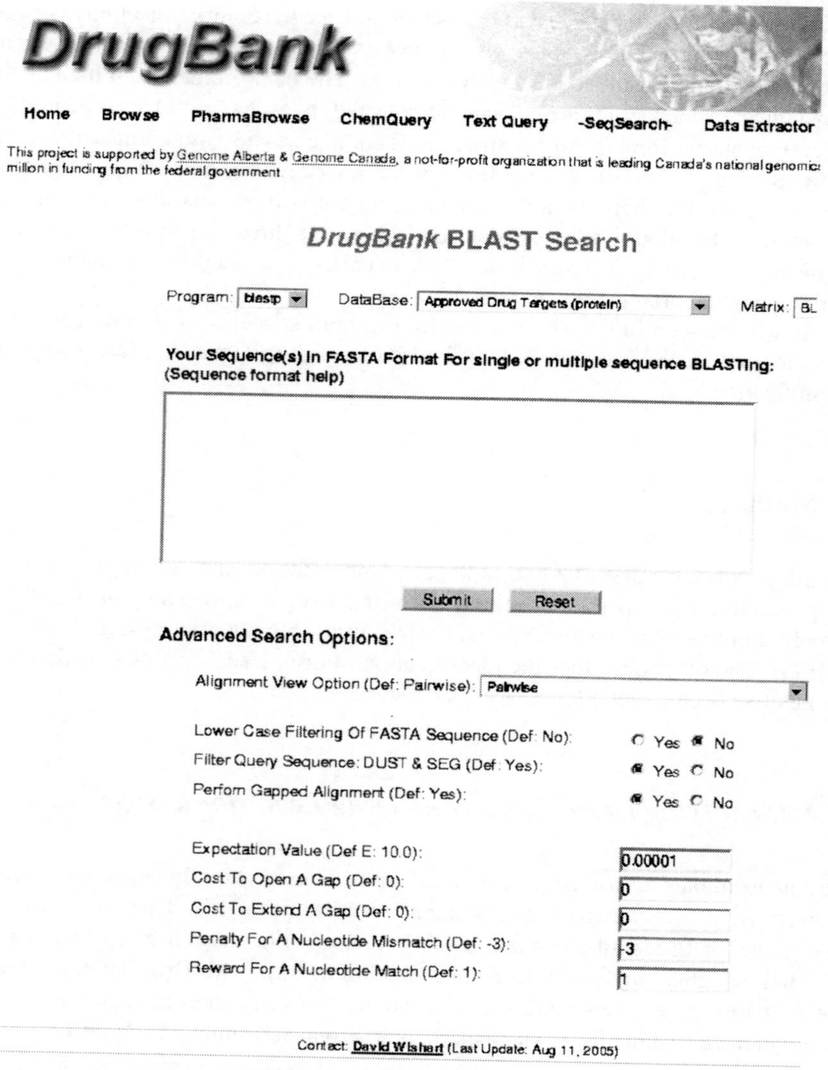

DrugBank

Home Browse PharmaBrowse ChemQuery Text Query -SeqSearch- Data Extractor

This project is supported by Genome Alberta & Genome Canada, a not-for-profit organization that is leading Canada's national genomic
million in funding from the federal government

DrugBank BLAST Search

Program: blastp DataBase: Approved Drug Targets (protein) Matrix: BL

Your Sequence(s) in FASTA Format For single or multiple sequence BLASTing:
(Sequence format help)

Submit Reset

Advanced Search Options:

Alignment View Option (Def: Pairwise): Pairwise

Lower Case Filtering Of FASTA Sequence (Def: No): Yes No
Filter Query Sequence: DUST & SEG (Def: Yes): Yes No
Perform Gapped Alignment (Def: Yes): Yes No

Expectation Value (Def E: 10.0): 0.00001
Cost To Open A Gap (Def: 0): 0
Cost To Extend A Gap (Def: 0): 0
Penalty For A Nucleotide Mismatch (Def: -3): -3
Reward For A Nucleotide Match (Def: 1): 1

Contact: David Wishart (Last Update: Aug 11, 2005)

Fig. 1 Screen shot of the DrugBank BLAST search page

buttons, as well as pull-down menus offering a choice **Programs** (BLASTP or BLASTN), **Databases** (14 choices), and scoring **Matrices** (BLOSUM or PAM). In almost all cases, users can leave everything (except the **Database** selection) in the default position. A unique feature of the **SeqSearch** program is its capacity to handle multiple FASTA-formatted sequences. This allows users to BLAST multiple sequences—or even entire proteomes.

3. For this example, we are looking for potential drug targets to a newly isolated retrovirus. To obtain the set of sequences to paste into the **SeqSearch** text box, launch a new browser window and go to: http://cpicanada.org/bioinfo2007/. Click on the **Virus** hyperlink. A list of 16 viral sequences should be visible. Select all 16 sequences by clicking and dragging through the window with your mouse. Copy the sequences (using the **Copy** option on your browser or using Ctrl + C).
4. Now click on the **SeqSearch** browser window to activate it and paste the sequences into the **SeqSearch** text box by clicking your mouse in the text box and using the **Paste** option on your browser (or Ctrl + V). You have now pasted 16 different protein sequences from the newly sequenced retrovirus. Use the scroll bars on the right side of the text box to see whether all 16 sequences are there.
5. Press the **Submit** button. Within a few seconds, the BLAST search for all 16 input sequences should be completed. The program will return a concatenated, text-based BLAST summary for each of the 16 proteins that were submitted. The top portion of the SeqSearch output consists of a summary of the submitted sequences. Below that is the BLAST result for the first sequence (Sequence 1, with 231 residues). The output should indicate ***** No hits found ******. Scrolling further down, the output for Sequence 2 should be seen for Enfuvirtide (a peptide drug) and further down is the output for Sequence 3 (a series of reverse transcriptase inhibitors—see Fig. 2).
6. Click on the hyperlink listed beside the word Nevirapine (users may select any one of the many hyperlinks in this list). This should take you to the DrugCard for Nevirapine or Viramune. This page describes the drug and its mode of action in detail and it suggests that Nevirapine may also be able to target this viral protein target.
7. Scroll down further through the **SeqSearch** output page and look for other sequences that exhibit hits to known DrugBank compounds and for drugs that would be likely to work on these protein targets. You should find that most of the 16 proteins in this virus seem to be potential drug targets and that multiple existing drugs could be effective against it.
8. For each of the proteins listed in the output from Step 5, click on the corresponding drug hyperlink(s). Scroll down the DrugCard page that is displayed (Fig. 3) and click on the field titled **PDB File Calculated Text** and download the PDB text file of the drug lead (Nevirapine, in this case). You may also obtain additional drug leads by going to the top of each DrugCard page and clicking on the button located on the top right corner, called **Find Similar Structures**. This will generate a table of chemically similar drugs that may exhibit potential

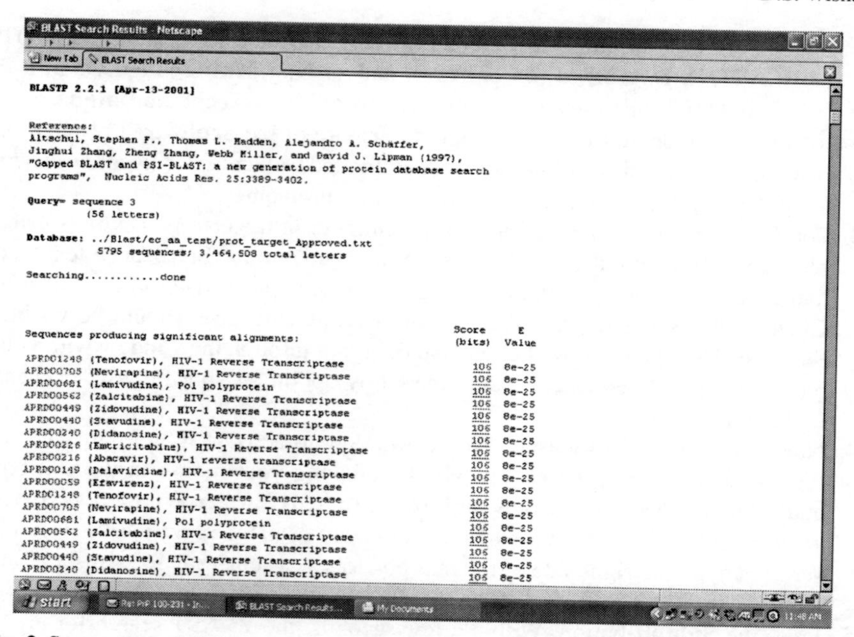

Fig. 2 Screen shot of the output from the DrugBank BLAST search using the 16 viral protein sequences belonging to a novel retrovirus

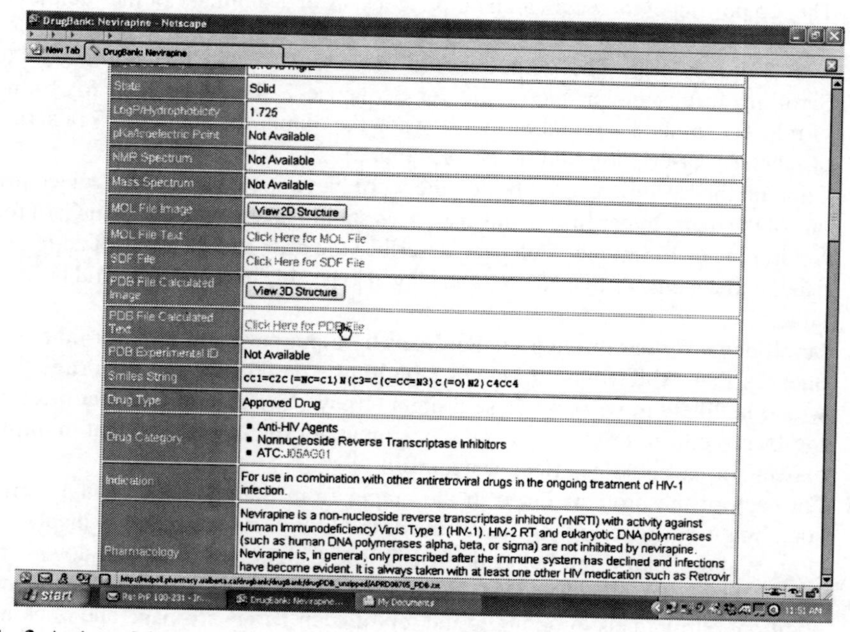

Fig. 3 A view of the tabular output found in the DrugCard for Viramune

activity against these viral proteins. Download these compounds as well. You should now have a large collection of PDB files (i.e., 3D structures) of possible drug leads for each of the unique proteins associate with the virus (see **Note 2**).

9. Scrolling up from the **PDB File Calculated Text** field, you will locate the **PubChem ID** field. Click on the Substance hyperlink (197039) listed in this field. This will take the user to the specific **PubChem Substance** page for that compound (Nevirapine in this case). PubChem is the largest online, open access collection of chemical compounds on the web. You can use PubChem to locate dozens or even hundreds of novel compounds that may have structural similarities to your compound (or drug lead) of interest. On the **PubChem Substance** page, you will find the **Structure Search** hyperlink beside the color diagram of the chemical's structure (Fig. 4). Clicking on this will take the user to PubChem's structure search tool. Go to the **PubChem Chemical Structure Search** located at the lower half of the page, because this is the type of search you wish to perform. Click on the **Search** button to activate the structure search tool. Users have several options to search for similar compounds (with different percent identity thresholds), identical compounds, matching chemical formulas, or matching substructures. The default of "Similar compounds, score ≥90%" is usually adequate. Note that the 2D structure files obtained from these PubChem

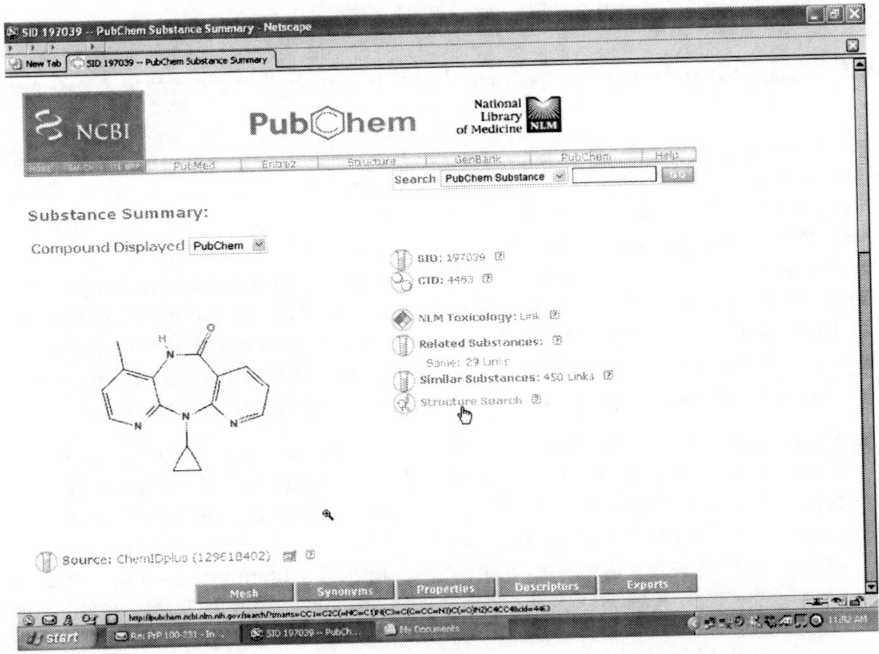

Fig. 4 A screen shot of the PubChem Substance page for Nevirapine or Viramune. The Structure Search link is on the bottom of the list located beside the structure

searches (SDF files) must be converted to 3D PDB files using a conversion and automated 3D modeling program, such as Cactus or MolConverter.

10. To perform docking or virtual screening experiments, it will be necessary to generate 3D structures of each of the protein targets identified through Steps 5 and 6. For many of the proteins identified in this exercise it is possible to generate a 3D homology model using either Modeller (3) or Swiss-Model (see Chap. 11). Swiss-Model users may go to the Swiss-Model website (http://swissmodel.expasy.org//SWISS-MODEL.html), paste in a sequence, provide an email address, and submit. An accurate 3D structure model, if it is possible to build, should be emailed to you within 30 minutes. Once you have done this for all of the proteins that can be modeled (not all will have 3D homologs) you should have a large collection of PDB files (i.e., 3D structures) of the key drug targets for this virus.
11. Use these two sets of structures (one for the drug leads, the other for the drug targets) to initiate a virtual screening run or to attempt to dock selected compounds into their corresponding protein targets using the methods described in Chapters 18 and 19. It is expected that the whole process (lead compound identification, lead compound structure generation, target protein structure generation) might take 1 to 2 hours, depending on how thorough one wishes to be (see **Note 3**).

3.2 Identifying Drug Targets and Drug Leads for Prostate Cancer

Prostate cancer is the second most common type of cancer in men in North America. It is responsible for more male deaths than any other cancer except lung cancer. It is a disease that generally strikes men older than the age of 50 years, however many factors beyond age, including genetics and diet, have been implicated in its development. In this example, we show how a large list of candidate target proteins can be easily obtained and then quickly reduced. From this list, we show how potential drug candidates or (anti)metabolites may be identified using DrugBank and the HMDB. We also demonstrate how the necessary coordinate files can be obtained to conduct rational drug design efforts via docking and virtual screening.

1. Start your local web browser and go to the UniProt website at http://www.pir.uniprot.org/. In the **Text Search** box at the top of the page (Fig. 5) type in the term “prostate cancer” and press the return key. A list of nearly 200 proteins that are associated with prostate cancer should be returned in a few seconds. UniProt is not necessarily the best source for identifying known disease genes, although it is very fast and the data are very reliable. Users could consider using the Prostate Gene Database or PGDB (http://www.ucsf.edu/pgdb/), a general PubMed/Entrez literature search or a search through PolySearch [25], or data obtained through a microarray experiment. Nevertheless, the UniProt list being used here is not only adequate for this example, it is also easy to manipulate and assess.

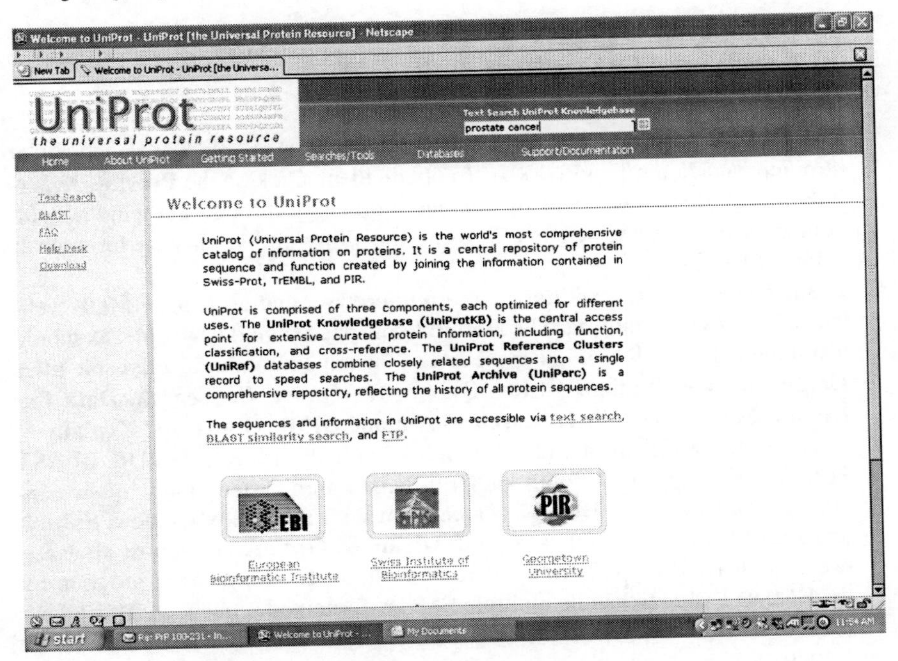

Fig. 5 Screen shot of the UniProt Knowledge Base search page

2. Scanning down the list of proteins, select those proteins that seem to be: 1) enzymes; 2) soluble proteins; and 3) able to bind or act on relatively unique small molecules. The reason for these selection criteria is that if one wants to develop a small molecule drug, the drug target should exhibit some propensity to bind a small molecule. Furthermore, if one wants to perform docking or virtual screening studies, the protein structure needs to be known or at least modeled. Because 99% of all proteins in the PDB are of soluble proteins or soluble fragments, the need for soluble protein targets is obviously very important. Readers may select the proteins of interest by clicking on the check boxes on the left side of the list. The complete list can be accessed by using the arrows at the top of the page. Once the desired proteins have been selected, users may save the sequences by clicking on the **FASTA** box on the upper right side of the screen under the **Save Options** tab. A reasonably good list of candidate protein targets that fit these three criteria is given below:

   (a) β-1, 3-galactosyltransferase 2
   (b) FK506-binding protein 5
   (c) Glutathione peroxidase 1
   (d) Nicotinamide mononucleotide adenylyltransferase 2
   (e) NADPH oxidase 5
   (f) Retinol dehydrogenase 11
   (g) 3-oxo-5-α-steroid 4-dehydrogenase 2

(h) Uridine-cytidine kinase 1
(i) α-methyl-acyl-CoA racemase
(j) Androgen receptor

The FASTA sequence file for all 10 proteins is available for download at http://cpicanada.org/bioinfo2007/. To obtain them, click on the **Prostate** hyperlink. Select all 10 sequences by clicking a dragging through the window with your mouse. Copy the sequences (using the **Copy** option on your browser or using Ctrl + C).

3. Launch a new window within your current browser and go to the HMDB website at http://www.hmdb.ca. The HMDB home page should be visible, as should a simple menu bar located near the top of the page with the 12 clickable titles **Home**, **Browse**, **Biofluids**, **ChemQuery**, **TextQuery**, **SeqSearch**, **Data Extractor**, **MS Search**, **NMR Search**, **Download**, **HML Home**, and **Explain**.
4. Click on the **SeqSearch** link. A window with the subtitle HMDB BLAST Search should appear. As with the DrugBank search system, the window contains a standard online BLAST search form with a text box window, **Submit** and **Reset** buttons, as well as several **Advanced Options**. In almost all cases, users can leave everything in the default position. The **SeqSearch** program for the HMDB is able to handle multiple FASTA-formatted sequences. This allows users to BLAST multiple sequences—such as those that might be obtained from a microarray experiment.
5. Now click on the **SeqSearch** browser window to activate it and paste the sequences into the **SeqSearch** text box by clicking your mouse in the text box and using the **Paste** option on your browser (or Ctrl + V). You have now pasted 10 different protein sequences that are potential drug/metabolite targets. Use the scroll bars on the right side of the text box to see whether all 10 sequences are there.
6. Press the **Submit** button. Within a few seconds, the BLAST search for all 10 input sequences should be completed. The program will return a concatenated, text-based BLAST summary for each of the 10 proteins that were submitted. The top portion of the SeqSearch output consists of a summary of the submitted sequences. Below that is the BLAST result for the first sequence (Sequence 1, with 422 residues) should be seen (Fig. 6).
7. Click on the hyperlink HMDB04968 listed beside the word Tetrahexosylceramide.... This should take you to the MetaboCard for tetrahexosylceramide (d18:1/25:0). This page describes the lipid, its structure, its metabolic importance, and the enzymes that act on it. For this particular example, more than 20 ceramides are listed, along with several UDP sugars. Although the ceramides may not be ideal drug candidates in terms of cost and lipophilicity, this result does suggest some candidate compounds that may prove to be either agonists or antagonists.
8. Scroll down further through the **SeqSearch** output page and look for other sequences that exhibit hits to known human metabolites and for metabolites that would be likely to work on these protein targets. You should find that most

Fig. 6 Screen shot of the output from a BLAST search against the HMDB using the 10 protein sequences identified as potential prostate cancer drug targets

of the proteins in this list seem to have interesting or unusual metabolites that could serve as the starting points to develop agonists or antagonists. Some of these (the androgens) have long been used as drug leads to treat prostate cancer. Among the most interesting novel compounds are the phytanic acids and pristanic acids identified for α-methyl-acyl-CoA racemase.

9. For each of the proteins listed in the output from Step 5, click on the corresponding HMDB hyperlink(s). Scroll down the MetaboCard page that is displayed (Fig. 7) and click on the field titled **PDB File Calculated Text** and download the PDB text file of the metabolite (or drug lead). You may also obtain additional drug leads by going to the top of each MetaboCard page and clicking on the button located on the top right corner called **Find Similar Structures**. This will generate a table of chemically similar compounds that may exhibit potential activity against these target proteins. Download these compounds as well. You should now have a large collection of PDB files (i.e., 3D structures) of possible drug leads for each of the prostate cancer-associated proteins.
10. Users may also want to use DrugBank and PubChem (as described in Sect. 3.1) to identify additional drug leads. Indeed, these efforts would prove to be very fruitful for this particular example. To generate models for the protein targets, we suggest that users follow Steps 10 and 11, as described in Sect. 3.1. This will allow them to complete the necessary steps required to set up docking and virtual screening efforts. It is expected that the whole process (target identification, lead compound identification, lead compound structure generation, and target protein structure generation) might take 2 to 3 hours.

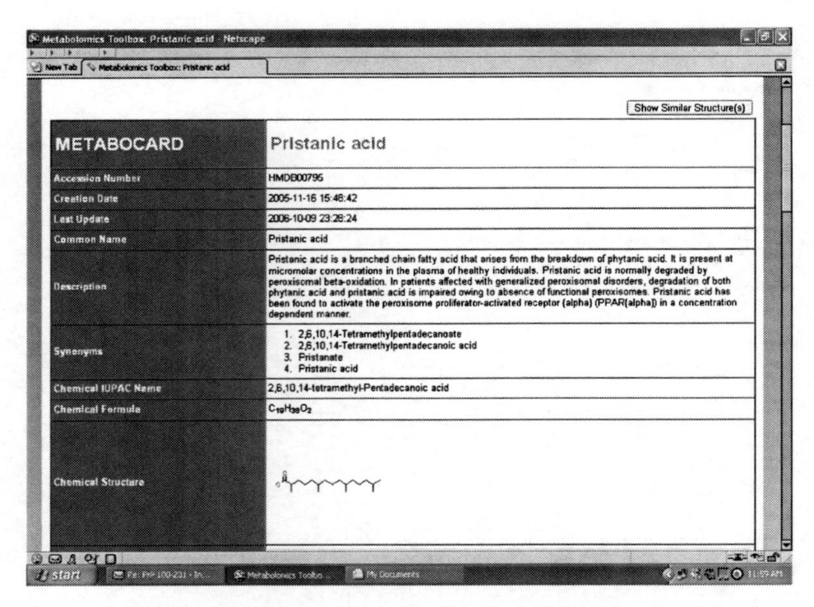

Fig. 7 Screen shot of the MetaboCard for pristanic acid. The hyperlinks for the PDB structure files and PubChem Substance identifications are located further down the card

4 Notes

1. The examples given in Sects. 3.1 and 3.2 are realistic but somewhat simplified compared with what might be necessary for "real life" drug discovery. In particular, the identification of drug targets always requires some critical assessment of the usefulness and viability of the drug target or drug lead. This typically requires a good deal of library research and additional experimentation. For instance, one must determine whether the drug target(s) should be inhibited (therefore, requiring an antagonist) or activated (therefore, requiring an agonist). As a general rule, the development of antagonists is generally much easier than agonists. Likewise, it is usually a good idea to determine whether the putative drug target has been previously identified and whether experimental lead compounds have already be explored. Even if a drug target appears viable, one should take particular care to determine whether the protein is essential, unique, or conditionally expressed for the associated disease or condition. Non-essential, non-unique, or continuously expressed proteins are generally not good drug targets. Likewise, proteins with generally weak affinities (i.e., most carbohydrate-binding proteins) or poor turnover rates often turn out to be poor drug targets.
2. The selection of drug leads also requires some careful consideration. Although DrugBank, HMDB, and PubChem can offer many useful suggestions, they are not the only sources for drug leads. Surveys through the literature or careful searches through specialized drug-screening databases can often yield very useful ideas. Once a collection of drug leads has been identified, it is usually prudent

to assess the suitability of the compound as a drug. Drug compounds must not be too soluble, too lipophilic, too unstable, or too toxic. These requirements are closely related to their physicochemical properties, which are also related to their absorption, distribution, metabolism, excretion, and toxcity (ADMET). ADMET prediction is becoming increasingly common in early stage drug discovery, drug screening, and drug design. Indeed, many computational chemists would argue that ADMET prediction is something that should *always* be done in the early phases of drug-lead selection. Fortunately, there are now a number of software packages, online servers, and standardized rules (Lipinski's Rule of Five) to determining the likely success or drug-likeness that a compound might have.

Examples of freely available ADMET prediction servers include the Actelion Property Prediction server (Google "Actelion Property Explorer") and the Pre-ADME web server (http://preadmet.bmdrc.org/preadmet/index.php). The Actelion Property Explorer is a web-enabled Java applet that allows users to draw chemical structures and then rapidly calculates various drug-related properties, including toxicity risks (mutagenicity, tumorigenicity, irritancy, and reproductive effect), solubility, logP, molecular weight, drug-likeness, and overall drug score. Pre-ADME is a web server that uses three classes of predictors, a molecular descriptors calculation, a drug likeness predictor, and an ADME predictor. The molecular descriptor calculator can predict nearly 1,000 molecular properties, including constitutional, topological, physicochemical, and geometrical descriptors, many of which are needed for ADME prediction. The drug likeness predictor uses Lipinski's rules (Rule of Five) and lead-like rules in its predictions. The ADME predictor is unique and can predict cell permeability, plasma protein binding, and skin permeability using an artificial neural network.

3. Perhaps the most important point to remember for each of the methods outlined here is that one is generating computer-based predictions. There is no guarantee that any of these predictions (drug targets or drug leads) will turn out to yield a viable therapeutic or even an interesting lead compound. As with any prediction in life science, one must always be prepared to thoroughly test the predictions using in vitro assays and animal models. In many cases, the computer predictions will turn out to be wrong. In rare cases, the initial predictions may prove to be very promising. Nevertheless, the results from any well-constructed wet-bench experiments can and should be used to help guide subsequent steps involved in the computational design, docking, and selection of drug leads.

References

1. Geldenhuys, W.J., Gaasch, K.E., Watson, M., Allen, D.D., and Van der Schyf, C.J. (2006) Optimizing the use of open-source software applications in drug discovery. *Drug Discov. Today*. **11**, 127–132.
2. Kirkpatrick, D.L., Watson, S., and Ulhaq, S. (1999) Structure-based drug design: combinatorial chemistry and molecular modeling. *Comb. Chem. High Throughput Screen.* **2**, 211–221.

3. Fiser, A. and Sali, A. (2003) Modeller: generation and refinement of homology-based protein structure models. *Methods Enzymol.* **374**, 461–491.
4. Mohan, V., Gibbs, A.C., Cummings, M.D., Jaeger, E.P., and DesJarlais, R.L. (2005) Docking: successes and challenges. *Curr. Pharm. Des.* **11**, 323–333.
5. Jain, A.N. (2004) Virtual screening in lead discovery and optimization. *Curr. Opin. Drug Discov. Devel.* **7**, 396–403.
6. Alvarez, J.C. (2004) High-throughput docking as a source of novel drug leads. *Curr. Opin. Chem. Biol.* **8**, 365–370.
7. Taylor, R.D., Jewsbury, P.J., and Essex, J.W. (2002) A review of protein-small molecule docking methods. *J. Comput. Aided Mol. Des.* **16**, 151–166.
8. Shoichet, B.K. and Kuntz, I.D. (1993) Matching chemistry and shape in molecular docking. *Protein Eng.* **6**, 723–732.
9. Goodsell, D.S., Morris, G.M., and Olson, A.J. (1996) Automated docking of flexible ligands: applications of AutoDock. *J. Mol. Recognit.* **9**, 1–5.
10. Jones, G., Willett, P., Glen, R.C., Leach, A.R., and Taylor, R. (1997) Development and validation of a genetic algorithm for flexible docking. *J. Mol. Biol.* **267**, 727–748.
11. Friesner, R.A., Banks, J.L., Murphy, R.B., Halgren, T.A., Klicic, J.J., Mainz, D.T., Repasky, M.P., Knoll, E.H., Shelley, M., Perry, J.K., Shaw, D.E., Francis, P., and Shenkin, P.S. (2004) Glide: a new approach for rapid, accurate docking and scoring. 1. Method and assessment of docking accuracy. *J. Med. Chem.* **47**, 1739–1749.
12. Wishart, D.S., Knox, C., Guo, A.C., Shrivastava, S., Hassanali, M., Stothard, P., Chang, Z., and Woolsey, J. (2006) DrugBank: a comprehensive resource for in silico drug discovery and exploration. *Nucleic Acids Res.* **34**(Database issue):D668–672.
13. Wheeler, D.L., Barrett, T., Benson, D.A., Bryant, S.H., Canese, K., Chetvernin, V., Church, D.M., DiCuccio, M., Edgar, R., Federhen, S., Geer, L.Y., Helmberg, W., Kapustin, Y., Kenton, D.L., Khovayko, O., Lipman, D.J., Madden, T.L., Maglott, D.R., Ostell, J., Pruitt, K.D., Schuler, G.D., Schriml, L.M., Sequeira, E., Sherry, S.T., Sirotkin, K., Souvorov, A., Starchenko, G., Suzek, T.O., Tatusov, R., Tatusova, T.A., Wagner, L., and Yaschenko, E. (2006) Database resources of the National Center for Biotechnology Information. *Nucleic Acids Res.* **34**(Database issue):D173–180.
14. Wishart, D.S., Metabolism and Metabolic Disease Resources on the Web, Chapter 3.1 (2006) in, *The Metabolic and Molecular Bases of Inherited Disease—OMMBID* (Scriver, C.R., Beaudet, A.L., Sly, W.S., Valle, D., Vogelstein, B. eds), McGraw-Hill, New York. www.ommbid.com
15. Sweetman, S. (2004) *Martindale: The Complete Drug Reference*, 34th edition, Pharmaceutical Press, New York, NY.
16. Krogh, A., Larsson, B., von Heijne, G., and Sonnhammer, E.L. (2001) Predicting transmembrane protein topology with a hidden Markov model: application to complete genomes. *J. Mol. Biol.* **305**, 567–580.
17. Gardy, J.L., Laird, M.R., Chen, F., Rey, S., Walsh, C.J., Ester, M., and Brinkman, F.S. (2005) PSORTb v.2.0: expanded prediction of bacterial protein subcellular localization and insights gained from comparative proteome analysis. *Bioinformatics*. **21**, 617–623.
18. Zhang, R., Ou, H.Y., and Zhang, C.T. (2004) DEG: a database of essential genes. *Nucleic Acids Res.* **32**(Database issue):D271–272.
19. Ihlenfeldt, W.D., Voigt, J.H., Bienfait, B., Oellien, F., and Nicklaus, M.C. (2002) Enhanced CACTVS browser of the Open NCI Database. *J. Chem. Inf. Comput. Sci.* **42**, 46–57.
20. Hamosh, A., Scott, A.F., Amberger, J.S., Bocchini, C.A., and McKusick, V.A. (2005) Online Mendelian Inheritance in Man (OMIM), a knowledgebase of human genes and genetic disorders. *Nucleic Acids Res.* **33**(Database issue), D514–517.
21. Darvasi, A. (2005) Dissecting complex traits: the geneticists' "Around the world in 80 days". *Trends Genet.* **21**, 373–376.
22. Clarke, P.A., te Poele, R., and Workman, P. (2004) Gene expression microarray technologies in the development of new therapeutic agents. *Eur. J. Cancer*. **40**, 2560–2591.

23. Rebhan, M., Chalifa-Caspi, V., Prilusky, J., and Lancet, D. (1998) GeneCards: a novel functional genomics compendium with automated data mining and query reformulation support. *Bioinformatics*. **14**, 656–664.
24. Hu, Y., Hines, L.M., Weng, H., Zuo, D., Rivera, M., Richardson, A., and LaBaer, J. (2003) Analysis of genomic and proteomic data using advanced literature mining. *J. Proteome Res.* **2**, 405–412.
25. Wishart, D.S. (2005) Bioinformatics in drug development and assessment. *Drug Metab. Rev.* **37**, 279–310.
26. Hopkins, A.L. and Groom, C.R. (2002) The druggable genome *Nat. Rev. Drug Discov.* **1**, 727–730.
27. Needleman, S.B. and Wunsch, C.D. (1970) A general method applicable to the search for similarities in the amino acid sequence of two proteins. *J. Mol. Biol.* **48**, 443–453.
28. Altschul, S.F., Madden, T.L., Schaffer, A.A., Zhang, J., Zhang, Z., Miller, W., and Lipman, D.J. (1997) Gapped BLAST and PSI-BLAST: a new generation of protein database search programs. *Nucleic Acids Res.* **25**, 3389–3402.
29. Weininger, D. (1988) SMILES, a chemical language and information system. 1. Introduction to methodology and encoding rules. *J. Chem. Inf. Comput. Sci*. **28**, 31–36.
30. Ullman, J.R. (1976) An algorithm for sub-graph isomorphism, *J. ACM* **23**, 31–42.

Chapter 18
Receptor Flexibility for Large-Scale *In Silico* Ligand Screens
Chances and Challenges

B. Fischer, H. Merlitz, and W. Wenzel

Summary An important contribution to today's computer-aided drug design is the automated screening of large compound databases against structurally resolved protein receptors targets. The introduction of ligand flexibility has, by now, become a standardized procedure. In contrast, a general approach to treat target degrees of freedom is still to be found, a consequence of the extreme increase of computational complexity, which comes along with the relaxation of protein degrees of freedom.

In this chapter, we discuss in some detail both benefits and present limitations of target flexibility for high-throughput *in silico* database screens. Among the benefits are an improved diversity of binding modes, which allows one to identify a wider class of drug candidates. The limitations are related to a diminishing docking accuracy and an increased number of false hits. Using the thymidine kinase receptor and ten known inhibitors as an example, we describe in detail how target flexibility was implemented and how it affected the screening performance.

Keywords: FlexScreen · Side chain flexibility · Simulation · Thymidine kinase · Virtual screening

1 Introduction

The key–lock principle, primarily focused only on geometric criteria [1], is the starting point for rational drug design. If either one ligand of the enzymatic process is known, or the x-ray crystallographic structure of a binding site of the protein has been determined, then a blueprint of a potential drug candidate, a pharmacophore model, can be constructed and molecules can be designed that share a certain similarity with that blueprint. This strategy has been applied during the last two decades in many successful drug design projects [2]. Current efforts focus on increasing the reliability of the predictions, which, to a large degree, depends on the viability of the underlying pharmacophore model.

From: Methods in Molecular Biology, vol. 443, Molecular Modeling of Proteins
Edited by Andreas Kukol © Humana Press, Totowa, NJ

With increasing computational power, a new strategy for computer-aided drug design has become the focus of interest [3] in recent years: the screening of large virtual compound databases to a receptor of known structure. The chemical and pharmaceutical industry have compiled large libraries of chemical compounds that were synthesized over the years, and it was realized that it would save time and costs to start a drug design process from molecules that can be readily purchased instead of being designed from scratch. In the following sections, we discuss the ingredients and some applications of *FlexScreen*, a tool for virtual database screening that has been developed in recent years at the Karlsruhe Research Center [4].

2 Theory: The Screening Tool *FlexScreen*

There are two major ingredients to an *in silico* screening method: 1) a *scoring function* that approximates the binding energy (ideally the affinity) of the receptor–ligand complex as a function of the conformation of this complex, and 2) an efficient *optimization method* that is able to locate the binding mode of a given ligand to the receptor as the global optimum of the scoring function. In a database screen, all ligands are assigned an optimal score, which is then used to sort the database to select suitable ligands for further investigations.

2.1 The Scoring Function

For virtual database screening, it is mandatory to model the receptor–ligand system as accurately as possible to obtain a precise representation of the key–lock principle. This is best achieved using a scoring function of atomistic resolution. *FlexScreen* contains an interaction-based scoring function, which is constructed from first principles, although major approximations are presently required to keep the computations feasible:

$$S = \sum_{Protein} \sum_{Ligand} \left(\frac{R_{ij}}{r_{ij}^{12}} - \frac{A_{ij}}{r_{ij}^{6}} + \frac{q_i q_j}{r_{ij}} \right) + \sum_{h-bonds} \cos \Theta_{ij} \left(\frac{\tilde{R}_{ij}}{r_{ij}^{12}} - \frac{\tilde{A}_{ij}}{r_{ij}^{10}} \right). \quad (1)$$

Equation 1 contains the empirical Pauli repulsion (first term), the Van de Waals attraction (second term), the electrostatic potential (third term) and an angular dependent hydrogen bond potential (terms four and five). The Lennard-Jones parameters R_{ij} and A_{ij} were taken from OPLSAA [5], the partial charges q_i were computed with Insight II and esff force field, and the hydrogen bond parameters $\tilde{R}_{ij}$ and $\tilde{A}_{ij}$ were taken from AutoDock [6]. This force field lacks solvation terms to model entropic or hydrophobic contributions. The omission of such terms has been argued to be appropriate for constricted receptor pockets in which all ligands with high affinity displace essentially all water molecules. In a database screen, such a scoring function may, therefore, preserve the correct *ranking* of the compounds, whereas the absolute scale of their binding energy is far from being accurate.

2.2 The Global Optimization Engine

As another important constituent of a screening tool, the global minimum of the potential energy surface (PES) of the protein–ligand complex has to be determined. For systems that contain high-dimensional conformational spaces, the *stochastic approach* to global optimization has turned out to be a most suitable tool [7]. Here, the conformation is modified in a random walk [8], and new conformations are accepted using a statistical method based on thermodynamic principles. Consequently, these methods are not exact and their results exhibit a statistical noise, the error of which has to be analyzed to judge the final accuracy of the database screen [9]. Among this class of methods, the stochastic tunneling (STUN) technique [10] has been proven to be especially effective for the receptor–ligand problem [11]. In STUN, the dynamical process explores not the original, but a transformed PES, which dynamically adapts and simplifies during the simulation. For the simulations reported here, the following transformation was applied:

$$E_{STUN} = \ln\left(x + \sqrt{x^2 + 1}\right). \tag{2}$$

Here, $x = \gamma(E - E_0)$, E is the energy of the present conformation and E_0 is the best energy found so far (in this particular simulation cycle). The problem-dependent transformation parameter γ controls the steepness of the transformation. The general idea of this approach is to flatten the PES in all regions that lie significantly above the best estimate for the minimal energy (E_0). Even at low temperatures, the dynamics of the system becomes diffusive at energies $E \gg E_0$ independent of the relative energy differences of the high-energy conformations involved. The dynamics of the conformation on the untransformed PES then seems to "tunnel" through energy barriers of arbitrary height, whereas low metastable conformations are still well resolved. Applied to receptor–ligand docking, this mechanism ensures that the ligand can reorient through sterically forbidden regions in the receptor pocket.

3 Method: Flexible Docking to Thymidine Kinase

The following docking simulations were performed using the thymidine kinase (TK) receptor as an example. This enzyme has long been a focus of pharmaceutical research because of its role in reproduction of the herpes simplex virus. Since then, it has emerged as a useful benchmark system in rational drug design, because not just one, but ten active inhibitors are known and the x-ray structures of their binding modes have been identified [12]. When mixing these ten inhibitors into a database of randomly selected compounds, the screening tool should be able to identify as many of them as possible as being "good" ligands, i.e., it should assign a high rank to these benchmark ligands. The present receptor is of particular interest, because the measured target conformations of the various complexes are significantly different, as we have discussed elsewhere [13].

The following project, using *FlexScreen* as a docking tool, could be partitioned into the following basic steps, which will be discussed in more detail in the upcoming paragraphs:

1. Preparation of the ligand database and the ten benchmark ligands.
2. Preparation of the docking site.
3. Database screen using a rigid receptor.
4. Identification of important side chains for individual binding modes.
5. Database screen with selected receptor degrees of freedom.
6. Evaluation of the selective power of different screens.

3.1 Preparation of the Ligands

First, 10,000 compounds were randomly chosen from the open NCI database [14]. Because, at this stage, no partial charges were assigned to these compounds, we used the Insight II package with ESFF forcefield [15] and an automated script to evaluate partial charges for each ligand atom. The compounds were then stored in Sybyl mol2 format, which is the standard format used by *FlexScreen*. The flexible bonds of the ligands are identified by *FlexScreen* using a set of simple standard rules. The ten benchmark ligands were taken out of the original files of the protein database (PDB) [16] and prepared in a similar manner using Insight II (see **Note 1**).

3.2 Preparation of the Docking Site

For our analysis, the uncomplexed x-ray TK receptor structure (1e2h [17]) was taken from the PDB database, partial charges were assigned using Insight II, and the structure was stored as a mol2 file. The uncomplexed structure was chosen to avoid any conformational bias, created by the ligand closely interacting with flexible side chains inside the binding pocket. *FlexScreen* creates grids for the Coulomb potential and lists of nearest Lennard-Jones interaction sites, to speed up the docking simulations (see **Note 2**).

3.3 Screen Using a Rigid Receptor

The database screen was carried out using the following strategy. The total number of simulation steps is divided into three partitions (cascadic approach [9]). We start with a population of 100 different conformations, for which we do short docking simulations with 7,500 steps. The energetic best five conformations are selected for further simulations on the next stage with 30,000 simulation steps each. Finally, the two best energetic conformations are again refined with 75,000 simulation steps. In each further stage of the cascade, the side group and ligand movement decreases until it matches, in the final stage, more or less the rotational and translational displacement of a free Brownian diffusion at 300 K. All ligands that score better than the energy of 10,000 kJ/mol (with our scoring function) are defined as docked and are potential candidates for further cascadic stages.

In this screen, 3,291 database ligands attained a stable conformation with negative binding affinity within the receptor pocket. The resulting ranks for the ten inhibitors during this screen are summarized in Table 1 (second column). The ligands dhbt and hpt were ranked with a very high affinity. The ligands hmtt and mct docked poorly. Hmtt has barely reached a negative binding energy, whereas mct has never been bound. The majorities of the benchmark ligands were energetically more or less close to each other but did not score especially well. Repeating the docking simulations for these ligands did not substantially improve their rank in the database, eliminating the possibility of statistical fluctuations of the docking algorithm as the source for this difficulty. This enrichment rate is comparable to the results of other scoring functions that were previously investigated for this system, but the overall performance is very disappointing [12] (see **Note 3**).

Table 1 Ranking of the TK inhibitors in a screen of 10,000 randomly chosen ligands of the NCI database

Inhibitor	Rigid	6 flex	6 flex + SO_4
Acv	221	55	38
Ahiu	1,454	1,315	794
Dhbt	2	2	1
Dt	308	172	45
Hmtt	3,117	2,934	2,076
Hpt	8	13	7
Idu	612	97	23
Mct	nd	4,180	4,054
Pcv	437	71	25
Gcv	187	14	4
Score	5,225	6,576	7,063

The top row designates the docking receptor model that was used in the screen. Nd, not docked

3.4 Identification of Important Side Chains

It is immediately clear that a relaxation of all side chains that are located near the binding site is not feasible with today's computational resources, a consequence of the combinatorial explosion of the conformation space. Instead, those side chains that would play an important role in the binding modes have to be identified and partially released.

For this particular prestudy, the x-ray structure of TK in complex with the ligand hmtt [18] was used. The side chain GLN125 was quickly identified as an important hotspot for the binding motif, forming two hydrogen bonds with hmtt. When comparing different receptor structures with one another, GLN125 turned out to be highly flexible; its conformation was significantly changed with the ligand with which it was interacting. To simulate such a system, three chemical bonds of GLN125 (among others) were made flexible to allow the ligands to find their individual binding motifs.

Figure 1 shows two final conformations of a docking simulation for the two ligands dt (left) and gcv (right). Similar to the measured x-ray structures, in our docking study, the side chain GLN125 has also changed its conformation. Dt, using the same binding mode as hmtt, did not modify the side chain orientation significantly compared with the x-ray side chain orientation. On the other hand, when gcv was docked into our flexible receptor structure, the side chain was moved to form the two important hydrogen bonds with the ligand, as can be seen in the right figure.

In the following, we analyze the dihedral angle population of the side chain GLN125 for the final receptor–ligand conformations after a docking run of the ten known active substances [12]. For each of the active compounds, 32 independent simulations were performed, and those two conformations with lowest energies were selected. For each dihedral angle, the number of occurrences at the different

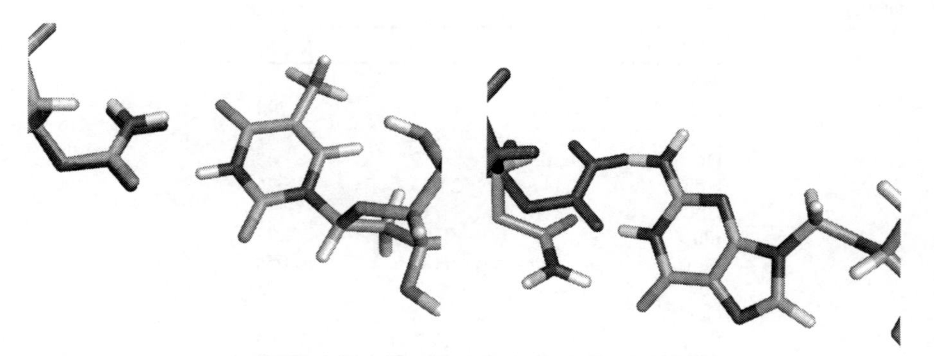

Fig. 1 The x-ray structure (*dark grey*) along with the simulated conformation of GLN125 and the docked ligand. *Left*: dt (deoxythymidine) docked into 1e2n. The side chain movement is insignificant because the binding pattern of this side chain matches to this ligand. *Right*: gcv (ganciclovir) is docked to 1e2n. Because the original crystal binding motif of GLN125 did not allow the ligand to form its individual interaction pattern, a new side chain conformation was energetically more favorable

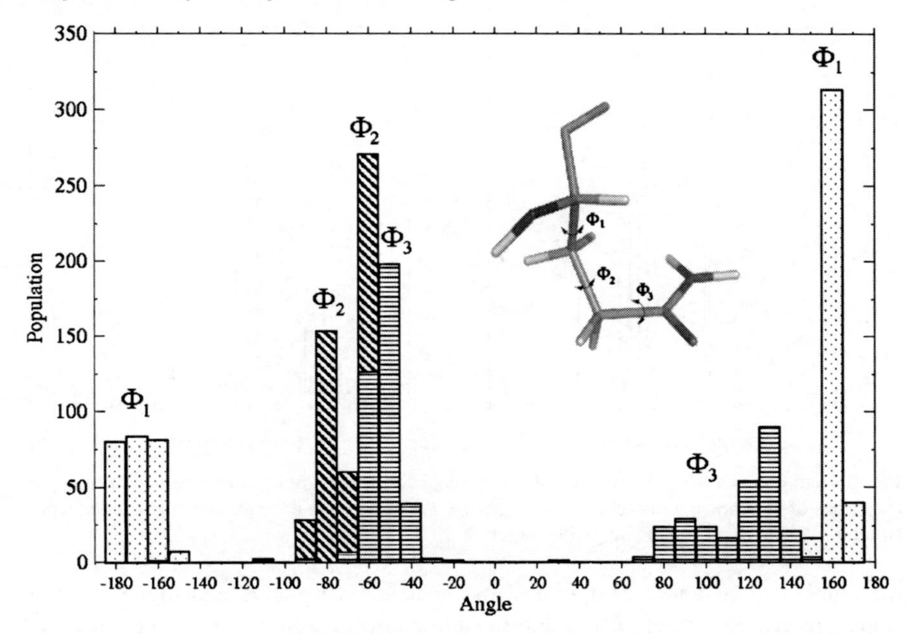

Fig. 2 Histogram of the three dihedral angles of GLN125. For 640 final conformations, the number of occurrences for each dihedral is plotted

angles is plotted in Fig. 2, which also contains a representation of the side chain with its three changing dihedral angles (Φ_1, Φ_2, and Φ_3).

The histogram does, in fact, show more than just three peaks for the different dihedrals. The dihedrals Φ_1 and Φ_2 formed two separate peaks, but the dihedral distribution of Φ_3 separated into two heaps. To allow for individual binding modes for these two ligands, this side chain had to flip around by 180 degrees, as was also observed in the x-ray structure.

In a similar manner, other side chains whose flexibility would contribute to increase the diversity of the database screen could be identified. For the following screen, four side chains with a total of six receptor degrees of freedom were introduced into the structure 1e2h, namely, dihedral rotations of the amino acids GLN125(3), TYR101(1), ARG222(1), and HIS58(1) (the numbers in parentheses denote the respective numbers of flexible bonds) (see **Note 4**).

3.5 Flexible Receptor Screen

Each step in the stochastic search now consisted of an additional random rotation for each side chain. The results of this screen are summarized in Fig. 3 (right panel), and the scores of the individual inhibitors are listed in the column labeled "6 flex" in Table 1. Now all ten ligands achieved a negative binding energy. As expected,

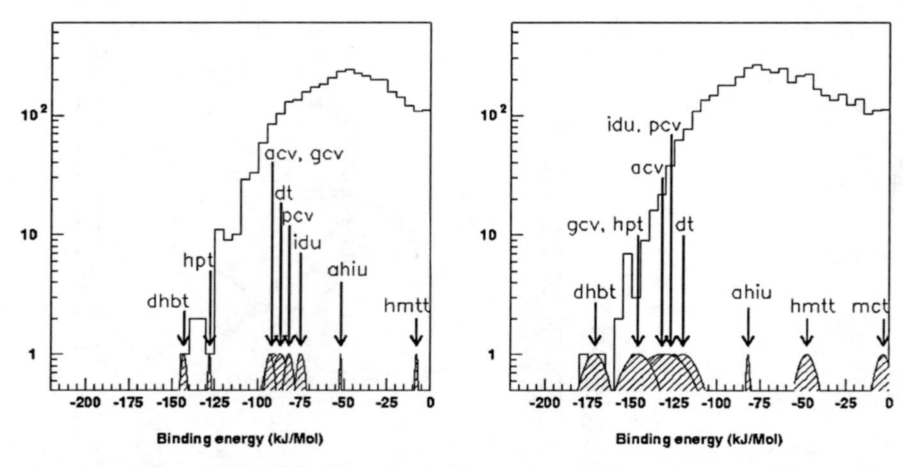

Fig. 3 Histogram of the binding energies of the docked ligands after a screen to the TK receptor (positions of the known inhibitors are highlighted). *Left*: Screen to the rigid uncomplexed receptor structure. *Right*: Screen using a flexible target

the number of database compounds that achieved a negative and higher binding energy increased as well, because a flexible conformation of the receptor reduced the bias of the screen against a specific binding pattern. Because 4,251 compounds had now got a negative binding energy (compared with 3,291 ligands of the rigid-receptor run), the diversity of the docking tool had increased by roughly 30%. At the same time, the specificity had decreased because the "lock" now allowed for a broader class of "keys" to fit into it. It was also observed that the accuracy of the flexible receptor screen was lower than that of the rigid receptor screen (with the same number of function evaluations), because the number of degrees of freedom increased. In Fig. 3, the docking inaccuracy is proportional to the width of the cone of the corresponding ligand.

The second flexible receptor screen, summarized in Fig. 4, includes additional a cofactor (SO_4) that was present in all investigated x-ray receptor pockets, but left out during the previous run. Because its position seemed to be almost invariant in the observed x-ray structures, it was kept fixed during the simulation. As a result, 4,078 of 10,000 ligands now docked to the target, a little less than before because of sterical restrictions. Comparing Fig. 4 with the flexible screen of Fig. 3 shows that the binding energies of the ten active compounds are not altered very much with the additional cofactor. It seems that the interaction with SO_4 could be neglected for such a database screen (see **Note 5**).

3.6 Comparison: Rigid Screen Versus Flexible Screens

To quantitatively compare different screens against the same ligand database, it is sensible to assign an overall score to each screen, which rates its performance [19]. We computed such a "score" for the entire screen from the ranks of the docked

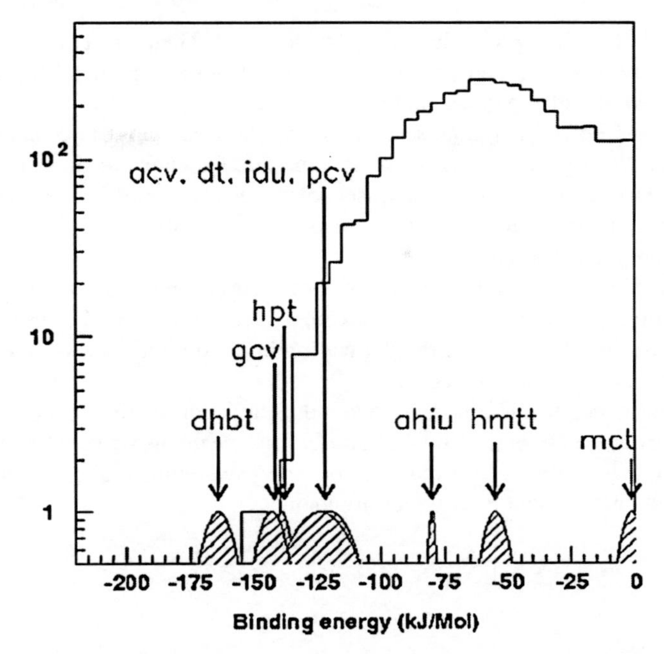

Fig. 4 Histogram of the binding energies of the docked ligands after a screen to the TK receptor (positions of the known inhibitors are highlighted). Screen using a flexible target including the cofactor SO_4

known inhibitors among the $N = 1{,}000$ best ligands (uppermost 10%). This score is computed as the sum of $N - P$, where P is the rank of the known inhibitor and shown in the bottom row of Table 1. An inhibitor ranking in the top of the screen contributes a score of 1,000 to the sum, a badly ranked inhibitor, comparatively little. Because the best N ligands are evaluated, screens that dock many known inhibitors with moderate rank may have comparable scores with screens that perform perfectly for one inhibitor, but fail for all others.

As expected, the rigid receptor screen with a score of 5,225 displayed the poorest performance among all screens because, here, the individual binding patterns of the inhibitors could not be supported. The flexible receptor screens performed much better. With a score of 6,576 and 7,063 for the screen with the additional cofactor, these results indicate that the increase in diversity of the technique had outbalanced the decrease in specificity, leading to an overall better docking performance of the screening tool.

4 Conclusions

The development of methods for high-throughput *in silico* screening has come a long way in the last decade. Fueled by both method development and the availability of computers that are more powerful, these methods have been established as a standard tool in modern preclinical drug development. Their success rate, measured by their ability to select medium affinity ligands from a large database of ligands now rivals that of wet-screening methods.

Much work, however, remains to be done. The treatment of receptor flexibility and induced fit remains one of the systematic challenges to these methods, as does the flexible treatment of semiconserved water molecules in the receptor. For the treatment of side chain flexibility, *FlexScreen* now offers a viable approach at moderate computational cost.

Open questions concern the development of accurate and transferable force fields and the treatment of large scale, i.e., backbone, receptor motion. For these reasons, it is not surprising that some of the known inhibitors do not dock well, even when side chain flexibility is accounted for.

With continued development of methods that address the points mentioned above, however, we are optimistic that further improvements will continue the trend of increased reliability and accuracy of *in silico* screening tools, which make an increasing impact in modern drug development.

5 Notes

1. As a matter of fact, there exists no semiclassical electrostatic force field that is generally acknowledged as being accurate for a widely diverse class of ligands. On the other hand, fully quantum mechanical *ab initio* methods are still too slow to be applicable to ten thousands of compounds. Hence, we have to accept inaccuracies of the scoring as a result of these force field deficiencies. Another quantum effect that has to be neglected is related to torsional potentials of semiflexible rotational bonds. In the present model, bonds are either rigid or freely rotatable (apart from steric restrictions).
2. Compared with the ligand case, the assignment of partial charges to the side chains inside the binding pocket is less prone to errors, because force fields are fairly well established for amino acids. However, many binding sites contain metal ions, so that binding energies that are more accurate would ask for the inclusion of quantum effects. Another unsolved problem is the inclusion of water molecules, which mediate hydrogen bonds between receptor and ligand. Their exact positions would depend on the ligand's properties, so that these water molecules should be made movable. Unfortunately, the excessive number of degrees of freedoms of such a system is prohibitive considering present computational resources.
3. In a previous study [20], we compared different database screens to different rigid receptor structures. In these cases, a high specificity of the receptor to its complexed ligands could be observed. The only ligands that scored well were those that were similar in their structure to the ligand with which the receptor formed the complex. This "memory effect" is a straight consequence of the key–lock principle. Such a high degree of selectivity is spurious, however, because

the natural receptor, the "key," contains a certain degree of flexibility to accommodate a variety of ligand structures. The rigidity of the model is, thus, causing a lack of diversity of the screen, and, consequently, the majority of potentially good drug candidates are rejected in such a simulation. The introduction of target degrees of freedom delivers an important tool to recover the diversity of the screening method.

4. In the more general case of screening projects without any known binding modes, the following strategy seems suitable. Starting with the rigid receptor, a subset of the database is docked and the hotspots (i.e., those side chains that formed the majority of high affinity bonds) are monitored. Side chains that carry most of these hotspots are defined as being important and are subsequently made flexible.
5. This could well be the result of lacking water molecules, which could mediate interactions between the ligands and the cofactor.

Acknowledgements We thank the Fond der Chemischen Industrie, the BMBF, the Deutsche Forschungsgemeinschaft (grant WE 1863/11-1), and the Kurt Eberhard Bode Stiftung for financial support.

References

1. Fischer, E. (1894) Einfluss der Konfiguration auf die Wirkung der Enzyme. *Ber. Dtsch. Chem. Ges.* **27**:2985–2993
2. Cramer III, R.D., Patterson, D.E. and Bunce, J.D. (1998) Comparative molecular field analysis (comfa). 1. Effect of shape on binding of steroids to carrier proteins. *J. Am. Chem. Soc.* **110**:5959–5967
3. Abagyan, R. and Totrov, M. (2001) High-throughput docking for lead generation. *Curr. Opin. Chem. Biol.* **5**:375–382
4. Merlitz, H., Burghardt, B. and Wenzel, W. (2003) Application of the stochastic tunneling method to high throughput database screening. *Chem. Phys. Lett.* **370**:68–73
5. Jorgensen, W.L. and McDonald, N.A. (1997) Development of an all-atom force field for heterocycles. Properties of liquid pyridine and diazenes. *Theochem-J. Mol. Struct.* **424**:145–155
6. Morris, G.M., Goodsell, D.S., Halliday, R., Huey, R., Hart, W.E., Belew, R.K. and Olson, A.J. (1998) Automated docking using a lamarckian genetic algorithm and an empirical binding free energy function. *J. Comput. Chem.* **19**:1639–1662
7. Kirkpatrick, S., Gelatt, C.D. and Vecchi, M.P. (1983) Optimization by simulated annealing. *Science* **220**:671–680
8. Metropolis, N. and Stanislaw, U. (1949) The Monte Carlo method. *JASA* **44**:335–341
9. Merlitz, H., Herges, T. and Wenzel, W. (2004) Fluctuation analysis and accuracy of a large-scale in silico screen. *J. Comp. Chem.* **25**:1568–1575
10. Wenzel, W. and Hamacher, K. (1999) Stochastic tunneling approach for global optimization of complex potential energy landscapes. *Phys. Rev. Lett.* **82**:3003–3007
11. Merlitz, H. and Wenzel, W. (2002) Comparison of stochastic optimization methods for receptor-ligand docking. *Chem. Phys. Lett.* **362**:271–277
12. Bissantz, C., Folkerts, G. and Rognan, D. (2000) Protein-based virtual screening of chemical databases. 1. Evaluation of different docking/scoring combinations. *J. Med. Chem.* **43**:4759–4767
13. Merlitz, H. and Wenzel, W. (2004) Impact of receptor flexibility on in silico screening performance. *Chem. Phys. Lett.* **390**:500–505

14. Milne, G.W.A., Nicklaus, M.C., Driscoll, J.S., Zaharevitz, D. and Wang, S. (1994) National cancer institute drug information system 3d database. *J. Chem. Inf. Comput. Sci.* **34**:1219–1224
15. Shi, S., Yan, L., Yang, Y., Fisher-Shaulsky, J. and Thacher, T. (2003) An extensible and systematic force field, ESFF, for molecular modeling of organic, inorganic, and organometallic systems. *J. Comput. Chem.* **24**:1059–1076
16. Bernstein, F.C., Koetzle, T.F., Williams, G.J.B., Meyer, E.F., Brice, Jr. M.D., Rodgers, J.R., Kennard, O., Shimanouchi, T. and Tasumi, M. (1977) The Protein Data Bank: A computer-based archival file for macromolecular structures. *J. Mol. Biol.* **112**:535–542
17. Vogt, J., Perozzo, R., Pautsch, A., Prota, A., Schelling, P., Pilger, P., Folkerts, G., Scapozza, L., and Schulz, G.E. (2000) Nucleoside binding site of herpes simplex type 1 thymidine kinase analyzed by x-ray crystallography. *Proteins* **42**:545–553
18. Wurth, C., Kessler, U., Vogt, J., Schulz, G.E., Folkers, G. and Scapozza, L. (2001) The effect of substrate binding on the conformation and structural stability of herpes simplex virus type 1 thymidine kinase. *Protein Sci.* **10**:60–73
19. Knegtel, R.M.A. and Wagnet, M. (1999) Efficacy and selectivity in flexible database docking. *Proteins* **37**:334–345
20. Fischer, B., Merlitz, H. and Wenzel, W. (2005) Increasing diversity in in silico screening with target flexibility. *CompLife* 186–197

Chapter 19
Molecular Docking

Garrett M. Morris and Marguerita Lim-Wilby

Summary Molecular docking is a key tool in structural molecular biology and computer-assisted drug design. The goal of ligand–protein docking is to predict the predominant binding mode(s) of a ligand with a protein of known three-dimensional structure. Successful docking methods search high-dimensional spaces effectively and use a scoring function that correctly ranks candidate dockings. Docking can be used to perform virtual screening on large libraries of compounds, rank the results, and propose structural hypotheses of how the ligands inhibit the target, which is invaluable in lead optimization. The setting up of the input structures for the docking is just as important as the docking itself, and analyzing the results of stochastic search methods can sometimes be unclear. This chapter discusses the background and theory of molecular docking software, and covers the usage of some of the most-cited docking software.

Keywords: AutoDock · Computer-assisted drug design · DOCK · FlexX · GOLD · ICM · Molecular recognition · Protein–ligand docking

1 Introduction

The field of molecular docking has emerged during the last three decades driven by the needs of structural molecular biology and structure-based drug discovery. It has been greatly facilitated by the dramatic growth in availability and power of computers, and the growing ease of access to small molecule and protein databases [1–4]. The goal of automated molecular docking software is to understand and predict molecular recognition, both structurally, finding likely binding modes, and energetically, predicting binding affinity. Molecular docking is usually performed between a small molecule and a target macromolecule. This is often referred to as ligand–protein docking, but there is growing interest in protein–protein docking. In this chapter, we will focus on ligand–protein docking, and use the more generic term

From: Methods in Molecular Biology, vol. 443, Molecular Modeling of Proteins
Edited by Andreas Kukol © Humana Press, Totowa, NJ

"target" to refer to the protein, DNA, or RNA macromolecule to which a much smaller molecule (or "ligand") is being docked.

Molecular docking has a wide variety of uses and applications in drug discovery, including structure–activity studies, lead optimization, finding potential leads by virtual screening, providing binding hypotheses to facilitate predictions for mutagenesis studies, assisting x-ray crystallography in the fitting of substrates and inhibitors to electron density, chemical mechanism studies, and combinatorial library design.

Virtual screening on the basis of molecular descriptors and physicochemical properties of (in)active ligands has great usefulness in finding hits and leads through library enrichment for screening [5], a strategy that is also well-used for reducing and enriching the library of ligands for molecular docking; there are recent reports that ligand shape-matching does as well as, if not better than, docking [6]. However, molecular docking when used as the final stage in virtual screening helps to provide a three-dimensional (3D), structural hypotheses of how a ligand interacts with its target.

Given the limitations of space, and in the interests of fairness, we are not able to survey the details of specific docking tool, except where illustrative, and we, therefore, refer the reader to the documentation provided with each of these tools. Instead, we aim to provide an overview comparing and contrasting the methodologies of the most cited [7] docking tools, namely AutoDock [8–10] http://autodock.scripps.edu; DOCK [11, 12] http://dock. compbio.ucsf.edu/; FlexX [13] http://www.biosolveit.de/FlexX; GOLD [14, 15] http://www.ccdc.cam.ac.uk/products/life_sciences/gold/; and ICM [16] http://www.molsoft.com/docking.html.

2 Theory

There are a number of excellent reviews of molecular docking methods [7, 17] and a large number of publications comparing the performance of a variety of molecular docking tools [18–29], often for virtual screening. It should be stressed that comparing docking methods is difficult [28], and because there is evidence that some docking methods do better with certain classes of target than others, the reader is encouraged to try several docking methods to determine which one(s) work best for their target of interest. The process of taking a known crystal structure of a complex of the target of interest, separating the ligand, and then docking the ligand back into the *apo*-form of the target is known as "re-docking." The reader should compare the ability of their chosen docking methods and parameters to re-dock a variety of ligands to the target of interest. Success is often measured in terms of root mean square deviation (RMSD) of the Cartesian coordinates of the atoms of the ligand in the docked and crystallographic conformations; a docking is generally regarded as successful if this is less than the somewhat arbitrary threshold of 2 Å; there are alternative measures of success, such as whether the correct ligand–target interactions are recovered.

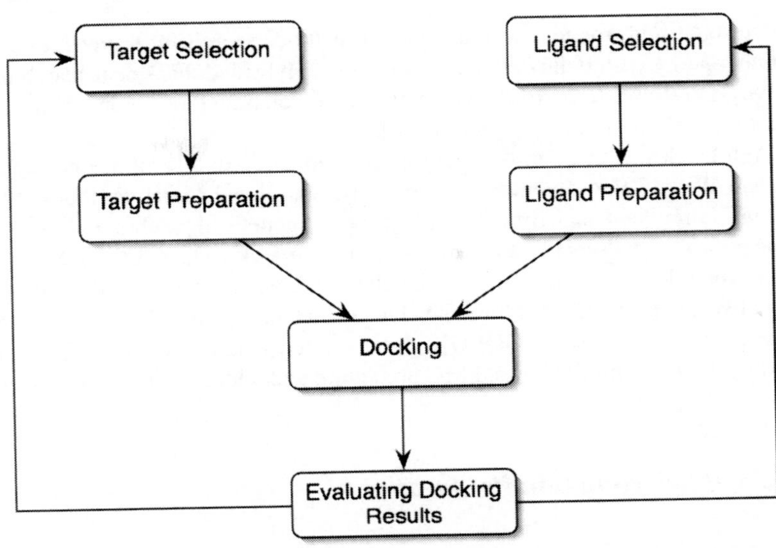

Fig. 1 A typical docking workflow. This flowchart shows the key steps common to all docking protocols. The 3D structures for the target macromolecule and the small molecule must first be chosen, and then each structure must be prepared in accordance with the requirements of the docking method being used. Following the docking, the results must be analyzed, selecting the binding modes with the best scores

Figure 1 shows the key steps in docking that are common to all protocols. Docking involves finding the most favorable binding mode(s) of a ligand to the target of interest. The binding mode of a ligand with respect to the receptor can be uniquely defined by its state variables. These consist of its position (x-, y-, and z-translations), orientation (Euler angles, axis-angle, or a quaternion), and, if the ligand is flexible, its conformation (the torsion angles for each rotatable bond). Each of these state variables describes one degree of freedom in a multidimensional search space, and their bounds describe the extent of the search. Rigid body docking is faster than treating the ligand as flexible, because the size of the search space is much smaller, but if the conformation of the ligand is not correct, then there will be a lower probability of finding a complementary fit.

All docking methods require a scoring function to rank the various candidate binding modes, and a search method to explore the state variables. Scoring functions can be empirical, force field based, or knowledge based, whereas search methods fall into two major categories: *systematic* and *stochastic*. Systematic search methods sample the search space at predefined intervals, and are deterministic. Stochastic search methods iteratively make random changes to the state variables until a user-defined termination criterion is met, so the outcome of the search varies; Sousa et al. discuss these classes of algorithms in more detail [7]. Search methods can also be classified by how broadly they explore the search space, as either *local* or *global*. Local search methods tend to find the nearest or local minimum energy to the current conformation, whereas global methods search for the best or global minimum energy within the defined search space. Hybrid global–local search methods have been shown to perform even better than global methods alone, being more efficient and able to find lower energies [8].

In AutoDock 4, for example, there is the choice of two local search methods (Solis and Wets [30] and Pattern Search [31]); two global search methods: Monte Carlo (MC) simulated annealing (SA) [32], and the genetic algorithm (GA) [33–35]; and one hybrid global–local search method, the Lamarckian GA (LGA) [8]. DOCK uses a systematic search method to match chemical features between the ligand and the negative image of the binding site. FlexX matches ligand features with complementary interaction sites. GOLD's global search method is a GA. The search method of ICM combines a biased MC procedure and a local energy minimization.

2.1 Target Selection and Preparation

Ideally, the target structure should be experimentally determined, usually by either x-ray crystallography or nuclear magnetic resonance. Docking has been performed successfully against homology models [36–39], although the reliability of the docking results depends heavily on the quality and bias of the homology model.

In some cases, the biologically relevant form of the target structure—the biological unit—is a multimer, which means that the appropriate symmetry-related molecules must also be included in the target structure. The online database Binding MOAD, for example, provides target structures as a biological unit [2] suitable for docking studies.

Many docking tools do not allow the target to be flexible, although this is a very important aspect of molecular recognition [40]. A target may adopt different conformations in the unbound and bound states, and with different classes of ligands; examples of different degrees of structural change on ligand binding are given in [41]. To tackle this and other problems, molecular dynamics has found an increasing number of applications in conjunction with molecular docking. These range from preparing the target before docking, to accounting for receptor flexibility, solvent effects, and induced fit, to calculating binding free energies and ranking docked ligands [42]. The so-called "relaxed complex method" developed in the laboratory of McCammon [43] generates snapshots from molecular dynamics simulations [44–46] of the *apo* form of the target, and then applies AutoDock to dock the ligand of interest; the technique effectively takes into account induced fit, and has been applied to develop novel inhibitors of HIV integrase [47].

In all x-ray crystal structures, there is a range of certainty with which atomic positions are defined. This is quantified by the temperature factors (also known as B-values) assigned to each atom in the PDB file. It is possible in some molecule viewers to color the atoms by B-value, which can visually indicate regions with more structural ambiguity. For a given x-ray crystal structure, atomic positions that may be suspect are those with 1) B-values higher than their surroundings, or 2) incomplete side chains (some atom positions are not assigned by the crystallographer). Furthermore, in certain crystal structures, alternate locations of atoms may be observed: in such cases, both alternatives must be tested.

To accelerate the scoring calculation, some docking methods precalculate grid maps to represent the receptor when calculating interaction energies with a ligand. A set of grid maps for a given receptor can be reused for docking of a library of ligands, also saving time. In general, grid maps are not transferable from one docking tool to another. For AutoDock, a grid map needs to be computed for each atom type in the ligand or set of ligands being docked, in addition to electrostatic potential and desolvation grid maps.

2.2 Ligand Selection and Preparation

The type of ligands chosen for docking will depend on the goal: for lead discovery, crude filters such as net charge, molecular weight, polar surface area, solubility, commercial availability, and price-per-compound can be applied to reduce the number of molecules to be docked. For lead optimization, filters such as similarity thresholds, pharmacophores, synthetic accessibility, and absorption, distribution, metabolism, excretion, and toxicology (ADME-Tox) properties are additionally applied. For focused lead optimization, a custom library of analogs that are related to the lead compound(s) is often constructed for docking, to inform and prioritize medicinal chemistry efforts [48]. Refer to Chap. 17 for more information regarding filtering libraries of compounds.

AutoDock uses a united-atom model for the ligand and receptor, in which only polar hydrogens are present. It also requires partial atomic charges to be assigned to the ligand. The AutoDock scoring functions were calibrated using Gasteiger charges [49] on the ligand, thus, to use the scoring functions correctly, the ligand must be assigned Gasteiger partial charges. It should be noted that alternative charge calculation methods for ligands have been successfully used in AutoDock [24].

Most docking tools treat ligands flexibly, with the exception of ring conformations. In general, the more rotatable bonds in a ligand, the more difficult and time consuming the docking will tend to be. This is because the size of the search space increases exponentially with the number of torsions. More highly branched torsion trees lead to more difficult searches than do linear torsion trees. Rotation of conjugated bonds, such as in amides, carbamates, ureas, etc., should be limited. One strategy to explore ring flexibility is to perform conformational analysis on any ring-containing ligands before docking [50–52]; another strategy is to compute the conformations of flexible ring systems during the docking, as can be done using FlexX [53] with either CORINA [54, 55] or Confort [56].

2.3 Docking

Molecular docking involves computationally exploring a search space that is defined by the molecular representation used by the method, and ranking candidate solutions to determine the best binding mode. Thus, docking requires both a search method and a scoring function.

Search methods can be divided into two main categories: *systematic* and *stochastic*. In the former case, the outcome of the search is deterministic, but the quality of the solution depends on the granularity of sampling of the search space. Stochastic methods rely on an element of randomness, therefore, the outcome varies. Systematic search methods are commonly used in rigid protein–rigid protein docking, where there are only six degrees of freedom, in programs such as DOT [57], GRAMM [58, 59], and ZDOCK [60]. Stochastic search methods are more suitable for higher-dimensional problems, such as flexible ligand–protein docking. Stochastic search methods include MCSA [10], GAs [8, 14, 15], and hybrid global–local search methods [8].

Scoring functions can be empirical, knowledge based, or molecular mechanics-based, see [17] for a review. In addition, some docking strategies use one scoring function during the docking, and a different one postdocking to rerank the results; such retrospective scoring, however, cannot affect the efficiency and accuracy of the primary scoring function [61].

The AutoDock scoring function is based on the molecular mechanics force field AMBER [62], with two additional terms: one to model the desolvation free energy change on binding, which is based on atomic solvation parameters [63]; and one empirical term to model the loss of conformational entropy on binding [8, 63]. The AutoDock scoring function in version 3 and later was inspired by the work of Böhm [64]. The individual contributions to the total energy of binding, namely van der Waals, hydrogen bonding, electrostatic, desolvation, and number of rotatable bonds in the ligand, were treated as independent variables. These were used to train a linear regression model given the observed free energy of binding, using a training set that included 30 protein–ligand complexes for AutoDock 3 [8] and 188 complexes for AutoDock 4 [63].

2.4 Evaluating Docking Results

Regardless of the ligand–protein docking tool used, docking results should be evaluated by considering the chemical complementarity between ligand and protein. Are all possible hydrogen bond donors and acceptors in the ligand satisfied? Are the charged groups in the ligand interacting with oppositely charged side chains in the receptor, or are they buried in hydrophobic pockets? Are hydrophobic groups in the ligand buried in hydrophobic pockets in the receptor?

Furthermore, the parameters chosen for the docking can be judged by the docking tool's ability to reproduce the binding mode of a ligand to protein, when the structure of the ligand–protein complex is known. The criterion usually used is the all-atom RMSD between the docked position and the crystallographically observed binding position of the ligand, and success is typically regarded as being less than 2 Å.

When docking using stochastic methods, it is recommended that the experiment be run at least 50 times with different initial conditions. The similarity of the predicted binding modes can be assessed by computing a matrix of pairwise RMSD values, and clustering docked conformations according to an RMSD threshold, typically 2 Å. If all of the dockings cluster into one family, this indicates that the search parameters were sufficient for each docking to converge. If there is no clustering at all, then the dockings should be repeated but with increased sampling: either increasing the number of iterations per search, increasing the number of searches, or, if the method is population based, increasing the population size.

If the scoring function were perfect, the docked conformation with the lowest energy would always correspond to the crystallographically observed binding mode, assuming that there are no bad contacts in the crystal structure. This is not always the case, and sometimes a different binding mode is observed significantly more often than the lowest energy binding mode. Furthermore, current docking methods will tend to find the binding mode with the lowest possible interaction energy for a given ligand: this score does not necessarily indicate whether the ligand even binds. There has been growing interest in developing methods to distinguish binders from nonbinders. One of the earliest reports that used docking to successfully discriminate binders from nonbinders [65] considered a simple metric that combined the mean binding energy for all of the conformations in the cluster, and the total number of conformationally distinct clusters found out of 100 dockings. The more clusters and the weaker the mean energy, the less likely the ligand was to bind. By building on statistical mechanical foundations, new methods are emerging that estimate the contributions of translational and rotational entropy to binding affinity, by approximating the configurational entropy using the sizes of the clusters [66, 67].

3 Methods

No matter which docking method is selected, the user needs to prepare the appropriate input files. This will depend on the docking method used, and in particular, on the molecular representation used in that method. To assist the user in setting up and in postdocking analysis, many docking programs include auxiliary tools, scripts, and graphical user interfaces (GUI); Table 1 summarizes some of these.

Docking methods that do not use a force field, such as FlexX and GOLD, do not require partial charges to be assigned to the atoms in the ligand and receptor molecules. AutoDock and UCSF DOCK, on the other hand, use an AMBER-derived force field and, therefore, require partial atomic charges. The AutoDock 3 scoring function was calibrated using Kollman united-atom partial charges on the macromolecule, unlike AutoDock 4, which uses Gasteiger PEOE charges for both ligand and macromolecule. It is important to note that other AutoDock users

Table 1 Ligand input requirements for the most commonly cited docking software

Docking tool	Auxiliary tools	File format	Hydrogen atoms	Partial charges
AutoDock 4	AutoGrid, ADT, BDT	mol2, PDBQT	United atom	Gasteiger PEOE[1]
DOCK 6	Chimera, Grid, Docktools, Nchemgrids, Sphgen, ANTECHAMBER	mol2	Explicit or united atom	AM1-BCC, Gasteiger
FlexX 2	FlexV	mol2, SD	United atom	Formal charge only
GOLD 3	GOLD Front End, SILVER	mol2, SD[2]	Explicit	None
ICM 3.4	ICM-Pro, ICM-VLS	mol2, SD	Explicit	MMFF, ICM

[1]Alternative partial charge calculations can be used (for example, AMSOL [68, 69] with the AM1-CM2 Hamiltonian; see [24])
[2]PDB format is also possible, but not recommended

have investigated the use of alternative partial charges on the ligand: e.g., Evans and Neidle concluded that the best charges to be used in AutoDock 3 for virtual screening of DNA minor groove binders came from calculations using AMSOL [68, 69] with the AM1-CM2 Hamiltonian for nonpolar organic solvent [24].

AutoDock is distributed with a GUI called AutoDockTools (ADT; see http://autodock.scripps.edu/resources/adt). ADT helps to prepare the ligand and receptor input files, and to set up the AutoGrid and AutoDock calculations. BDT [70] is an alternative preparatory tool to ADT (see http://www.quimica.urv.cat/~pujadas/BDT/) that helps in setting up virtual screening runs with AutoDock, and in setting up collections of AutoGrid maps for blind docking and also in combining grid maps to incorporate structural variability in the receptor. AutoDock's AutoGrid program and DOCK's Grid program precompute the necessary grid maps that describe the chemical potential at regular intervals around the target. In addition, DOCK's Sphgen program uses spheres to create the required "negative image" of the binding site.

The key stages in docking are: 1) target selection and preparation, 2) ligand selection and preparation, 3) docking setup, and 4) evaluating docking results; these are discussed in the following sections.

3.1 Target Selection and Preparation

1. Gather structures of the target, ideally with bound ligands, from internal and external sources. Good publicly available sources include the Protein Data Bank [4], http://www.rcsb.org/pdb; ReLiBase [1], http://relibase.ccdc.cam.ac.uk; and Binding MOAD [2, 71] http://www.bindingmoad.org. See **Note 1**.

2. Discard any structures that lack the biologically necessary cofactors, if any are required for biological activity. Structures that are incomplete or missing side chains should also be disregarded.
3. If there is more than one target structure, overlay them by superimposing the key residues in the binding site or region of interest using a least-squares superimposition method. SwissPdbViewer [72], a freely available tool from http://www.expasy.org/spdbv, offers several superimposition options under its "Fit" menu, such as "Magic Fit" and "Fit molecules (from selection)". Note also that SwissPdbViewer can also automatically reconstruct incomplete side chains.
4. Identify the extent of the structural variability and select a representative structure (see **Note 2**).
5. Add all hydrogen atoms to the target at the desired pH; under physiological conditions at pH 7.2, the following residues have ionized side chains: arginine, lysine, aspartic acid, and glutamic acid. This defines the formal charges (see **Note 3**). Each histidine side chain can be either neutral or positively charged at physiological pH. If it is neutral, either the delta or the epsilon nitrogen can be protonated (see **Note 4**).
6. The atomic assignments of imidazole rings in histidine and amido groups in asparagine and glutamine side chains can be ambiguous; tools such as REDUCE and its web interface, MOLPROBITY [73, 74] can evaluate 180° flips of these groups to optimize the hydrogen-bond network, and add hydrogen atoms appropriately.
7. Remove all water molecules, except those that are integral to your binding hypothesis(see **Note 5**).
8. If the representative target structure is complexed with a ligand, remove the ligand.
9. Calculate the partial charges, if required by the docking tool (see Table 1). Some tools may use a dictionary of amino acid partial charges to simply assign the charges. If there are any cofactors in the target structure, it will be necessary to compute the appropriate partial charges if required by the docking method.
10. When using AutoDock, merge nonpolar hydrogens, because it uses a united-atom representation (see Table 1 and **Note 6**).
11. AutoDock uses grid maps that must be calculated using AutoGrid. Each map describes a 3D grid of interaction energies with the target, one for each atom type in the ligand (see **Note 7**).

3.2 Ligand Selection and Preparation

Most docking tools require a 3D structure for each ligand, including explicit hydrogens. Depending on the source of the ligands—real molecules, molecules that have yet to be synthesized, or vendor libraries—the steps required to process the molecules will vary. The following steps exemplify how to obtain these structures, and how to process them for use in AutoDock.

1. ZINC is one of the largest collections of commercially available compounds; it is well curated and has 4.6 million compounds (http://blaster.docking.org/zinc; see also [3]). It is particularly useful for molecular docking because it provides 3D structures in SYBYL MOL2 formats, and is also free of charge. Subsets of compounds can be created by composing a query that specifies constraints on both molecular properties and two-dimensional (2D) molecular topology.
2. Ligands in the form of SMILES strings [75] can be converted into full 3D atomic coordinates, including hydrogens, using tools such as CORINA [54, 55] or ZINC [3].
3. Ligands in 2D SD format [76] can be converted into full 3D atomic coordinates using CORINA [54, 55] or Ghemical [77–79]. Ghemical can be used to sketch the ligand in 3D and then perform energy minimization, molecular dynamics, or conformational search to identify low energy conformations. PRODRG [80–82] can take PDB format, MDL MOL files, or even ASCII–text drawings of the molecule, instead of SD format. PRODRG is available as a standalone executable or as a web service, where the user can sketch the molecule in 2D and then convert the molecule into 3D; PRODRG is convenient for AutoDock 2.4 and 3, because it outputs PDBQ format.
4. It is important that the protonation, tautomeric, and stereoisomeric forms of the ligand be correct, otherwise subsequent calculations will be highly suspect. The enumeration of all possible ligand tautomers can be achieved with such programs as QUACPAC (Open Eye) [83], TAUTOMER (Molecular Networks), and LigPrep (Schrödinger).
5. When preparing ligands for AutoDock, the GUI AutoDockTools (ADT) can be used to set up the necessary input files. The first step for AutoDock is to calculate Gasteiger partial charges [49] and assign AutoDock atom types to each atom in the ligand (see **Notes 8–10**).
6. Define the "root" of the torsion tree and the rotatable bonds interactively using ADT. The "Ligand > Torsion Tree > Detect Root..." option automatically examines all the rotatable bonds in the ligand and chooses the atom that is nearest to the center of the torsion tree. The "Ligand > Torsion Tree > Choose Torsions..." option displays all rotatable bonds as green or magenta, indicating that they are active or inactive, respectively. Clicking on these bonds toggles whether they are active or not. Make sure any conjugated bonds are not rotatable.
7. AutoDock 4 requires the ligand to be in PDBQT format, which is very similar to PDB format, but also includes the partial atomic charge and the AutoDock atom type for each atom. The ligand should be saved using the "Ligand > Output > Save as PDBQT..." option in ADT.

3.3 Docking

1. Define the search space. There are two possibilities, depending on how much is known about the binding site:
   (a) If there is no previous information regarding the location of the binding site, then the translational search space should encompass the entire surface of the receptor. This is known as "blind docking," and is possible with AutoDock [84]. If the docking tool cannot encompass the whole target, then probable sites such as cavities large enough to contain the ligand(s) should be investigated separately; the third-party tool BDT [70] can be used to set up staggered grid boxes for AutoGrid.
   (b) If there is previous information, such as ligands with known binding modes, active site residues, or mutagenesis data, then the search space can be reduced to focus on the region of interest, thus, simplifying the search problem.
2. Set the target to be docked to, using the ADT menu item "Docking > Macromolecule > Set Rigid Filename..."
3. Select the search method (if there is more than one), and set the appropriate parameters. AutoDock offers MCSA, a traditional GA, and a hybrid global–local search method called LGA. The best search algorithm was shown to be LGA [8], therefore, we recommend this for most dockings (see **Note 11**).
4. Save the input parameter file for the docking tool, if necessary. For AutoDock, use the "Docking > Output > Lamarckian GA..." option in ADT to save an AutoDock docking parameter file (DPF) set up to perform LGA dockings.

3.4 Evaluating Docking Results

When evaluating the results of dockings, there are two main criteria to consider: 1) how well did the binding mode predicted by the docking match known structural data, where available; and 2) how well did the docking rank the ligands? If the method's scoring function is designed to predict binding affinities, how well did it match experimental binding data?

To answer the first criteria, a crystal structure of the complex of the ligand bound to the target must be known, and then the RMSD between the docked and the "reference" crystallographic binding mode of the ligand can be calculated; success is usually counted as RMSD less than 2 Å. To answer the second criteria, inhibition constants, or K_i values, must be known for the ligands and the target system.

When the search method used is stochastic, it is important to consider how often a given binding mode was predicted across all the dockings that were run. This is usually achieved using conformational clustering, building families of related conformations using RMSD tolerances to decide whether two conformations belong in the same cluster.

1. Read in all of the docked conformations into the docking analysis tool. For AutoDock, use ADT with the menu option "Analyze > Dockings > Open..." for one docking log (DLG) (see **Note 12**).
2. It is useful to view the dockings in the context of the target, therefore, if necessary, load the structure of the target. In ADT, use "Analyze > Macromolecule > Open..." to read in the target PDBQT structure used to compute the AutoGrid maps.
3. Perform conformational cluster analysis on the dockings to assess the level of agreement in the results. In ADT, use "Analyze > Clusterings > Recluster...", and type in a list of RMSD tolerances in angstroms separated by spaces. This performs clustering for each RMSD tolerance value, grouping the docked conformations accordingly.
4. Display the conformational clustering as a histogram, and visually inspect each cluster. In ADT, use "Analyze > Clusterings > Show..." and then choose the RMSD tolerance value. This displays a histogram of number of docked conformations in the cluster, versus the energy of the most tightly binding conformation in that cluster. The histogram is interactive, thus, clicking on a histogram bar sets up the "play" buttons in the "Conformation Player" window to play through the conformations in that cluster. This window has buttons to play forward and backward, and to step through the conformations one at a time.
5. It is possible to examine AutoDock-docked conformations in more detail using the "Conformation Player" in ADT, by clicking on the "&" button. This displays a panel in which it is possible to show more information regarding the current docking, by clicking on the "Show Info" check-button. It is also possible to monitor which hydrogen bonds are formed between the ligand and the target using the "Build H-bonds" check-button. The atoms in the ligand can be colored by a color scale that goes from dark blue to green to yellow to orange to red, indicating more favorable to less favorable interaction energies, using the "Color by" option; "vdw" colors by van der Waals or H-bond plus desolvation free energy, "elec_stat" colors by electrostatic interaction energy, and "total" colors by the total interaction energy; and "atom" returns to the default color-by-atom coloring.
6. If the docking results do not cluster into at least one significantly populated cluster, with an RMSD tolerance of between 2 and 3 Å, this is an indication that the dockings did not search for long enough. In AutoDock and ADT, increase the number of energy evaluations used in the LGA, and rerun the dockings. To get decent statistics, it is advisable to repeat the docking at for at least 50 runs. See Sect. 3.3, Step 3.
7. If the docked conformations are too far from the target structure, make sure that the AutoGrid grid box is centered on or near the target; the grid box can be visualized in ADT using the "Analyze > Grids > Open...", then choosing one of the grid map files. The x-, y-, and z-axes are color-coded red, green, and blue, respectively. The energy values in the grid map can be isocontoured by dragging the blue solid triangle on the "IsoValue" slider, with lower energy values indicating pockets of tighter binding affinity, and higher-resolution isocontours can be plotted using a "Sampling" value of 1 instead of the default value, 3.

4 Notes

1. It is preferable to use only high-resolution structures where available, ideally better than 2.5 Å.
2. As an alternative to Steps 1 to 3 in Sect. 3.1, *Target Selection and Preparation*, a representative structure or "leader" for a 90% homology family of structures is already precalculated and available from Binding MOAD [2] http://www.bindingmoad.org.
3. In AutoDockTools, use the "Edit > Hydrogens > Add" then choose "All Hydrogens"; all hydrogens are required for the initial Gasteiger partial charge calculation, but the nonpolar hydrogens will be merged later on.
4. AutoDockTools offers a tool to help set the desired protonation state of each His side chain, under the "Edit > Hydrogens > Edit Histidine Sidechains" menu. Which protonation state a His adopts will depend on its environment in the target.
5. Consolv [85], freely available from http://www.bch.msu.edu/labs/kuhn/software.html, "predicts whether water molecules bound to the surface of a protein are likely to be conserved or displaced in other, independently solved crystallographic structures of the same protein."
6. AutoDockTools calculates the partial charges and merges the nonpolar hydrogens automatically when the user selects the "Grid > Macromolecule > Choose…" menu items.
7. AutoGrid requires the target to be saved in PDBQT format, and it requires a Grid Parameter File (GPF). To save the receptor, use the "File > Save > Write PDBQT…" option. Set which types of grid maps should be calculated using either "Grid > Set Map Types > Directly…" or "Grid > Set Map Types > Choose Ligand…". To set up the location and grid spacing of the grid maps, use "Grid > Grid Box…". Finally, to save the GPF, use "Grid > Output > Save GPF…".
8. Note that ADT can read in a ligand with partial charges using SYBYL mol2 format. Use "File > Read Molecule…" and change the "Files of type" button to "MOL2 files (*.mol2)".
9. If the ligand is missing hydrogen atoms, then they must be added before calculating the Gasteiger charges. After selecting the ligand in ADT, use the menu option "Edit > Hydrogens > Add". It is very important to consider the tautomeric and ionization states when adding hydrogens, or use one of the tools in Step 4 in Sect. 3.2.
10. AutoDock atom types are assigned automatically in ADT by choosing the "Ligand > Input > Choose…" option. This command will also merge the nonpolar hydrogens, making the ligand suitable for use with the united-atom force field in AutoDock.
11. It is important to make sure that the number of energy evaluations is increased from the default value of 250,000 if the ligand has any rotatable bonds. Use the ADT menu option "Docking > Search Parameters > Genetic Algorithm…" and change the "Maximum number of energy evaluations" to at least 2,500,000.

It is also possible to increase the "Number of GA runs" in the panel from the default value 10. One other important parameter is the "Population Size"; the default is 150, although Hetenyi et al. showed that larger values up to 300 can improve the efficiency of the search [84]. Note also that this panel works for both the traditional and LGA search methods.

12. Alternatively, if the same ligand has been docked to the same target, but separate runs of AutoDock have produced uniquely named DLG files, as is the case when running dockings in parallel on computational clusters, use the "Analyze > Dockings > Open All..." option to read in all the dockings.

Acknowledgements This manuscript is TSRI publication number 18829. This work was supported by the National Institutes of Health (NIH) grant R01-GM069832.

References

1. Hendlich, M. (1998) Databases for protein-ligand complexes. *Acta Crystallogr D Biol Crystallogr*, **54**(Pt 6 Pt 1): 1178–1182.
2. Hu, L., Benson, M.L., Smith, R.D., Lerner, M.G., and Carlson, H.A. (2005) Binding MOAD (Mother Of All Databases). *Proteins*, **60**(3): 333–340.
3. Irwin, J.J. and Shoichet, B.K. (2005) ZINC—a free database of commercially available compounds for virtual screening. *J Chem Inf Model*, **45**(1): 177–182.
4. Berman, H.M., Westbrook, J., Feng, Z., Gilliland, G., Bhat, T.N., Weissig, H., Shindyalov, I.N., and Bourne, P.E. (2000) The Protein Data Bank. *Nucleic Acids Res*, **28**(1): 235–242.
5. Pozzan, A. (2006) Molecular descriptors and methods for ligand based virtual high throughput screening in drug discovery. *Curr Pharm Des*, **12**(17): 2099–2110.
6. Hawkins, P.C., Skillman, A.G., and Nicholls, A. (2007) Comparison of shape-matching and docking as virtual screening tools. *J Med Chem*, **50**(1): 74–82.
7. Sousa, S.F., Fernandes, P.A., and Ramos, M.J. (2006) Protein-ligand docking: Current status and future challenges. *Proteins*, **65**(1): 15–26.
8. Morris, G.M., Goodsell, D.S., Halliday, R.S., Huey, R., Hart, W.E., Belew, R.K., and Olson, A.J. (1998) Automated docking using a Lamarckian genetic algorithm and an empirical binding free energy function. *J Comput Chem*, **19**: 1639–1662.
9. Morris, G.M., Goodsell, D.S., Huey, R., and Olson, A.J. (1996) Distributed automated docking of flexible ligands to proteins: parallel applications of AutoDock 2.4. *J Comput Aided Mol Des*, **10**(4): 293–304.
10. Goodsell, D.S. and Olson, A.J. (1990) Automated docking of substrates to proteins by simulated annealing. *Proteins*, **8**(3): 195–202.
11. Ewing, T.J.A. and Kuntz, I.D. (1997) Critical evaluation of search algorithms for automated molecular docking and database screening. *J Comput Chem*, **18**(9): 1175–1189.
12. Kuntz, I.D., Blaney, J.M., Oatley, S.J., Langridge, R., and Ferrin, T.E. (1982) A geometric approach to macromolecule-ligand interactions. *J Mol Biol*, **161**(2): 269–288.
13. Rarey, M., Kramer, B., Lengauer, T., and Klebe, G. (1996) A fast flexible docking method using an incremental construction algorithm. *J Mol Biol*, **261**(3): 470–489.
14. Jones, G., Willett, P., Glen, R.C., Leach, A.R., and Taylor, R. (1997) Development and validation of a genetic algorithm for flexible docking. *J Mol Biol*, **267**(3): 727–748.
15. Jones, G., Willett, P., and Glen, R.C. (1995) Molecular recognition of receptor sites using a genetic algorithm with a description of desolvation. *J Mol Biol*, **245**(1): 43–53.
16. Abagyan, R.A., Totrov, M.M., and Kuznetzov, D.A. (1994) ICM—a new method for protein modeling and design: applications to docking and structure prediction from the distorted native conformation. *Journal of Computational Chemistry*, **15**: 488–506.

17. Taylor, R.D., Jewsbury, P.J., and Essex, J.W. (2002) A review of protein-small molecule docking methods. *J Comput Aided Mol Des*, **16**(3): 151–166.
18. Bissantz, C., Folkers, G., and Rognan, D. (2000) Protein-based virtual screening of chemical databases. 1. Evaluation of different docking/scoring combinations. *J Med Chem*, **43**(25): 4759–4767.
19. Friesner, R.A., Banks, J.L., Murphy, R.B., Halgren, T.A., Klicic, J.J., Mainz, D.T., Repasky, M.P., Knoll, E.H., Shelley, M., Perry, J.K., Shaw, D.E., Francis, P., and Shenkin, P.S. (2004) Glide: a new approach for rapid, accurate docking and scoring. 1. Method and assessment of docking accuracy. *J Med Chem*, **47**(7): 1739–1749.
20. Halgren, T.A., Murphy, R.B., Friesner, R.A., Beard, H.S., Frye, L.L., Pollard, W.T., and Banks, J.L. (2004) Glide: a new approach for rapid, accurate docking and scoring. 2. Enrichment factors in database screening. *J Med Chem*, **47**(7): 1750–1759.
21. Kellenberger, E., Rodrigo, J., Muller, P., and Rognan, D. (2004) Comparative evaluation of eight docking tools for docking and virtual screening accuracy. *Proteins*, **57**(2): 225–242.
22. Kontoyianni, M., McClellan, L.M., and Sokol, G.S. (2004) Evaluation of docking performance: comparative data on docking algorithms. *J Med Chem*, **47**(3): 558–565.
23. Perola, E., Walters, W.P., and Charifson, P.S. (2004) A detailed comparison of current docking and scoring methods on systems of pharmaceutical relevance. *Proteins*, **56**(2): 235–249.
24. Evans, D.A. and Neidle, S. (2006) Virtual screening of DNA minor groove binders. *J Med Chem*, **49**(14): 4232–4238.
25. Evers, A., Hessler, G., Matter, H., and Klabunde, T. (2005) Virtual screening of biogenic amine-binding G-protein coupled receptors: comparative evaluation of protein- and ligand-based virtual screening protocols. *J Med Chem*, **48**(17): 5448–5465.
26. Cummings, M.D., DesJarlais, R.L., Gibbs, A.C., Mohan, V., and Jaeger, E.P. (2005) Comparison of automated docking programs as virtual screening tools. *J Med Chem*, **48**(4): 962–976.
27. Cotesta, S., Giordanetto, F., Trosset, J.Y., Crivori, P., Kroemer, R.T., Stouten, P.F., and Vulpetti, A. (2005) Virtual screening to enrich a compound collection with CDK2 inhibitors using docking, scoring, and composite scoring models. *Proteins*, **60**(4): 629–643.
28. Cole, J.C., Murray, C.W., Nissink, J.W., Taylor, R.D., and Taylor, R. (2005) Comparing protein-ligand docking programs is difficult. *Proteins*, **60**(3): 325–332.
29. Vigers, G.P. and Rizzi, J.P. (2004) Multiple active site corrections for docking and virtual screening. *J Med Chem*, **47**(1): 80–89.
30. Solis, F.J. and Wets, R.J.-B. (1981) Minimization by random search techniques. *Mathematical Operations Research*, **6**: 19–30.
31. Conn, A.R., Gould, N.I.M., and Toint, P.L. (1991) A globally convergent augmented Lagrangian pattern search algorithm for optimization with general constraints and simple bounds. *SIAM Journal on Numerical Analysis*, **28**(2): 545–572.
32. Kirkpatrick, S., C. D. Gelatt, J., and Vecchi, M.P. (1983) Optimization by simulated annealing. *Science*, **220**(4598): 671–680.
33. Holland, J.H., *Adaptation in natural and artificial systems*. 1992, Cambridge, MA: The MIT Press. 211.
34. Goldberg, D.E., *Genetic Algorithms in Search, Optimization and Machine Learning*. 1st ed. 1989, Boston, MA: Addison-Wesley Longman Publishing Co., Inc. 372.
35. Michalewicz, Z., *Genetic Algorithms + Data Structures = Evolution Program*. 3rd ed. 1996, London, UK: Springer-Verlag. 387.
36. de Graaf, C., Oostenbrink, C., Keizers, P.H., van der Wijst, T., Jongejan, A., and Vermeulen, N.P. (2006) Catalytic site prediction and virtual screening of cytochrome P450 2D6 substrates by consideration of water and rescoring in automated docking. *J Med Chem*, **49**(8): 2417–2430.
37. Diller, D.J. and Li, R. (2003) Kinases, homology models, and high throughput docking. *J Med Chem*, **46**(22): 4638–4647.
38. Evers, A. and Klabunde, T. (2005) Structure-based drug discovery using GPCR homology modeling: successful virtual screening for antagonists of the alpha1A adrenergic receptor. *J Med Chem*, **48**(4): 1088–1097.

39. Shoichet, B.K., McGovern, S.L., Wei, B., and Irwin, J.J. (2002) Lead discovery using molecular docking. *Curr Opin Chem Biol*, **6**(4): 439–446.
40. Murray, C.W., Baxter, C.A., and Frenkel, A.D. (1999) The sensitivity of the results of molecular docking to induced fit effects: application to thrombin, thermolysin and neuraminidase. *J Comput Aided Mol Des*, **13**(6): 547–562.
41. Gunasekaran, K. and Nussinov, R. (2007) How different are structurally flexible and rigid binding sites? Sequence and structural features discriminating proteins that do and do not undergo conformational change upon ligand binding. *J Mol Biol*, **365**(1): 257–273.
42. Alonso, H., Bliznyuk, A.A., and Gready, J.E. (2006) Combining docking and molecular dynamic simulations in drug design. *Med Res Rev*, **26**(5): 531–568.
43. Lin, J.H., Perryman, A.L., Schames, J.R., and McCammon, J.A. (2002) Computational drug design accommodating receptor flexibility: The relaxed complex scheme. *J Am Chem Soc*, **124**(20): 5632–5633.
44. Case, D.A., Pearlman, D.A., Caldwell, J.W., Cheatham, T.E., III, Ross, W.S., Simmerling, C.L., Darden, T.A., Merz, K.M., Stanton, R.V., Cheng, A.L., Vincent, J.J., Crowley, M., Ferguson, D.M., Radmer, R.J., Seibel, G.L., Singh, U.C., Weiner, P.K., and Kollman, P.A. (1997) *AMBER 5* University of California: San Francisco.
45. Case, D.A., Pearlman, D.A., Caldwell, J.W., Cheatham, T.E., III, J., W., Ross, W.S., C., S., Darden, T., Merz, K.M., Stanton, R.V., Cheng, A., Vincent, J.J., Crowley, M., V., T., Gohlke, R.R., Duan, Y., Pitera, J., Massova, I., Seibel, G.L., Singh, C., Weiner, P., and Kollman, P.A. (2002) *AMBER 7* University of California: San Francisco.
46. Phillips, J.C., Braun, R., Wang, W., Gumbart, J., Tajkhorshid, E., Villa, E., Chipot, C., Skeel, R.D., Kale, L., and Schulten, K. (2005) Scalable molecular dynamics with NAMD. *J Comput Chem*, **26**: 1781–1802.
47. Schames, J.R., Henchman, R.H., Siegel, J.S., Sotriffer, C.A., Ni, H., and McCammon, J.A. (2004) Discovery of a novel binding trench in HIV integrase. *J Med Chem*, **47**(8): 1879–1881.
48. Gastreich, M., Lilienthal, M., Briem, H., and Claussen, H. (2006) Ultrafast *de novo* docking: combining pharmacophores and combinatorics. *J Comput Aided Mol Des*, in press.
49. Gasteiger, J. and Marsili, M. (1978) A new model for calculating atomic charges in molecules. *Tetrahedron Lett.*, **34**: 3181–3184.
50. Mulakala, C., Nerinckx, W., and Reilly, P.J. (2006) Docking studies on glycoside hydrolase Family 47 endoplasmic reticulum alpha-(1 → 2)-mannosidase I to elucidate the pathway to the substrate transition state. *Carbohydrate Research*, **341**(13): 2233–2245.
51. Laederach, A. and Reilly, P.J. (2005) Modeling protein recognition of carbohydrates. *Proteins-Structure Function and Bioinformatics*, **60**(4): 591–597.
52. Rockey, W.M., Laederach, A., and Reilly, P.J. (2000) Automated docking of alpha-(1 → 4)- and alpha-(1 → 6)-linked glucosyl trisaccharides and maltopentaose into the soybean beta-amylase active site. *Proteins-Structure Function and Genetics*, **40**(2): 299–309.
53. Rarey, M., Kramer, B., and Lengauer, T. (1995) Time-efficient docking of flexible ligands into active sites of proteins. *Proc Int Conf Intell Syst Mol Biol*, **3**: 300–308.
54. Gasteiger, J. and Sadowski, J. (1992) *CORINA* 3.4, Molecular Networks GmbH: Erlangen, Germany, http://www.molecular-networks.com/online_demos/corina_demo.html.
55. Gasteiger, J., Rudolph, C., and Sadowski, J. (1992) Automatic generation of 3D-atomic coordinates for organic molecules. *Tetrahedron Comput. Methodol.*, **3**: 537–547.
56. Pearlman, R.S. and Balducci, R. *Confort: A Novel Algorithm For Conformational Analysis*. in *National Meeting of the American Chemical Society*. 1998. New Orleans, LA.
57. Ten Eyck, L.F., Mandell, J., Roberts, V.A., and Pique, M.E., *Surveying Molecular Interactions With DOT*. 1995.
58. Vakser, I.A. (1997) Evaluation of GRAMM low-resolution docking methodology on the hemagglutinin-antibody complex. *Proteins*, **Suppl 1**: 226–230.
59. Tovchigrechko, A. and Vakser, I.A. (2006) GRAMM-X public web server for protein-protein docking. *Nucleic Acids Res*, **34**(Web Server issue): W310–314.
60. Chen, R. and Weng, Z. (2002) Docking unbound proteins using shape complementarity, desolvation, and electrostatics. *Proteins*, **47**(3): 281–294.

61. Mohan, V., Gibbs, A.C., Cummings, M.D., Jaeger, E.P., and DesJarlais, R.L. (2005) Docking: successes and challenges. *Curr Pharm Des*, **11**(3): 323–333.
62. Cornell, W.D., Cieplak, P., Bayly, C.I., Gould, I.R., Merz, J., Kenneth M., Ferguson, D.M., Spellmeyer, D.C., Fox, T., Caldwell, J.W., and Kollman, P.A. (1995) A second generation force field for the simulation of proteins, nucleic acids, and organic molecules. *J. Am. Chem. Soc.*, **117**: 5179–5197.
63. Huey, R., Morris, G.M., Olson, A.J., and Goodsell, D.S. (2007) A semiempirical free energy force field with charge-based desolvation. *J Comput Chem*, **28**(6): 1145–1152.
64. Böhm, H.-J. (1994) The development of a simple empirical scoring function to estimate the binding constant for a protein-ligand complex of known three-dimensional structure. *J Comput Aided Mol Des*, **8**(3): 243–256.
65. Rosenfeld, R.J., Goodsell, D.S., Musah, R.A., Morris, G.M., Goodin, D.B., and Olson, A.J. (2003) Automated docking of ligands to an artificial active site: augmenting crystallographic analysis with computer modeling. *Journal of Computer-Aided Molecular Design*, **17**(8): 525–536.
66. Ruvinsky, A.M. and Kozintsev, A.V. (2005) New and fast statistical-thermodynamic method for computation of protein-ligand binding entropy substantially improves docking accuracy. *Journal of Computational Chemistry*, **26**(11): 1089–1095.
67. Ruvinsky, A.M. and Kozintsev, A.V. (2006) Novel statistical-thermodynamic methods to predict protein-ligand binding positions using probability distribution functions. *Proteins-Structure Function and Bioinformatics*, **62**(1): 202–208.
68. Cramer, C.J. and Truhlar, D.G. (1992) AM1-SM2 and PM3-SM3 parameterized SCF solvation models for free energies in aqueous solution. *J Comput Aided Mol Des*, **6**(6): 629–666.
69. Hawkins, G.D., Giesen, D.J., Lynch, G.C., Chambers, C.C., Rossi, I., Storer, J.W., Li, J., Zhu, T., Thompson, J.D., Winget, P., Lynch, B.J., Rinaldi, D., Liotard, D.A., Cramer, C.J., and Truhlar, D.G. (2007) *AMSOL* 7.1, Department of Chemistry and Supercomputer Institute, University of Minnesota: Minneapolis, Minnesota, http://comp.chem.umn.edu/amsol/.
70. Vaque, M., Arola, A., Aliagas, C., and Pujadas, G. (2006) BDT: an easy-to-use front-end application for automation of massive docking tasks and complex docking strategies with AutoDock. *Bioinformatics*, **22**(14): 1803–1804.
71. Smith, R.D., Hu, L., Falkner, J.A., Benson, M.L., Nerothin, J.P., and Carlson, H.A. (2006) Exploring protein-ligand recognition with Binding MOAD. *J Mol Graph Model*, **24**(6): 414–425.
72. Guex, N. and Peitsch, M.C. (1997) SWISS-MODEL and the Swiss-PdbViewer: an environment for comparative protein modeling. *Electrophoresis*, **18**(15): 2714–2723.
73. Davis, I.W., Murray, L.W., Richardson, J.S., and Richardson, D.C. (2004) MOLPROBITY: structure validation and all-atom contact analysis for nucleic acids and their complexes. *Nucleic Acids Res*, **32**(Web Server issue): W615–619.
74. Lovell, S.C., Davis, I.W., Arendall, W.B., 3rd, de Bakker, P.I., Word, J.M., Prisant, M.G., Richardson, J.S., and Richardson, D.C. (2003) Structure validation by Calpha geometry: phi,psi and Cbeta deviation. *Proteins*, **50**(3): 437–450.
75. Weininger, D., *Daylight Theory Manual*. 2006, Daylight Chemical Information Systems, Inc.
76. Dalby, A., Nourse, J.G., Hounshell, W.D., Gushurst, A.K.I., Grier, D.L., Leland, B.A., and Laufer, J. (1992) Description of Several Chemical Structure File Formats Used by Computer Programs Developed at Molecular Design Limited. *J Chem Inf Comput Sci*, **32**: 244–255.
77. Acton, A., Banck, M., Bréfort, J., Cruz, M., Curtis, D., Hassinen, T., Heikkilä, V., Hutchison, G., Huuskonen, J., Jensen, J., Liboska, R., and Rowley, C. (2006) *Ghemical* 2.00, Department of Chemistry, University of Kuopio: Kuopio, Finland, http://www.uku.fi/~thassine/projects/ghemical/.
78. Hassinen, T. and Peräkylä, M. (2001) New energy terms for reduced protein models implemented in an off-lattice force field. *J Comput Chem*, **22**(12): 1229–1242.
79. Hassinen, T., Hutchison, G., Cruz, M., Banck, M., Rowley, C., and Curtis, D. (2007) *Ghemical-GMS* 2.10, Department of Chemistry, University of Iowa.: Iowa City, IA, http://www.uiowa.edu/~ghemical/ghemical.shtml.
80. van Aalten, D. and Oswald, S. (2007) *PRODRG* 2, University of Dundee: Dundee, Scotland, http://davapc1.bioch.dundee.ac.uk/programs/prodrg/.

81. van Aalten, D.M., Bywater, R., Findlay, J.B., Hendlich, M., Hooft, R.W., and Vriend, G. (1996) PRODRG, a program for generating molecular topologies and unique molecular descriptors from coordinates of small molecules. *J Comput Aided Mol Des*, **10**(3): 255–262.
82. Schuttelkopf, A.W. and van Aalten, D.M. (2004) PRODRG: a tool for high-throughput crystallography of protein-ligand complexes. *Acta Crystallogr D Biol Crystallogr*, **60**(Pt 8): 1355–1363.
83. Skillman, A.G. *QUACPAC* OpenEye Scientific Software: Santa Fe, NM, http://www.eyesopen.com/products/applications/quacpac.html.
84. Hetenyi, C. and van der Spoel, D. (2002) Efficient docking of peptides to proteins without prior knowledge of the binding site. *Protein Sci*, **11**(7): 1729–1737.
85. Raymer, M.L., Sanschagrin, P.C., Punch, W.F., Venkataraman, S., Goodman, E.D., and Kuhn, L.A. (1997) Predicting conserved water-mediated and polar ligand interactions in proteins using a K-nearest-neighbors genetic algorithm. *J Mol Biol*, **265**(4): 445–464.

Index

Lightning Source UK Ltd.
Milton Keynes UK
UKOW030258230313

208010UK00019B/273/P